Biocontrol Mechanisms of Endophytic Microorganisms

Biocontrol Mechanisms of Endophytic Microorganisms

Edited by

Radhakrishnan E.K.

School of Biosciences, Mahatma Gandhi University, Kottayam, India; Inter University Centre for Organic Farming and Sustainable Agriculture (IUCOFSA), Mahatma Gandhi University, Kottayam, India

Ajay Kumar

Agriculture Research Organization, Ministry of Agriculture and Rural Development Volcani Center, Rishon LeZion, Israel

Aswani R

School of Biosciences, Mahatma Gandhi University, Kottayam, India

ELSEVIER

ACADEMIC PRESS

An imprint of Elsevier

Academic Press is an imprint of Elsevier
125 London Wall, London EC2Y 5AS, United Kingdom
525 B Street, Suite 1650, San Diego, CA 92101, United States
50 Hampshire Street, 5th Floor, Cambridge, MA 02139, United States
The Boulevard, Langford Lane, Kidlington, Oxford OX5 1GB, United Kingdom

British Library Cataloguing-in-Publication Data
A catalogue record for this book is available from the British Library

Library of Congress Cataloging-in-Publication Data
A catalog record for this book is available from the Library of Congress

ISBN: 978-0-323-88478-5

For Information on all Academic Press publications
visit our website at https://www.elsevier.com/books-and-journals

Publisher: Andre G. Wolff
Acquisitions Editor: Linda Versteeg-Buschman
Editorial Project Manager: Megan Ashdown
Production Project Manager: Stalin Viswanathan
Cover Designer: Christian J. Bilbow

Typeset by MPS Limited, Chennai, India

Contents

List of contributors

Aswani R
School of Biosciences, Mahatma Gandhi University, Kottayam, India

Javeria Ayub
Department of Biosciences, COMSATS University Islamabad, Islamabad, Pakistan; Department of Microbiology, Balochistan University of Information Technology, Engineering and Management Sciences Quetta, Quetta, Pakistan

Jasim Basheer
Department of Cell Biology, Centre of the Region Haná for Biotechnological and Agricultural Research, Faculty of Science, Palacký University, Olomouc, Czech Republic

Joorie Bhattacharya
Genetic Gains, International Crops Research Institute for the Semi-Arid Tropics, Hyderabad, India; Department of Genetics, Osmania University, Hyderabad, India

Ashok Bhattacharyya
Department of Plant Pathology, Assam Agricultural University, Jorhat - 785013, Assam, India

Musfirah Bint-e-Mansoor
Department of Biosciences, COMSATS University Islamabad, Islamabad, Pakistan

Robin Chandra Boro
Department of Agricultural Biotechnology, Assam Agricultural University, Jorhat - 785013, Assam, India

Puneet Singh Chauhan
Microbial Technology Division, Council of Scientific and Industrial Research-National Botanical Research Institute (CSIR-NBRI), India; Academy of Scientific and Innovative Research (AcSIR), India

Elizabeth Cherian
Department of Botany, CMS College Kottayam, Kottayam, India

Vijaykant Dixit
Microbial Technology Division, Council of Scientific and Industrial Research-National Botanical Research Institute (CSIR-NBRI), India

Sougata Ghosh
Department of Microbiology, School of Science, RK University, Rajkot, India

Rashmi Gondi
Symbiosis School of Biological Sciences, Symbiosis International (Deemed University), Pune, India

Alisha Gupta
Department of Genetics, University of Delhi, South Campus, India

Fauzia Yusuf Hafeez
Department of Biosciences, COMSATS University Islamabad, Islamabad, Pakistan

Dibya Jyoti Hazarika
Department of Agricultural Biotechnology, Assam Agricultural University, Jorhat, India; DBT - North East Centre for Agricultural Biotechnology, Assam Agricultural University, Jorhat, India

Priya Jaiswal
Microbial Technology Division, Council of Scientific and Industrial Research-National Botanical Research Institute (CSIR-NBRI), India; Academy of Scientific and Innovative Research (AcSIR), India

Merilin Kakoti
Department of Agricultural Biotechnology, Assam Agricultural University, Jorhat, India

Jun Kang
School of Life Sciences, Tianjin University, Tianjin, P.R. China

Sristi Kar
Microbial Technology Division, Council of Scientific and Industrial Research-National Botanical Research Institute (CSIR-NBRI), India

Ajay Kumar
Postharvest Science, Agriculture Research Organization, Volcani Centre, Israel

Deepak Kumar
Department of Botany, Institute of Science, Banaras Hindu University, India

Sushil Kumar
Department of Biotechnology, Anand Agricultural University, Anand, India

Sijo Mathew
Department of Botany, Government College Kottayam, Kottayam, India

Shashank Kumar Mishra
Microbial Technology Division, Council of Scientific and Industrial Research-National Botanical Research Institute (CSIR-NBRI), India; Academy of Scientific and Innovative Research (AcSIR), India

Sankalp Misra
Microbial Technology Division, Council of Scientific and Industrial Research-National Botanical Research Institute (CSIR-NBRI), India; Academy of Scientific and Innovative Research (AcSIR), India

Zakira Naureen
Department of Biological Sciences and Chemistry, College of Arts and Science, University of Nizwa, Nizwa, Oman

Rahul Nitnavare
Division of Plant and Crop Sciences, School of Biosciences, University of Nottingham, Nottingham, United Kingdom; Department of Plant Sciences, Rothamsted Research, Harpenden, United Kingdom

Poonam Patel
Department of Biotechnology, Anand Agricultural University, Anand, India

Amrita Patil
Symbiosis School of Biological Sciences, Symbiosis International (Deemed University), Pune, India

Radhakrishnan E.K.
School of Biosciences, Mahatma Gandhi University, Kottayam, India

Meenakshi Raina
Government Degree College (Boys), India

Reedhu Raj
Department of Botany, Government Victoria College, Palakkad, India

Prasanna Rajan
Department of Botany, Government College Kottayam, Kottayam, India

Vinay Rale
Symbiosis School of Biological Sciences, Symbiosis International (Deemed University), Pune, India

A. Remakanthan
Department of Botany, University College, Thiruvananthapuram, Thiruvananthapuram, India

Sunil D. Saroj
Symbiosis School of Biological Sciences, Symbiosis International (Deemed University), Pune, India

Mythili Sathiavelu
School of Biosciences and Technology, Vellore Institute of Technology, Vellore, India

Gayathri Segaran
School of Biosciences and Technology, Vellore Institute of Technology, Vellore, India

Saranya Shankar
School of Biosciences and Technology, Vellore Institute of Technology, Vellore, India

Aishwarya Shankhapal
Genetic Gains, International Crops Research Institute for the Semi-Arid Tropics, Hyderabad, India

Pooja Singh
Symbiosis Centre for Waste Resource Management, Symbiosis International (Deemed University), India

Ayesha T. Tahir
Department of Biosciences, COMSATS University Islamabad, Islamabad, Pakistan

Roshmi Thomas
Sanatana Dharma College, Alappuzha, India

Introduction

From their initial discovery, over a 100 years ago, the evolution of the lines of work on endophytes has taken different turns. Some earlier workers realized that certain European forest species were hosts to a plethora of different fungal species but never really asked serious questions about what roles these organisms maybe playing in the biology of the host plant as simply finding and identifying those seemed to satisfy their curiosity. Then, by the middle of the last century, the term "endophyte" began taking a serious negative connotation as workers in New Zealand and the United States showed an unequivocal relationship between the *Epichloe* sp. endophytes in fescue grasses and certain disease manifestations in livestock grazing on these infested plants. The literature on this association became large and the work was comprehensive in virtually all respects. Endophytes were looked on as the bane of agriculture. Nevertheless, certain beneficial microbial endophytic interactions had been known and recognized for many years such as mycorrhizal fungi and various species of rhizobium, but these seemed to be viewed as isolated cases of beneficial host–plant microbe interactions and taken as exceptional instances of microbe–plant cooperation.

I entered this field in the early 1990s, which was 20 years after the anticancer drug taxol was discovered in the bark of a Pacific yew tree. This unique compound was then shown to have an extremely unique mode of action, which was the inhibition of the depolymerization of tubulin, which was lethal to rapidly dividing cancer cells. Taxol turned out to be the world's first billion-dollar anticancer drug. We realized that it might be possible that certain endophytic fungi, associated with yew tree species, may also be making taxol. This fact was published in an article in *Science* in 1993. Since then, many endophytic fungi have been shown to produce taxol. This work opened the floodgates for the imagination as others began to think of endophytes in a different and more compellingly and useful manner, namely, as novel sources of useful natural bioactive products.

Then, about 14 years ago, the National Institutes of Health instituted a huge initiative on the nature, role, and importance of the human microbiome. It was learned that the human gut contains more than 300 different microbial species, and in an ever-revealing manner, new information is constantly forthcoming on the biology of the microbiome and its role in the entire human condition. Obviously, the literature on the human gut led to the fact that plants also have a microbiome and endophytes are the major players in this scene. As human survival is totally dependent on the ability of man to grow and maintain plants; it behooves us to learn and understand all aspects of plant life. Thus the American Academy of Microbiology (AAM) released a special publication *How Microbes Can Help Feed the World.*

The piece examines the intimate relationships between microbes and plant agriculture including why plants need microbes, what types of microbes they need, and how they interact. A series of recommendations was made, including greater investment in plant microbiome research, to take on one or more grand challenges such as the characterization of the complete microbiome of one important crop plant and the establishment of formal processes for moving scientific discoveries from the laboratory to the field. Finally, a number of scientific challenges were listed and based on the current state of knowledge in this important field. All of the same questions apply to all of the other 350,000 plant species in the world and the work to be done is enormous.

Now, literally hundreds of endophytes have been studied for their novel bioactive natural products and other potentially useful aspects of their biology. This includes the role that endophytes may play in allowing their host plants to withstand the rigors of their environment, in crop productivity, and in crop protection. On top of this, serious works on the details of endophyte biology have dramatically increased starting with a complete examination of the entire endophyte population of a particular plant, to the seasonal variation in the endophyte populations and the specific roles that these organisms may play in the life cycle of the plant. It seems to be the case that all plants have endophytic populations that have the potential to be found in virtually all tissues and organs of the plant. In many cases, their identity and their exact role in the plant remains a mystery, but the possibilities for discovery are enormous.

Accepting many of the major challenges relating the novel and important recent advances in endophyte biology, the authors of this text—Radhakrishnan E.K (Mahatma Gandhi University, Kottayam, India), Aswani R. (Mahatma Gandhi University), and Ajay Kumar (Agricultural Research Organization, Volcani Center, Israel)—have selected a series of important and current topics in endophyte biology ranging from the colonization, diversity, and distribution of the endophytic community in plants to the techniques that are currently being applied to the identification of these microorganisms. Also, topics on the biocontrol mechanisms as well as antimicrobial metabolites associated with these microbes are included. Virtually all aspects of the role that endophytes may play in plants and in the induction of defense mechanisms with consideration for plant disease management and the potential use of endophytes in disease control are addressed. To me, it seems as if this text is an extremely useful contribution to the advanced literature on the importance as well as future challenges to be met in dealing with the impact of endophyte biology and its potential utility to mankind. This text addresses many of the topics and challenges that were set forth in the AAM publication as described earlier and, thus, is extremely timely.

Gary Strobel
Department of Plant Sciences, Montana State University, Bozeman, MT, United States

CHAPTER

Colonization, diversity, and distribution of endophytic microbial communities in different parts of plants

1

Jasim Basheer

Department of Cell Biology, Centre of the Region Haná for Biotechnological and Agricultural Research, Faculty of Science, Palacký University, Olomouc, Czech Republic

1.1 Introduction

Associations of plants and microbes are critical for them to overcome many challenges posed by the natural environment through their beneficial interactions (Afzal, Shinwari, Sikandar, & Shahzad, 2019; Santoyo, Moreno-Hagelsieb, del Carmen Orozco-Mosqueda, & Glick, 2016). The microbes can be divided into different groups based on their association with the plants as epiphytic (found externally on the leaf and stem surfaces); rhizospheric (found associated with roots in the rhizosphere region); or endophytic (which live inside the plant tissues) (Compant, Clément, & Sessitsch, 2010; John, Kumar, & Ge, 2020; Tichá, Illésová, & Hrbáčková, 2020). These associations have numerous mutualistic benefits to each other by providing help in mitigating various biotic and abiotic stresses (Miliute, Buzaite, Baniulis, & Stanys, 2015). Among these different classes of microbes, we are mainly focusing on the endophytic microbial communities being the most important contributor of plant growth.

First description of endophytes was done by a German botanist named Heinrich Friedrich Link in 1809 suggesting the presence of microbes inside plants. However, for a long time, the term "endophytae" was used only to describe pathogenic fungi due to their prevalence in diseased plants, thus, to the belief that healthy plants are sterile. But later in 19th century, reports of Galippe proved the presence of microbes inside the plants which pointed to study the beneficial role of soil microbes and their migration from soil to the plants (Hardoim, van Overbeek, & Berg, 2015). The term "endophyte" was coined by De Bary in 1866 using two Greek words "endon" which means "within," and "phyton" which means "plant" for defining microbes that grow inside plants which was modified later by other researchers from their specific observations (Chanway, 1997). One of the significant observations was the isolation of pure culture of nitrogen fixing bacteria (later classified as *Rhizobium leguminosarum*) from the root nodules of

Biocontrol Mechanisms of Endophytic Microorganisms. DOI: https://doi.org/10.1016/B978-0-323-88478-5.00008-0

Leguminosae plants by Martinus Willem Beijerinck in 1888 (Anyasi & Atagana, 2019). Another important one was from Albert Bernhard Frank who observed the mutualistic relationship between tree root and underground fungi which he later termed mycorrhiza (translates as fungus roots) where both parties benefited and, thus, the first observation of symbiosis (Domka, Rozpaądek, & Turnau, 2019). These observation enabled the researchers to study different aspects of these relationships which motivated Orlando Petrini in 1991 to modify the definition of endophytes to "all organisms inhabiting plant organs that at some time in their life cycle can colonize internal plant tissues without causing apparent harm to their host" and is evolving since then by updated observations (Petrini, 1991).

The importance of endophytes increases since nearly all species of the plants existing on the surface of the earth are colonized by one or more endophytic microbes (Ryan, Germaine, & Franks, 2008). Each one of these microbes thrives in the unique microenvironments based on the host's metabolite capacity, making the endophytes a treasure trove of biosynthetic potency (Jasim, Benny, Sabu, Mathew, & Radhakrishnan, 2016). Due to the diverse biosynthetic capabilities, endophytes are reported to support the plant in growth enhancement and resistance to different stress factors (Chanway, Shishido, & Nairn, 2000; Cipollini, Rigsby, & Barto, 2012; Hardoim, van Overbeek, & van Elsas, 2008; Mei & Flinn, 2010; Rosenblueth & Martínez-Romero, 2006). Endophytes mostly originate from the rhizosphere and colonize the plant in some stages of the plant growth and share almost all the growth-promoting properties exhibited by rhizosphere microbes, thus, considered a subclass of rhizosphere bacteria by some researchers.

1.2 The rhizosphere as a microbial contributor

The rhizosphere is the area of soil that is in close contact with the plant root which has a dynamic role in maintaining the health and physiology of the plants and the microbiome associated with it (Singh, Singh, & Kumar, 2017; Singh, Singh, Singh, & Kumar, 2018; Kumar, Droby, Singh, Singh, & White, 2020). The process of colonization is highly complex which involves a cascade of processes from both host plant and microbial counterpart. Plants initiate the process of selection and attraction by communication with its associated microbes using chemical molecules present in the root exudates (de Weert et al., 2002; Rosenblueth & Martínez-Romero, 2006). The main components present in the root exudates are proteins, amino acids, and organic acids that play the key role in the process of selecting the microbial communities associated with the host plant. Oxalates are found to have such use in different studies which concludes that the levels of oxalates can even dictate the concentration of microbes attached to the plant. Microbes with defective oxalate utilization mechanism were found to have significant loss in colonization capability when studied in *Burkholderia phytofirmans* PsJN (Esmaeel, Miotto, & Rondeau, 2018).

Microbes use compounds from the exudates as quorum sensing molecules for the specific attachment to the host plant. Compounds with quorum sensing capability play an important role in the endophyte colonization and extending their support to the host. *B. phytofirmans* PsJN which has a knockdown quorum sensing gene was found to have lost the capability of colonizing and growth enhancement in *Arabidopsis thaliana* which confirms the importance of these compounds in the process (Esmaeel, Miotto, & Rondeau, 2018). *N*-Acyl-homoserine lactones (AHLs) are commonly found in Gram-negative bacteria and cyclic peptides are found in Gram-positive bacteria which act as quorum sensing molecules. The action of bacterial AHLs to trigger specific responses was the first to report in *Phaseolus vulgaris* and *Medicago truncatula* (Joseph & Phillips, 2003; Mathesius, Mulders, & Gao, 2003). Early responses in plants to these signals are currently believed to have significance either in recognizing pathogens to prepare themselves for the attack or to welcome mutualistic microbes for colonizing. There is a significant difference in the accumulation of over 150 different proteins as a response to AHLs *Sinorhizobium meliloti* (symbiotic bacteria) and *Pseudomonas aeruginosa* (pathogenic bacteria). AHL (with C4 and C6 side chains) produced by *Serratia liquefaciens* MG1 was found to have induction of systemic resistance proteins after inoculation on the roots of tomato plants sensing the presence of pathogens (Schikora, Schenk, & Hartmann, 2016). These reports confirm that plants are likely to be involved in quorum sensing for recognizing the difference between pathogens and beneficial microbes (Fig. 1.1). It is also demonstrated that some plant extracts have the capability of quorum quenching which helps them to protect against pathogens. There are reports which suggest that the capacity of endophytes to synthesize has LuxR homologs from the evidence that LuxR-LuxI type quorum sensing gene pairs isolated from several microbes of endophytic origin confirming their role in plant microbe crosstalk (Kandel, Joubert, & Doty, 2017).

1.3 Attachment and colonization of endophytes

Attachment of microbe to the plant surface is the first and foremost step in the process of microbial colonization. The chemo-attracted microbial population in the rhizosphere migrates to the plant root surface and initiates colonizing potential entry sites like site of emergence of lateral root or sites of other injuries. In the case of bacteria, their cells synthesize exopolysaccharides (EPS) which helps in the attachment onto the root surface. In *Gluconacetobacter diazotrophicus* Pal5, EPS was found to be an important factor for its attachment and colonization to the root surface. After attachment, EPS prevents the cells from oxidative burst caused as an immune response to the invasion. EPS purified from the same strain (Pal5) was even capable of inducing the colonization and biofilm production in mutant strains (Meneses, Gonçalves, & Alquéres, 2017). Reports of Balsanelli, Tuleski, anf de Baura, (2013) demonstrated that, in maize, *N*-acetyl glucosamine

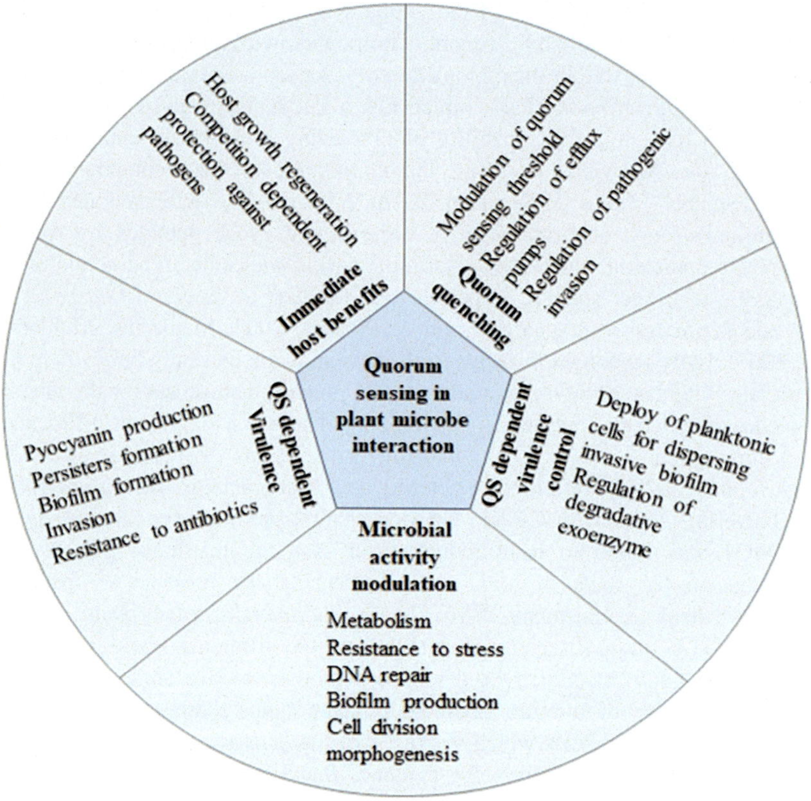

FIGURE 1.1

Different roles of Quorum sensing compounds played during plant microbe interactions.

residue of Lipopolysaccharides (LPS) binds to the root lectins which is crucial for the attachment and colonization of bacteria inside the roots. Endophytic fungi colonize the host either one of two methods: (1) vertical transmission in which the fungi in the maternal plants are transmitted to the offspring through the progeny seeds. The spores of the fungi present in the seeds will start to germinate and colonize once the seed starts to germinate (Gagic, Faville, & Zhang, 2018; Hodgson, de Cates, & Hodgson, 2014). (2) Horizontally transmitted ones mostly transmit via spores/hyphal fragmentation with the help of either biotic means by herbivores like insects or by abiotic means like wind and rain (Wiewióra, Żurek, & Pańka, 2015). The beginning of colonization after the attachment to the plant cell is the formation of the appressorium-like structures that help in the migration and colonization to the internal plant tissues (Esparza-Reynoso, Pelagio-Flores, & López-Bucio, 2020). During this process, unlike the pathogenic invasion, the cell integrity is not disturbed, which was proved by the microscopic observations of *Trichoderma* colonization in tomato roots (Yan & Khan, 2021).

1.4 Mechanisms involved in endophytic colonization

Endophytic microbes were highly successful in coevolving with their host plants which helped them to get equipped with all the necessary traits needed to internalize, colonize, and translocate into the intercellular spaces of the plant. Even though plant and endophytic interaction stays on the beneficial side, plants activate their immune system when sensing microbial presence (Zipfel & Oldroyd, 2017). They use innate immune responses for recognizing the signal molecules to trigger the defense mechanisms. The defense mechanism involves either microbe-associated molecular patterns (MAMPs) by recognizing the cell surface–localized pattern recognition receptors which activate the MAMP-triggered immunity; or by recognizing the molecules synthesized by microbes (effectors molecules) with the help of intracellular receptors which will activate the effector-triggered immunity (López-Gómez, Lara-Herrera, & Bravo-Lozano, 2012; Mendoza-Mendoza, Zaid, & Lawry, 2018; Zamioudis & Pieterse, 2012). Components of bacterial cell surfaces are distinct and unique from that of pathogens. MAMP was found to be triggered by the presence of bacterial flagellum (Butchart, Scharlemann, & Evans, 2012). However, studies conducted on flagellin-sensing system (flg22-Flagellin Sensing 2) derived from endophytic *B. phytofirmans* and pathogenic *P. aeruginosa* or *Xanthomonas campestris* in grapevine found to recognize the flagellin in a differential manner suggesting the recognition of endophytes (Trdá, Fernandez, & Boutrot, 2014). Studies of Chen, Marszałkowska, and Reinhold-Hurek (2020) explain the possibility of downregulated Mitogen-Activated Protein Kinase (MAPK) signaling pathway due to the slight upregulation of four different protein phosphatase 2C (PP2C) homologs during *Azoarcus* colonization (PGPB); however, in the case of Xoo infection, another PP2C homolog was downregulated leading to upregulated MAPK signaling leading to stronger PTI response when studied in detail in rice plants demonstrating its role in both beneficial and pathogenic interactions.

Another most important groups with host immune modulation are bacterial protein secretion systems (SSs) consist of large protein complexes that include Type I SS ~ Type VI SS, Sec and Tat in Gram-negative and Sec, Tat, secA2, Sortase, Injectosome, and Type VII SS in Gram-positive bacteria (Green & Mecsas, 2016; Tseng, Tyler, & Setubal, 2009). Among these, T3SS and T4SS are either present in low concentration or absent in endophytic bacteria, thus, have mild or no defense response, whereas present in pathogens leading to a stronger response (Green & Mecsas, 2016). There are lot of genomic and metagenomic studies concentrated on the abundance of T3SS and T4SS genes from endophytes, including *Herbaspirillum frisingense* GSF30(T), *G. diazotrophicus* PAI5, *Azoarcus* sp. BH72, *Klebsiella pneumoniae* 342, *Azospirillum* sp. B510, from diverse plants which lack the presence of T3SS genes; *Herbaspirillum* sp. lacks T4SS; and *Azoarcus* sp. strain BH72 lacks both T3SS and T4SS (Juhas, Van Der Meer, & Gaillard, 2009; Piromyou, Buranabanyat, & Tantasawat,

2011; Reinhold-Hurek & Hurek, 2011; Straub, Zabel, & Gilfillan, 2013). However, there are exceptions also found by having both the sets of genes present in *Bradyrhizobium* sp. SUTN9-2 isolated from *Aeschynomene americana* L., which was found crucial in colonization process (Piromyou, Buranabanyat, & Tantasawat, 2011).

Production of reactive oxygen species (ROS) is a nonspecific defense mechanism used by plants which will induce hypersensitive response thereby leading to programmed cell death against biotrophic pathogens (Apel & Hirt, 2004). Most of the endophytes protect themselves with the help of EPS, and the residual ROS is neutralized with the help of ROS-scavenging enzymes (Alquéres, Meneses, & Rouws, 2013; You & Chan, 2015). Quantification and diversity studies on the genes responsible to produce superoxide dismutase and glutathione reductase in metagenome of the endophytic bacterial communities in rice roots, *Enterobacter* sp. 638, *G. diazotrophicus*, etc. are significantly higher than free living and are found to be essential for colonization process (Liu, Carvalhais, & Crawford, 2017; Sessitsch, Hardoim, & Döring, 2012; Taghavi, van der Lelie, & Hoffman, 2010). It is also found that the transcript levels of these genes are upregulated when present inside the plants confirming their role in successful colonization (Liu, Carvalhais, & Crawford, 2017).

Another important mechanism influences the colonization of endophytes are the phytohormones because of their role in the plant defense signaling pathways. They have the capacity to regulate the structure of plant-associated microbiome involved in beneficial interactions, plant nutrition, and defense-related interactions (Liu, Carvalhais, & Crawford, 2017). Iniguez, Dong, & Carter, 2005) demonstrated the role of ethylene signaling pathway (ET) activation resulted in the suppression of colonization of both *K. pneumoniae* 342 (Kp342) (PGPB) and *Salmonella enterica* serovar *Typhimurium* (human enteric pathogen) in wild-type *M. truncatula* and when inoculated in ET insensitive *M. truncatula* leads to extensive colonization by Kp342. Similar studies are also conducted in the case of Jasmonic acid (JA), which suggests that its level is downregulated during colonization process to maintain plant favorable microbial densities inside the plant tissues. During early nodulation stages in *Lotus japonicus* suppression of JA-signaling pathway is observed by Nakagawa and Kawaguchi (2006). Similar results were also observed during the colonization of *Azospirillum brasilense* 245 on Arabidopsis roots where the JA-signaling was strongly downregulated (Spaepen, Versées, & Gocke, 2007). It is not the case in the case of rice varieties like Japonica and Indica rice cultivar where RT-PCR and proteomic analyses confirm the strong upregulation of markers like *OsJAR1* and *OsJAmyb* during the endophytic colonization of *Azoarcus olearius*, *Azospirillum* B510, and *G. diazotrophicus* (Drogue, Sanguin, & Borland, 2013). Even though we have strong indication that the different defense mechanisms in the host plants are differentially modulated, the total mechanism involved is not yet clear (Plett & Martin, 2018). Endophytic microbes isolated from different plants are summarized in Table 1.1.

Table 1.1 Consolidation of endophytes isolation from different hosts.

Host plant	Endophytic microbe	Reference
Colobanthus quitensis	*Aspergillus, Cadophora, Davidiella, Entrophospora, Fusarium, Geomyces, Gyoerffyella, Microdochium, Mycocentrospora, Phaeosphaeria*	Rosa, Almeida Vieira, Santiago, and Rosa (2010)
Dendrobium loddigesii	*Acremonium, Alternaria, Ampelomyces, Bionectria, Cercophora, Chaetomella, Cladosporium, Colletotrichum, Davidiella, Fusarium, Lasiodiplodia, Nigrospora, Paraconiothyrium, Pyrenochaeta, Sirodesmium, Verticillium, Xylaria*	Chen, Wang, and Li (2010)
Hevea brasiliensis	*Alternaria, Arthrinium, Cladosporium, Endomelanconiopsis, Entonaema, Fimetariella, Fusarium, Guignardia, Penicillium, Perisporiopsis, Pestalotiopsis, Nigrospora, Trichoderma, Umbelopsis*	Gazis and Chaverri (2010)
Magnolia liliifera	*Colletotrichum, Corynespora, Fusarium, Guignardia, Leptosphaeria, Phomopsis*	Promputtha, Hyde, McKenzie, Peberdy, & Lumyong, 2010
Theobroma cacao	*Acremonium, Arthrinium, Aspergillus, Clonostachys, Colletotrichum, Coniothyrium, Curvularia, Cylindrocladium, Fusarium, Gliocladium, Lasiodiplodia, Myrothecium, Paecilomyces, Penicillium, Pestalotiopsis, Phoma, Septoria, Talaromyces, Tolypocladium, Trichoderma, Verticillium*	Hanada, Pomella, and Costa (2010)
Theobroma grandiflorum	*Acremonium, Asteromella, Lasiodiplodia, Pestalotiopsis, Phoma*	
Acer truncatum	*Alternaria, Ascochytopsis, Bipolaris, Cladosporium, Clypeopycnis, Colletotrichum, Coniothyrium, Coprinellus, Cryptodiaporthe, Cyclothyrium, Diaporthe, Discula, Drechslera, Epicoccum, Fusarium, Geniculosporium, Gibberella, Glomerella, Guignardia, Helminthosporium, Leptosphaeria, Melanconis, Microdiplodia, Microsphaeropsis, Nigrospora, Paraconiothyrium, Phoma, Phomopsis, Podosordaria, Preussia, Pseudocercosporella, Sclerostagonospora, Septoria, Sirococcus, Xylaria*	Sun, Guo, and Hyde (2011)
Aquilaria sinensis	*Chaetomium, Cladosporium, Coniothyrium, Epicoccum, Fusarium, Hypocrea, Lasiodiplodia, Leptosphaerulina, Paraconiothyrium, Phaeoacremonium, Phoma, Pichia, Rhizomucor, Xylaria*	Cui, Guo, and Xiao (2011)
Dendrobium devonianum	*Acremonium, Arthrinium, Cladosporium, Fusarium, Glomerella, Leptosphaerulina, Phoma, Pestalotiopsis, Rhizopus, Trichoderma, Xylaria*	Xing, Chen, & Cui, 2011
Dendrobium thyrsiflorum	*Alternaria, Colletotrichum, Epicoccum, Fusarium, Glomerella, Leptosphaerulina, Pestalotiopsis, Phoma, Rhizopus, Xylaria*	
Ledum palustre	*Arthrinium, Fusarium, Lecythophora, Penicillium, Sordaria, Sphaeriothyrium*	Tejesvi, Kajula, Mattila, and Pirttilä (2011)
Lippia sidoides	*Alternaria, Colletotrichum, Corynespora, Curvularia, Drechslera, Fusarium, Guignardia, Microascus, Peacilomyces, Periconia, Phoma, Phomopsis*	de Siqueira, Conti, de Araújo, and Souza-Motta (2011)
Mansoa alliacea	*Alternaria, Aspergillus, Chaetomium, Curvularia, Fusarium, Penicillium, Phomopsis, Rhizoctonia, Stenella, Trichoderma*	Kharwar, Verma, & Mishra, 2011

(*Continued*)

Table 1.1 Consolidation of endophytes isolation from different hosts. *Continued*

Host plant	Endophytic microbe	Reference
Pinus halepensis	*Alternaria, Aureobasidium, Camarosporium, Chaetomium, Chalastospora, Davidiella, Diplodia, Epicoccum, Fusarium, Gremmeniella, Leptosphaeria, Lophodermium, Naemacyclus, Paraconiothyrium, Penicillium, Pestalotiopsis, Peziza, Phaeomoniella, Phaeosphaeria, Phoma, Phomopsis, Pleospora, Preussia, Pyronema, Sordaria, Trichoderma, Tryblidiopsis, Truncatella, Ulocladium, Xylaria*	Botella and Diez (2011)
Solanum cernuum	*Arthrobotrys, Bipolaris, Botryosphaeria, Candida, Cercospora, Colletotrichum, Coprinellus, Cryptococcus, Curvularia, Diatrypella, Edenia, Eutypella, Fusarium, Glomerella, Leptosphaeria, Mucor, Petriella, Phoma, Meyerozyma, Flavodon, Hapalopilus, Hohenbuehelia, Kwoniella, Oudemansiella, Phanerochaete, Phlebia, Phlebiopsis, Schizophyllum*	Vieira, Hughes, & Gil, 2012
Taxus globosa	*Alternaria, Aspergillus, Annulohypoxylon, Cercophora, Cochliobolus, Colletotrichum, Conoplea, Coprinellus, Daldinia, Hypocrea, Hypoxylon, Lecythophora, Letendraea, Massarina, Nigrospora, Penicillium, Phialophorophoma, Phoma, Polyporus, Sporormia, Trametes, Trichophaea, Xylaria, Xylomelasma*	Rivera-Orduña, Suarez-Sanchez, & Flores-Bustamante, 2011
Tylophora indica	*Alternaria, Chaetomium, Colletotrichum, Nigrospora, Thielavia*	Kumar, Kaushik, and Edrada-Ebel (2011)
Acer tataricum	*Alternaria, Cladosporium, Epicoccum, Fusarium, Neurospora, Penicillium, Phoma, Phomopsis, Trichoderma*	Qi, Jing, and Zhan (2012)
Cinnamomum camphora	*Alternaria, Arthrinium, Arthrobotrys, Aspergillus, Chaetomium, Chaetophoma, Cladosporium, Curvularia, Drechslera, Gliomastix, Humicola, Nigrospora, Penicillium, Periconia, Pestalotiopsis, Phacidium, Phomopsis, Phyllosticta, Stachybotrys, Trichoderma*	Kharwar, Maurya, & Verma, 2012
Echinacea purpurea	*Ceratobasidium, Cladosporium, Colletotrichum, Fusarium, Glomerella, Mycoleptodiscus*	Rosa, Tabanca, & Techen, 2012
Ginkgo biloba	*Alternaria, Cladosporium, Colletotrichum, Fusarium, Pestalotiopsis, Peyronellaea, Phoma, Phomopsis, Phyllosticta*	Thongsandee, Matsuda, and Ito (2012)
Holcoglossum flavescens	*Alternaria, Cladosporium, Didymella, Epulorhiza, Fusarium*	Tan, Chen, & Wang, 2012
Nyctanthes arbor-tristis	*Acremonium, Alternaria, Aspergillus, Chaetomium, Cladosporium, Colletotrichum, Drechslera, Humicola, Fusarium, Nigrospora, Penicillium, Phomopsis, Rhizoctonia*	Gond, Mishra, & Sharma, 2012
Opuntia ficus-indica	*Acremonium, Aspergillus, Cladosporium, Fusarium, Monodictys, Nigrospora, Penicillium, Pestalotiopsis, Phoma, Phomopsis, Tetraploa, Xylaria*	Bezerra, Santos, and Svedese (2012)
Panicum virgatum	*Alternaria, Ampelomyces, Aspergillus, Candida, Cladosporium, Colletotrichum, Epicoccum, Fusarium, Kretzschmaria, Monographella, Nemania, Nigrospora, Nodulisporium, Ophiosphaerella, Phaeosphaeria, Phoma, Phomopsis, Preussia, Schizosaccharomyces, Septoria, Stagonospora, Xylaria*	Kleczewski, Bauer, & Bever, 2012

(*Continued*)

Table 1.1 Consolidation of endophytes isolation from different hosts. *Continued*

Host plant	Endophytic microbe	Reference
Picea abies	*Acephala*, *Chalara*, *Cistella*, *Cladosporium*, *Entomocorticium*, *Fomitopsis*, *Lophodermium*, *Mollisia*, *Mycena*, *Neonectria*, *Ophiostoma*, *Phacidiopycnis*, *Phacidium*, *Phialocephala*, *Rhizoscyphus*, *Rhizosphaera*, *Sarea*, *Scleroconidioma*, *Sirococcus*, *Valsa*, *Xylomelasma*, *Zalerion*	Koukol, Kolařík, Kolářová, and Baldrian (2012)
Piper hispidum	*Alternaria*, *Bipolaris*, *Colletotrichum*, *Glomerella*, *Guignardia*, *Lasiodiplodia*, *Marasmius*, *Phlebia*, *Phoma*, *Phomopsis*, *Schizophyllum*	Orlandelli, Alberto, Rubin Filho, and Pamphile (2012)
Reynoutria japonica	*Alternaria*, *Arthrinium*, *Bionectria*, *Colletotrichum*, *Didymella*, *Glomerella*, *Nigrospora*, *Pestalotiopsis*, *Phoma*, *Phomopsis*, *Phyllosticta*, *Septoria*, *Xylaria*	Kurose, Furuya, and Tsuchiya (2012)
Sapindus saponaria	*Alternaria*, *Cochliobolus*, *Curvularia*, *Diaporthe*, *Phoma*, *Phomopsis*	García, Rhoden, Filho, Nakamura, & Pamphile, 2012
Stryphnodendron adstringens	*Alternaria*, *Arthrobotrys*, *Aspergillus*, *Botryosphaeria*, *Cladosporium*, *Colletotrichum*, *Coniochaeta*, *Cytospora*, *Diaporthe*, *Guignardia*, *Fimetariella*, *Massarina*, *Muscodor*, *Neofusicoccum*, *Nigrospora*, *Paraconiothyrium*, *Penicillium*, *Pestalotiopsis*, *Phomopsis*, *Preussia*, *Pseudofusicoccum*, *Sordaria*, *Sporormiella*, *Trichoderma*, *Xylaria*	Carvalho, Gonçalves, & Pereira, 2012
Tinospora sinensis	*Acremonium*, *Alternaria*, *Aspergillus*, *Botryosphaeria*, *Botrytis*, *Cladosporium*, *Chaetomium*, *Colletotrichum*, *Curvularia*, *Drechslera*, *Emericella*, *Fusarium*, *Guignardia*, *Humicola*, *Monilia*, *Nigrospora*, *Penicillium*, *Pseudofusicoccum*, *Trichoderma*, *Veronaea*	Mishra, Gond, & Kumar, 2012
Trichilia elegans	*Cordyceps*, *Diaporthe*, *Phomopsis*	Rhoden, Garcia, & Rubin Filho, 2012
Vitis vinifera	*Absidia*, *Alternaria*, *Aspergillus*, *Aureobasidium*, *Botrytis*, *Cladosporium*, *Epicoccum*, *Fusarium*, *Mortierella*, *Mucor*, *Penicillium*, *Pithomyces*, *Rhizopus*, *Trichoderma*, *Umbelopsis*, *Zygorhynchus*	Pancher, Ceol, & Corneo, 2012
Cannabis sativa	*Aspergillus*, *Chaetomium*, *Eupenicillium*, *Penicillium*	Kusari, Kusari, Spiteller, and Kayser (2013)
Glycine max	*Alternaria*, *Ampelomyces*, *Annulohypoxylon*, *Arthrinium*, *Cercospora*, *Chaetomium*, *Cladosporium*, *Cochliobolus*, *Colletotrichum*, *Curvularia*, *Davidiella*, *Diaporthe*, *Didymella*, *Epicoccum*, *Eutypella*, *Fusarium*, *Gibberella*, *Guignardia*, *Leptospora*, *Magnaporthe*, *Myrothecium*, *Nectria*, *Neofusicoccum*, *Nigrospora*, *Ophiognomonia*, *Paraconiothyrium*, *Phaeosphaeriopsis*, *Phoma*, *Phomopsis*, *Rhodotorula*, *Sporobolomyces*, *Stemphylium*, *Xylaria*	de Souza Sebastianes, Romão-Dumaresq, & Lacava, 2013
Jatropha curcas	*Alternaria*, *Chaetomium*, *Colletotrichum*, *Fusarium*, *Guignardia*, *Nigrospora*	Kumar and Kaushik (2013)
Kigelia africana	*Alternaria*, *Aspergillus*, *Botryodiplodia*, *Chaetomium*, *Colletotrichum*, *Curvularia*, *Drechslera*, *Fusarium*, *Mucor*, *Nigrospora*, *Nodulisporium*, *Penicillium*, *Pestalotiopsis*, *Phoma*, *Phomopsis*, *Rhizopus*, *Trichoderma*	Maheswari and Rajagopal (2013)
Panax ginseng	*Aspergillus*, *Cladosporium*, *Engyodontium*, *Fusarium*, *Penicillium*, *Plectosphaerella*, *Verticillium*	Wu, Yang, You, and Li (2013)

(*Continued*)

Table 1.1 Consolidation of endophytes isolation from different hosts. *Continued*

Host plant	Endophytic microbe	Reference
Stellera chamaejasme	*Acremonium*, *Alternaria*, *Aporospora*, *Ascochyta*, *Aspergillus*, *Bionectria*, *Botryotinia*, *Cadophora*, *Colletotrichum*, *Dothiorella*, *Emericellopsis*, *Eucasphaeria*, *Eupenicillium*, *Fusarium*, *Geomyces*, *Ilyonectria*, *Leptosphaeria*, *Mucor*, *Nectria*, *Neonectria*, *Paecilomyces*, *Paraphoma*, *Penicillium*, *Schizophyllum*, *Scytalidium*, *Sordaria*, *Sporormiella*	Jin, Yan, & Liu, 2013
Taxus x media	*Alternaria*, *Colletotrichum*, *Gibberella*, *Glomerella*, *Guignardia*, *Nigrospora*, *Phoma*, *Phomopsis*	Xiong, Wang, Hao, and Wang (2013)
Brassica napus	*Acremonium*, *Alternaria*, *Arthrinium*, *Aspergillus*, *Aureobasidium*, *Botrytis*, *Chaetomium*, *Clonostachys*, *Cryptococcus*, *Dioszegia*, *Dothidea*, *Dothiorella*, *Epicoccum*, *Fusarium*, *Guignardia*, *Hypoxylon*, *Leptosphaeria*, *Macrophomina*, *Nigrospora*, *Penicillium*, *Periconia*, *Phoma*, *Rhizoctonia*, *Rhizopus*, *Simplicillium*, *Sporidiobolus*, *Sporobolomyces*	Zhang, Xu, & Yang, 2014
Viola philippica	*Micromonospora violae*	
Pinus wallichiana	*Alternaria*, *Anthostomella*, *Aspergillus*, *Cadophora*, *Cladosporium*, *Cochliobolus*, *Coniochaeta*, *Coniothyrium*, *Epicoccum*, *Fimetariella*, *Fusarium*, *Geopyxis*, *Lecythophora*, *Leptosphaeria*, *Lophiostoma*, *Lophodermium*, *Microdiplodia*, *Neurospora*, *Nigrospora*, *Paraconiothyrium*, *Penicillium*, *Pestalotiopsis*, *Phoma*, *Phomopsis*, *Preussia*, *Pseudoplectania*, *Rachicladosporium*, *Rosellinia*, *Sclerostagonospora*, *Sordaria*, *Sporormiella*, *Therrya*, *Tricharina*, *Trichoderma*, *Thielavia*, *Tritirachium*, *Truncatella*, *Xylaria*	Qadri, Rajput, & Abdin, 2014
Alnus firma	*Bacillus* sp.	Shin, Shim, & You, 2012
Alyssum bertolonii	*Arthrobacter, Bacillus, Curtobacterium, Leifsonia, Microbacterium, Paenibacillus, Pseudomonas, Staphylococcus*	Barzanti, Ozino, and Bazzicalupo (2007)
Calystegia soldanella	*Acinetobacter, Arthrobacter, Chryseobacterium, Curtobacterium, Enterobacter, Microbacterium, Pantoea, Pedobacter, Pseudomonas, Stenotrophomonas*	Park, Jung, & Lee, 2005
Elymus mollis	*Acinetobacter, Arthrobacter, Chryseobacterium, Enterobacter, Exiguobacterium, Flavobacterium, Klebsiella, Pedobacter, Pseudomonas, Stenotrophomonas*	
Cannabis sativa	*Acinetobacter gyllenbergii, Acinetobacter nosocomialis, Acinetobacter parvus, Acinetobacter pittii, Bacillus anthracis, Chryseobacterium* sp., *Enterobacter asburiae, Enterococcus casseliflavus, Nocardioides albus, Nocardioides kongjuensis, Pantoea vagans, Planomicrobium chinense, Pseudomonas taiwanensis, Rhizobium radiobacter, Streptomyces eurocidicus, Xanthomonas gardneri*	Afzal, Iqrar, and Shinwari (2015)
Commelina communis	*Arthrobacter, Bacillus, Bacillus pumilus, Herbaspirillum, Microbacterium, Sphingomonas*	Sun, Zhang, & He, 2010
Elsholtzia splendens	*Acinetobacter calcoaceticus, Acinetobacter junii, Bacillus, Bacillus firmus, Bacillus megaterium, Burkholderia, Exiguobacterium aurantiacum, Micrococcus luteus, Moraxella, Paracoccus, Serratia marcescens*	

(*Continued*)

Table 1.1 Consolidation of endophytes isolation from different hosts. *Continued*

Host plant	Endophytic microbe	Reference
Cressa cretica, Salicornia brachiate, Suaeda nudiflora, Sphaeranthus indicus	*Acinetobacter, Arthrobacter, Bacillus, Kocuria, Oceanobacillus, Paenibacillus, Pseudomonas, Virgibacillus*	Sanjay, Purvi, Meghna, and Rao (2014)
Dodonaea viscosa	*Agrococcus terreus, Bacillus cereus, Bacillus idriensis, Bacillus simplex, Bacillus subtilis, Brevundimonas subvibrioides, Inquilinus limosus, Microbacterium trichothecenolyticum, Pseudomonas geniculata, Pseudomonas taiwanensis, Rhizobium huautlense, Streptomyces alboniger, Streptomyces caeruleatus, Xanthomonas sacchari, Xanthomonas translucens*	Afzal, Iqrar, Shinwari, and Yasmin (2017)
Halimione portulacoides	*Altererythrobacter, Hoeflea, Labrenzia, Marinilactibacillus, Microbacterium, Salinicola, Vibrio*	Fidalgo, Henriques, Rocha, Tacão, & Alves, 2016
Mammillaria fraileana (cactus)	*Azotobacter vinelandii, Bacillus megaterium, Enterobacter sakazakii, Pseudomonas putida*	Lopez, Bashan, and Bacilio (2011)
Miscanthus sinensis	*Pseudomonas koreensis*	Babu, Kim, and Oh (2013)
Pinus sylvestris	*Bacillus thuringiensis*	
Noccaea caerulescens	*Agreia, Arthrobacter, Bacillus, Kocuria, Microbacterium, Stenotrophomonas, Variovorax*	Visioli, D'Egidio, Vamerali, Mattarozzi, & Sanangelantoni, 2014
Pachycereus pringlei (cardon cactus)	*Acinetobacter, Bacillus, Citrobacter, Klebsiella, Paenibacillus, Pseudomonas, Staphylococcus*	Puente, Li, and Bashan (2009)
Pinus contorta (Lodgepole pine)	*Bacillus, Brevibacillus, Brevundimonas, Cellulomonas, Kocuria, Paenibacillus, Pseudomonas*	Bal, Anand, Berge, and Chanway (2012)
Thuja plicata (Red cedar)	*Arthrobacter, Bacillus, Paenibacillus, Pseudomonas, Streptoverticillium*	
Polygonum pubescens	*Rahnella* sp. *JN6*	He, Ye, & Yang, 2013
Prosopis strombulifera	*Achromobacter xylosoxidans, Bacillus licheniformis, Bacillus pumilus, Bacillus subtilis, Brevibacterium halotolerans, Lysinibacillus fusiformis, Pseudomonas putida*	Sgroy, Cassán, & Masciarelli, 2009
Salix caprea	*Bacillus, Frigoribacterium, Frondihabitans, Kocuria, Leifsonia, Massilia, Methylobacterium, Microbacterium, Ochrobactrum, Pedobacter, Plantibacter, Rhodococcus, Sphingomonas, Spirosoma, Subtercola*	Kuffner, De Maria, & Puschenreiter, 2010
Sedum alfredii Hance	*Burkholderia, Sphingomonas, Variovorax*	Zhang, Lin, & Zhu, 2013
Wild prairie plants	*Cellulomonas, Clavibacter, Curtobacterium, Microbacterium*	Zinniel, Lambrecht, & Harris, 2002
Limonium sinense	*Kineococcus endophytica, Streptomyces* sp., *Glutamicibacter halophytocola* sp. nov.	Bian, Feng, and Qin (2012), Feng, Wang, & Bai, 2017
Lupinus termis	*Actinoplanes missouriensis*	El-Tarabily (2003)
Triticum aestivum	*Streptomyces* sp., *Microbispora* sp., *Micromonospora* sp., *Nocardioides* sp.	Coombs and Franco (2003)
Curcuma phaeocaulis	*Streptomyces phytohabitans*	Bian, Qin, and Yuan (2012)
Solanum melongena	*Nonomuraea solani*	Wang, Zhao, and Liu (2013)
Glycine max	*Actinoplanes hulinensis, Streptomyces harbinensi, Wangella harbinensis*	Liu, Liu, Yu, Yan, & Qi, 2012, Jia, Liu, & Wang, 2013, Shen, Liu, & Wang, 2013
Oryza sativa (Thai jasmine rice plant)	*Actinoallomurus oryzae* sp.	Indananda, Thamchaipenet, & Matsumoto, 2011
Acacia auriculiformis	*Actinoallomurus acaciae, Streptomyces* sp., *Actinoallomurus coprocola, Amycolatopsis tolypomycina, Kribbella* sp., *Microbispora mesophila*	Bunyoo, Duangmal, Nuntagij, & Thamchaipenet, 2009, Thamchaipenet, Indananda, & Bunyoo, 2010

(*Continued*)

Table 1.1 Consolidation of endophytes isolation from different hosts. *Continued*

Host plant	Endophytic microbe	Reference
Piper nigrum	*Bacillus firmus, Paenibacillus dendritiformis, Pseudomonas hibiscicola, Bordetella avium, Bacillus flexus, Stenotrophomonas maltophilia, Bacillus licheniformis, Klebsiella pneumoniae, Enterobacter cloacae, Enterobacter* sp., *Klebsiella oxytoca*, and *Pantoea agglomeran*	Jasim, John Jimtha, Jyothis, and Radhakrishnan (2013)
Zingiber officinale	*Bacillus barbaricus, Pseudomonas putida, Stenotrophomonas maltophilia*, and *Staphylococcus pasteuri*	Jasim, Joseph, John, Mathew, & Radhakrishnan, 2014
Curcuma longa	*Paenibacillus favisporus* and *Paenibacillus* sp.	Aswathy, Jasim, Jyothis, and Radhakrishnan (2012)
Capsicum annuum	*Bacillus* sp.	Jasim, Mathew, and Radhakrishnan (2016)
Bacopa monnieri	*Bacillus* sp., *Bacillus mojavensis*, and *Aspergillus* sp.	Jasim, Benny, Sabu, Mathew, & Radhakrishnan, 2016, Jasim, Sreelakshmi, Mathew, and Radhakrishnan (2016), Jasim, Daya, & Sreelakshmi, 2017
Eucalyptus microcarpa	*Promicromonospora endophytica*	Kaewkla and Franco (2013)
Elettaria cardamomum	*Pantoea* sp., *Polaromonas* sp., *Pseudomonas* sp., and *Ralstonia* sp.	Jasim, Anish, Shimil, Jyothis, & Radhakrishnan, 2015
Camptotheca acuminata	*Blastococcus endophyticus, Plantactinospora endophytica*	Zhu, Zhao, and Zhao (2012), Zhu, Zhang, and Qin (2013)
Dracaena cochinchinensis	*Streptomyces* sp. *Nocardiopsis* sp., *Pseudonocardia* sp.	Salam, Khieu, & Liu, 2017
Aquilaria crassna	*Streptomyces javensis, Nonomuraea rubra, Actinomadura glauciflava, Pseudonocardia halophobica, Nocardia alba*	Nimnoi, Pongsilp, and Lumyong (2010)
Xylocarpus granatum	*Jishengella endophytica*	Xie, Wang, & Wang, 2011
Avicennia marina, Aegiceras corniculatum, Kandelia obovata, Bruguiera gymnorrhiza, Thespesia populnea	*Streptomyces* sp., *Curtobacterium* sp., *Mycobacterium* sp., *Micrococcus* sp., *Brevibacterium* sp., *Kocuria* sp., *Nocardioides* sp., *Kineococcus* sp., *Kytococcus* sp., *Marmoricola* sp., *Mycobacterium* sp. *Micromonospora*, sp., *Actinoplanes* sp., *Agrococcus* sp., *Amnibacterium* sp., *Brachybacterium* sp., *Citricoccus* sp., *Dermacoccus* sp., *Glutamicibacter* sp., *Gordonia* sp., *Isoptericola* sp., *Janibacter* sp., *Leucobacter* sp., *Nocardia* sp., *Nocardiopsis* sp., *Pseudokineococcus*, sp., *Sanguibacter* sp., *Verrucosispora* sp.	Jiang, Tuo, & Huang, 2018
Thespesia populnea	*Marmoricola endophyticus*	Jiang, Pan, & Li, 2017
Lobelia clavatum	*Pseudonocardia endophytica*	Chen, Qin, & Li, 2009
Elaeagnus angustifolia	*Micromonospora* sp., *Nonomuraea* sp., *Pseudonocardia* sp., *Planotetraspora* sp.	Chen, Zhang, and Zhang (2011)
Aloe arborescens	*Micrococcus aloeverae, Streptomyces zhaozhouensis*	He, Liu, & Zhao, 2014, Prakash, Nimonkar, and Munot (2014)
Psammosilene tunicoides	*Allostreptomyces psammosilenae*	Huang, Rao, & Salam, 2017
Centella asiatica	*Streptomyces* sp., *Wenchangensis, Actinoplanes brasiliensis, Couchioplanes caeruleus, Gordonia otitidis, Micromonospora schwarzwaldensis*	Ernawati, Solihin, and Lestari (2016)
Terminalia mucronata	*Micromonospora terminaliae*	Kaewkla, Thamchaipinet, and Franco (2017)

(*Continued*)

Table 1.1 Consolidation of endophytes isolation from different hosts. *Continued*

Host plant	Endophytic microbe	Reference
Jatropha curcas	*Jatrophihabitans endophyticus*, *Nocardioides panzhihuaensis*, *Nocardia endophytica*, *Kibdelosporangium phytohabitans*	Qin, Xing, & Jiang, 2011, Qin, Yuan, & Zhang, 2012, Qin, Feng, & Xing, 2015, Madhaiyan, Hu, & Kim, 2013
Sonneratia apetala	*Micromonospora sonneratiae*	Li, Tang, & Wei, 2013
Bruguiera sexangula	*Mangrovihabitans endophyticus*	Liu, Carvalhais, & Crawford, 2017
Ferula sinkiangensis	*Amycolatopsis* sp.	
Salicornia europaea	*Modestobacter roseus*	Qin, Bian, & Zhang, 2013
Tamarix chinensis	*Streptomyces halophytocola*	Lour, Qin, and Bian (2013)
Dendranthema indicum	*Glycomyces phytohabitans*, *Amycolatopsis jiangsuensis*	Xing, Liu, & Zhang, 2013, Xing, Qin, & Zhang, 2014
Costus speciosus	*Micromonospora costi*	Thawai (2015)
Glycyrrhiza uralensis	*Phytoactinopolyspora endophytica*	Li, Ma, and Mohamad (2015)
Salsola affinis	*Okibacterium endophyticum*, *Arthrobacter endophyticus*	Wang, Zhang, & Chen, 2015, Wang, Zhang, & Cheng, 2015
Anabasis elatior	*Frigoribacterium endophyticum*, *Labedella endophytica*	
Dysophylla stellata	*Rothia endophytica*	Xiong, Zhang, & Zhang, 2013
Anabasis aphylla L.	*Glycomyces anabasis*	Zhang, Wang, & Alkhalifah, 2018
Seaweed	*Streptomyces sp.*	Rasool and Hemalatha (2017)
Thalassia hemprichii	*Micromonospora* sp., *Saccharomonospora* sp., *Mycobacterium* sp., *Actinomycetospora lutea*, *Nonomuraea maheshkhaliensis*, *Verrucosispora sediminis*, *Nocardiopsis composta*, *Microbacterium esteraromaticum*, *Glycomyces arizonensis*, *Streptomyces* sp.	Wu (2012)

1.5 Diversity and distribution of endophytic microbial communities

Ecosystem and its processes are controlled by the diversity and interaction of different species and their collective bioactivities within a community. Most of these activities are closely associated with the microbial secondary metabolism and their contributions to the plants in terms of nutrients. Due to the absence of associated bacteria to a plant, it will be much susceptible to stress and pathogenic invasion (Timmusk, Paalme, & Pavlicek, 2011). Endophytic diversity of a plant is highly depended on different factors, including specificity of plant and microbes, condition and health of host plant, environmental and soil factors (Ding & Melcher, 2016; Shi, Yang, Zhang, Sun, & Lou, 2014). Studies done on the enumeration of endophytic communities on the poplar tree roots suggest a 10-fold lower concentration than the surrounding rhizosphere population which can be a result of selective recruitment by the plants (Gottel, Castro, & Kerley, 2011). These populations are mainly dominated by microbes belonging to both cultivable and uncultivable groups. The rhizosphere communities comprise Acidobacteria and Alphaproteobacteria; however, endophytic communities mainly include Actinobacteria, Alphaproteobacteria, Firmicutes, Gammaproteobacteria, and

Bacteroidetes (Edwards, Johnson, & Santos-Medellín, 2015; Ringelberg, Foley, & Reynolds, 2012). Among these groups the most abundant groups of endophytic microbes are *Pseudomonas*, *Bacillus*, *Burkholderia*, *Enterobacter*, *Serratia* (Deng, Zhao, & Kong, 2011; Kumar, Singh, & Yadav, 2016; Taghavi, Garafola, & Monchy, 2009; Taghavi, van der Lelie, & Hoffman, 2010; Weilharter, Mitter, & Shin, 2011).

Reports from different authors suggest that the role of medium in which the host plant grows can alter the endophytic communities present in mature plants. A higher number of Firmicutes and Cyanobacteria were found in plants grown in sand and Betaproteobacteria in plants grown in peat-based growing mix (Ferrando & Fernández Scavino, 2015). The reports of the presence of common endophytic communities in the different part of the plants can be a result of migration of these groups along the xylem region. However, it is overserved that the xylem is the less populated region in the plant due to the lesser availability of nutrients (Compant, Clément, & Sessitsch, 2010; Fürnkranz, Köster, & Chun, 2011; Kumar, Droby, Singh, Singh, & White, 2020). Endophytic microbial communities have been isolated in almost all part of the plant in different concentrations. Leguminous plants that are capable of harboring nodule forming rhizobial population are poorly colonized with common endophytic bacteria, but there are certain exceptions observed in soybean roots with *Pantoea*, *Serratia*, *Acinetobacter*, *Bacillus*, *Agrobacterium*, and *Burkholderia* (Li, Wang, Chen, & Chen, 2008).

Another example of endophytic association with rhizobium observed in legume *Lespedeza* sp. with *Rhizobium* and *Bradyrhizobium* bacteria has been found to have *Arthrobacter, Bacillus, Burkholderia, Dyella, Methylobacterium, Microbacterium*, and *Staphylococcus* (Palaniappan, Chauhan, Saravanan, Anandham, & Sa, 2010). There is evidence of gene transfer where *nifH* genes found in *Bradyrhizobium japonicum* were also isolated from endophytic *Bacillus* strains from the same plants. The oxic–anoxic transition zone influences on plant–endophyte interactions are well studied in rice because of its irrigation patterns, including their plant growth-promoting properties (Loaces, Ferrando, & Scavino, 2011). Endophytes from the leaves of different rice varieties obtained using multiphasic detection methods confirmed that 51% of the total isolates belong to *Pantoea ananatis* and *Pseudomonas syringae* (Ferrando, Mañay, & Scavino, 2012). Studies in sugarcane for the presence of diazotrophic endophytes resulted in the characterization of endophytes, including *Pseudomonas, Stenotrophomonas, Xanthomonas, Acinetobacter, Rhanella, Enterobacter, Pantoea, Shinella, Agrobacterium*, and *Achromobacter* (Taulé, Mareque, & Barlocco, 2012). The most important and prominent among them is *G. diazotrophicus* with the ability to colonize and fix nitrogen in the inner tissues of sugarcane (Boddey, Polidoro, Resende, Alves, & Urquiaga, 2001; Sevilla, Burris, Gunapala, & Kennedy, 2001). The occurrence of *Gluconacetobacter* in plants growing in diverse geographical distributions like sweet potato, coffee, tea, banana, carrot, radish, and rice increases the importance of the organism in farming practices (Saravanan, Madhaiyan, Osborne, Thangaraju, & Sa, 2008).

1.6 Future perspective

Plant-associated microbes are a vital part of the environment and are having significant role in plant growth and disease management, thus, a potential source of biofertilizers and biopesticides.

There are a huge number of studies which suggest the potential, plant colonizing, and growth-promoting properties of these microbes. However, there are only a few efforts have been made to channel them toward field applications, thus obtaining inconsistent results under field conditions. The understanding of the control and selection of endophytes by host plant and their complex dynamics need to be elucidated more at molecular level by studying the expressed genes and the proteome by both plant and the bacteria at the endosphere. The biggest challenge in obtaining this is the low concentration of bacterial population and difficulty in obtaining bacterial and plant transcriptome without cross contamination. Another important area in endophytic microbial treasure trove is the possible genes present in unculturable microbes with high potential in medicinal applications. Efforts must be made to combine different areas of research to get more insight in these areas with so much potential in agricultural and medicinal applications.

References

Afzal, I., Iqrar, I., & Shinwari, Z. K. (2015). Selective isolation and characterization of agriculturally beneficial endopytic bacteria from wild hemp using canola. *Pakistan Journal of Botany*, *47*, 1999–2008.

Afzal, I., Iqrar, I., Shinwari, Z. K., & Yasmin, A. (2017). Plant growth-promoting potential of endophytic bacteria isolated from roots of wild *Dodonaea viscosa* L. *Plant Growth Regulation*, *81*, 399–408. Available from https://doi.org/10.1007/s10725-016-0216-5.

Afzal, I., Shinwari, Z. K., Sikandar, S., & Shahzad, S. (2019). Plant beneficial endophytic bacteria: Mechanisms, diversity, host range and genetic determinants. *Microbiological Research*, *221*, 36–49. Available from https://doi.org/10.1016/j.micres.2019.02.001.

Alquéres, S., Meneses, C., Rouws, L., Rothballer, M, Baldani, I, Schmid, M, & Hartmann, A (2013). The bacterial superoxide dismutase and glutathione reductase are crucial for endophytic colonization of rice roots by *Gluconacetobacter diazotrophicus* PAL5. *Molecular Plant-Microbe Interactions*, *26*, 937–945. Available from https://doi.org/10.1094/MPMI-12-12-0286-R.

Anyasi, R. O., & Atagana, H. I. (2019). Endophyte: Understanding the microbes and its applications. *Pakistan Journal of Biological Sciences*, *22*, 154–167. Available from https://doi.org/10.3923/pjbs.2019.154.167.

Apel, K., & Hirt, H. (2004). Reactive oxygen species: Metabolism, oxidative stress, and signal transduction. *Annual Review of Plant Biology*, *55*, 373–399. Available from https://doi.org/10.1146/annurev.arplant.55.031903.141701.

Aswathy, A. J., Jasim, B., Jyothis, M., & Radhakrishnan, E. K. (2012). Identification of two strains of *Paenibacillus* sp. as indole 3 acetic acid-producing rhizome-associated endophytic bacteria from *Curcuma longa. 3 Biotech*, *3*, 219–224. Available from https://doi.org/10.1007/s13205-012-0086-0.

Babu, A. G., Kim, J. D., & Oh, B. T. (2013). Enhancement of heavy metal phytoremediation by *Alnus firma* with endophytic *Bacillus thuringiensis* GDB-1. *Journal of Hazardous Materials*, *250–251*, 477–483. Available from https://doi.org/10.1016/j.jhazmat.2013.02.014.

Bal, A., Anand, R., Berge, O., & Chanway, C. P. (2012). Isolation and identification of diazotrophic bacteria from internal tissues of *Pinus contorta* and *Thuja plicata*. *Canadian Journal of Forest Research. Journal Canadien de la Recherche Forestiere*, *42*, 808–813.

Balsanelli, E., Tuleski, T. R., de Baura, V. A., Yates, M. G., Chubatsu, L. S., de Oliveira Pedrosa, F., … Monteiro, R. A. (2013). Maize root lectins mediate the interaction with *Herbaspirillum seropedicae* via *N*-acetyl glucosamine residues of lipopolysaccharides. *PLoS One*, *8*. Available from https://doi.org/10.1371/journal.pone.0077001.

Barzanti, R., Ozino, F., Bazzicalupo, M., Gabbrielli, R, Galardi, F, Gonnelli, C, & Mengoni, A (2007). Isolation and characterization of endophytic bacteria from the nickel hyperaccumulator plant *Alyssum bertolonii*. *Microbial Ecology*, *53*, 306–316. Available from https://doi.org/10.1007/s00248-006-9164-3.

Bezerra, J. D. P., Santos, M. G. S., Svedese, V. M., Lima, D. M. M., Fernandes, M. J. S., Paiva, L. M., & Souza-Motta, C. M. (2012). Richness of endophytic fungi isolated from *Opuntia ficus-indica* Mill. (Cactaceae) and preliminary screening for enzyme production. *World Journal of Microbiology and Biotechnology*, *28*, 1989–1995. Available from https://doi.org/10.1007/s11274-011-1001-2.

Bian, G.-K., Feng, Z.-Z., Qin, S., Xing, K, Wang, Z, Cao, C.-L, … Jiang, J.-H (2012). *Kineococcus endophytica* sp. nov., a novel endophytic actinomycete isolated from a coastal halophyte in Jiangsu, China. *Antonie Van Leeuwenhoek*, *102*, 621–628. Available from https://doi.org/10.1007/s10482-012-9757-4.

Bian, G.-K., Qin, S., Yuan, B., Zhang, Y.-J, Xing, K, Ju, X.-Y, … Jiang, J.-H (2012). *Streptomyces phytohabitans* sp. nov., a novel endophytic actinomycete isolated from medicinal plant *Curcuma phaeocaulis*. *Antonie Van Leeuwenhoek*, *102*, 289–296. Available from https://doi.org/10.1007/s10482-012-9737-8.

Boddey, R. M., Polidoro, J. C., Resende, A. S., Alves, B. J. R., & Urquiaga, S (2001). Use of the 15N natural abundance technique for the quantification of the contribution of N2 fixation to sugar cane and other grasses. *Australian Journal of Plant Physiology*, *28*, 889–895.

Botella, L., & Diez, J. J. (2011). Phylogenic diversity of fungal endophytes in Spanish stands of *Pinus halepensis*. *Fungal Divers*, *47*, 9–18. Available from https://doi.org/10.1007/s13225-010-0061-1.

Bunyoo, C., Duangmal, K., Nuntagij, A., & Thamchaipenet, A. (2009). Characterisation of endophytic actinomycetes isolated from wattle trees (*Acacia auriculiformis* A. Cunn. ex Benth.) in Thailand. *Environmental Microbiology*, *2*, 155–163.

Butchart, S. H. M., Scharlemann, J. P. W., Evans, M. I. Quader, S., Aricò, S., Arinaitwe, J., Balman, M., Bennun, L. A., Bertzky, B., Besançon, C., Boucher, T. M., Brooks, T. M., Burfield, I. J., Burgess, N. D., Chan, S., Clay, R. P., Crosby, M. J., Davidson, N. C., de Silva, N., Devenish, C., Dutson, G. C. L., Fernández, D. F. D., Fishpool, L. D. C., Fitzgerald, C., Foster, M., Heath, M. F., Hockings, M., Hoffmann, M., Knox, D., Larsen, F. W., Lamoreux, J. F., Loucks, C., May, I., Millett, J., Molloy, D., Morling,

P., Parr, M., Ricketts, T. H., Seddon, N., Skolnik, B., Stuart, S. N., Upgren, A., Woodley, S.,. (2012). Protecting important sites for biodiversity contributes to meeting global conservation targets. *PLoS One*, *7*. Available from https://doi.org/10.1371/journal.pone.0032529.

Carvalho, C. R., Gonçalves, V. N., & Pereira, C. B.Johann, S., Galliza, I. V., Alves, T.M. A., Rabello, A., Sobral, M.E.G., Zani, C.L., Rosa, C.A., Rosa, L.H. (2012). The diversity, antimicrobial and anticancer activity of endophytic fungi associated with the medicinal plant *Stryphnodendron adstringens* (Mart.) Coville (Fabaceae) from the Brazilian savannah. *Symbiosis*, *57*, 95–107. Available from https://doi.org/10.1007/s13199-012-0182-2.

Chanway, C. P. (1997). Inoculation of tree roots with plant growth promoting soil bacteria: An emerging technology for reforestation. *Forest Science*, *43*, 99–112.

Chen, H., Wang, C. S., Li, M., et al. (2010). *A novel angiogenesis model for screening anti-angiogenic compounds: The chorioallantoic membrane/feather bud assay*. pp. 71–79. https://doi.org/10.3892/ijo.

Chanway, C. P., Shishido, M., Nairn, J., et al.Jungwirth, S., Markham, J., Xiao, G., Holl, F.B. (2000). Endophytic colonization and field responses of hybrid spruce seedlings after inoculation with plant growth-promoting rhizobacteria. *Forest Ecology and Management*, *133*, 81–88. Available from https://doi.org/10.1016/S0378-1127(99)00300-X.

Chen, H. H., Qin, S., & Li, J. Sanchez, E., Li, J., Berenson, A., Wirtschafter, E., Wang, J., Shen, J., Li, Z., Bonavida, B., Berenson, J.R. (2009). *Pseudonocardia endophytica* sp. nov., isolated from the pharmaceutical plant *Lobelia clavata*. *International Journal of Systematic and Evolutionary Microbiology*, *59*, 559–563. Available from https://doi.org/10.1099/ijs.0.64740-0.

Chen, M., Zhang, L., & Zhang, X. (2011). Isolation and inoculation of endophytic actinomycetes in root nodules of *Elaeagnus angustifolia*. *Modern Applied Science*, *5*, 264–267. Available from https://doi.org/10.5539/mas.v5n2p264.

Chen, X., Marszałkowska, M., & Reinhold-Hurek, B. (2020). Jasmonic acid, not salicyclic acid restricts endophytic root colonization of rice. *Frontiers in Plant Science*, *10*, 1–15. Available from https://doi.org/10.3389/fpls.2019.01758.

Cipollini, D., Rigsby, C. M., & Barto, E. K. (2012). Microbes as targets and mediators of allelopathy in plants. *Journal of Chemical Ecology*, *38*, 714–727. Available from https://doi.org/10.1007/s10886-012-0133-7.

Compant, S., Clément, C., & Sessitsch, A. (2010). Plant growth-promoting bacteria in the rhizo- and endosphere of plants: Their role, colonization, mechanisms involved and prospects for utilization. *Soil Biology & Biochemistry*, *42*, 669–678. Available from https://doi.org/10.1016/j.soilbio.2009.11.024.

Coombs, J. T., & Franco, C. M. M. (2003). Isolation and identification of actinobacteria from surface-sterilized wheat roots. *Applied and Environmental Microbiology*, *69*, 5603–5608. Available from https://doi.org/10.1128/AEM.69.9.5603-5608.2003.

Cui, J. L., Guo, S. X., & Xiao, P. G. (2011). Antitumor and antimicrobial activities of endophytic fungi from medicinal parts of *Aquilaria sinensis*. *Journal of Zhejiang University. Science. B*, *12*, 385–392. Available from https://doi.org/10.1631/jzus.B1000330.

de Siqueira, V. M., Conti, R., de Araújo, J. M., & Souza-Motta, C. M. (2011). Endophytic fungi from the medicinal plant Lippia sidoides Cham. and their antimicrobial activity. *Symbiosis*, *53*, 89–95. Available from https://doi.org/10.1007/s13199-011-0113-7.

de Souza Sebastianes, F. L., Romão-Dumaresq, A. S., & Lacava, P. T.Harakava, R., Azevedo, J.L., de Melo, I.S., Pizzirani-Kleiner, A.A. (2013). Species diversity of culturable endophytic fungi from Brazilian mangrove forests. *Current Genetics*, *59*, 153–166. Available from https://doi.org/10.1007/s00294-013-0396-8.

Deng, Z. S., Zhao, L. F., Kong, Z. Y., Yang, W. Q., Lindström, K., Wang, E. T., Wei, G. H. (2011). Diversity of endophytic bacteria within nodules of the *Sphaerophysa salsula* in different regions of Loess Plateau in China. *FEMS Microbiology Ecology*, *76*, 463–475. Available from https://doi.org/10.1111/j.1574-6941.2011.01063.x.

Ding, T., & Melcher, U. (2016). Influences of plant species, season and location on leaf endophytic bacterial communities of non-cultivated plants. *PLoS One*, *11*, 1–13. Available from https://doi.org/10.1371/journal.pone.0150895.

Domka, A. M., Rozpaądek, P., & Turnau, K. (2019). Are fungal endophytes merely mycorrhizal copycats? The role of fungal endophytes in the adaptation of plants to metal toxicity. *Frontiers in Microbiology*, *10*, 371.

Drogue, B., Sanguin, H., Borland, S., Picault, N., Prigent-Combaret, C., Wisniewski-Dyé, F. (2013). Host specificity of the plant growth-promoting cooperation between Azospirillum and rice. In Netw Plast Microbiol communities secret to success 12th Symp Bact Genet Ecol (BAGECO12), Ljubljana, Slovénie, 9–13 June 2013.

Edwards, J., Johnson, C., Santos-Medellín, C., Lurie, E., Podishetty, N. K., Bhatnagar, S., Eisen, J. A., Sundaresan, V., Jeffery, L. D. (2015). Structure, variation, and assembly of the root-associated microbiomes of rice. *Proceedings of the National Academy of Sciences of the United States of America*, *112*, E911–E920. Available from https://doi.org/10.1073/pnas.1414592112.

El-Tarabily, K. A. (2003). An endophytic chitinase-producing isolate of *Actinoplanes missouriensis*, with potential for biological control of root rot of lupin caused by *Plectosporium tabacinum*. *Australian Journal of Botany*, *51*, 257–266. Available from https://doi.org/10.1071/BT02107.

Ernawati, M., Solihin, D. D., & Lestari, Y. (2016). Community structures of endophytic actinobacteria from medicinal plant *Centella asiatica* L. Urban-based on metagenomic approach. *International Journal of Pharmacy and Pharmaceutical Sciences*, *8*, 292–297.

Esmaeel, Q., Miotto, L., Rondeau, M., Leclère, V., Clément, C., Jacquard, C., Sanchez, L., Barka, E. A. (2018). *Paraburkholderia phytofirmans* PsJN-plants interaction: From perception to the induced mechanisms. *Frontiers in Microbiology*, *9*, 1–14. Available from https://doi.org/10.3389/fmicb.2018.02093.

Esparza-Reynoso, S., Pelagio-Flores, R., & López-Bucio, J. (2020). Chapter 3—Mechanism of plant immunity triggered by *Trichoderma*. In V. K. Gupta, S. Zeilinger, H. B. Singh, I. B. T.-N. Druzhinina, & FD in MB and B (Eds.), *New and future developments in microbial biotechnology and bioengineering recent developments in trichoderma research* (pp. 57–73). Elsevier.

Feng, W. W., Wang, T. T., Bai, J. L., Ding, P., Xing, K., Jiang, J. H., Peng, X., Qin, S. (2017). *Glutamicibacter halophytocola* sp. nov., an endophytic actinomycete isolated from the roots of a coastal halophyte, *Limonium sinense*. *International Journal of Systematic and Evolutionary Microbiology*, *67*, 1120–1125. Available from https://doi.org/10.1099/ijsem.0.001775.

Ferrando, L., & Fernández Scavino, A. (2015). Strong shift in the diazotrophic endophytic bacterial community inhabiting rice (*Oryza sativa*) plants after flooding. *FEMS Microbiology Ecology*, *91*. Available from https://doi.org/10.1093/femsec/fiv104.

Ferrando, L., Mañay, J. F., & Scavino, A. F. (2012). Molecular and culture-dependent analyses revealed similarities in the endophytic bacterial community composition of leaves from three rice (*Oryza sativa*) varieties. *FEMS Microbiology Ecology*, *80*, 696–708. Available from https://doi.org/10.1111/j.1574-6941.2012.01339.x.

Fidalgo, C., Henriques, I., Rocha, J., Tacão, M., & Alves, A. (2016). Culturable endophytic bacteria from the salt marsh plant *Halimione portulacoides*: Phylogenetic diversity, functional characterization, and influence of metal(loid) contamination. *Environmental Science and Pollution Research*, *23*, 10200–10214. Available from https://doi.org/10.1007/s11356-016-6208-1.

Fürnkranz, A., Köster, I., Chun, K. R. J., Metzner, A., Mathew, S., Konstantinidou, M., Ouyang, F., Kuck, K. H. (2011). Cryoballoon temperature predicts acute pulmonary vein isolation. *Hear Rhythm*, *8*, 821–825. Available from https://doi.org/10.1016/j.hrthm.2011.01.044.

Gagic, M., Faville, M. J., Zhang, W., Forester, N. T., Rolston, M. P., Johnson, R. D., Ganesh, S., Koolaard, J. P., Easton, H. S., Hudson, D., Johnson, L. J., Moon, C. D., Voisey, C. R. (2018). Seed transmission of epichloë endophytes in lolium perenne is heavily influenced by host genetics. *Frontiers in Plant Science*, *871*, 1–16. Available from https://doi.org/10.3389/fpls.2018.01580.

García, A., Rhoden, S. A., Filho, C. J. R., Nakamura, C. V., & Pamphile, J. A. (2012). Diversity of foliar endophytic fungi from the medicinal plant *Sapindus saponaria* L. and their localization by scanning electron microscopy. *Biological Research*, *45*, 139–148. Available from https://doi.org/10.4067/S0716-97602012000200006.

Gazis, R., & Chaverri, P. (2010). Diversity of fungal endophytes in leaves and stems of wild rubber trees (*Hevea brasiliensis*) in Peru. *Fungal Ecology*, *3*, 240–254. Available from https://doi.org/10.1016/j.funeco.2009.12.001.

Gond, S. K., Mishra, A., Sharma, V. K., Verma, S. K., Kumar, J., Kharwar, R. N., Kumar, A. (2012). Diversity and antimicrobial activity of endophytic fungi isolated from *Nyctanthes arbor-tristis*, a well-known medicinal plant of India. *Mycoscience*, *53*, 113–121. Available from https://doi.org/10.1007/s10267-011-0146-z.

Gottel, N. R., Castro, H. F., Kerley, M., Yang, Z., Pelletier, D. A., Podar, M., Karpinets, T., Uberbacher, E. D., Tuskan, G. A., Vilgalys, R., Doktycz, M. J., Schadt, C. W. (2011). Distinct microbial communities within the endosphere and rhizosphere of *Populus deltoides* roots across contrasting soil types. *Applied and Environmental Microbiology*, *77*, 5934–5944. Available from https://doi.org/10.1128/AEM.05255-11.

Green, E. R., & Mecsas, J. (2016). Bacterial secretion systems: An overview. *Virulence Mechanisms of Bacterial Pathogens*, *4*, 213–239. Available from https://doi.org/10.1128/9781555819286.ch8.

Hanada, R. E., Pomella, A. W. V., Costa, H. S., Bezerra, J. L., Loguercio, L. L., & Pereira, J. O. (2010). Endophytic fungal diversity in *Theobroma cacao* (cacao) and *T. grandiflorum* (cupuaçu) trees and their potential for growth promotion and biocontrol of black-pod disease. *Fungal Biology*, *114*, 901–910. Available from https://doi.org/10.1016/j.funbio.2010.08.006.

Hardoim, P. R., van Overbeek, L. S., Berg, G., Pirttilä, A. M., Compant, S., Campisano, A., Döring, M., Sessitsch, A. (2015). The hidden world within plants: Ecological and evolutionary considerations for defining functioning of microbial endophytes. *Microbiology and Molecular Biology Reviews: MMBR*, *79*, 293–320. Available from https://doi.org/10.1128/mmbr.00050-14.

Hardoim, P. R., van Overbeek, L. S., & van Elsas, J. D. (2008). Properties of bacterial endophytes and their proposed role in plant growth. *Trends in Microbiology*, *16*, 463–471. Available from https://doi.org/10.1016/j.tim.2008.07.008.

He, H., Liu, C., Zhao, J., Li, W., Pan, T., Yang, L., Wang, X., Xiang, W. (2014). *Streptomyces zhaozhouensis* sp. nov., an actinomycete isolated from candelabra aloe (*Aloe arborescens* Mill). *International Journal of Systematic and Evolutionary Microbiology*, *64*, 1096–1101. Available from https://doi.org/10.1099/ijs.0.056317-0.

He, H., Ye, Z., Yang, D., Yan, J., Xiao, L., Zhong, T., Yuan, M., Cai, X., Fang, Z., Jing, Y. (2013). Characterization of endophytic *Rahnella* sp. JN6 from *Polygonum pubescens* and its potential in promoting growth and Cd, Pb, Zn uptake by *Brassica napus*. *Chemosphere*, *90*, 1960–1965. Available from https://doi.org/10.1016/j.chemosphere.2012.10.057.

Hodgson, S., de Cates, C., Hodgson, J., Morley, N. J., Sutton, B. C., Gange, A. C. (2014). Vertical transmission of fungal endophytes is widespread in forbs. *Ecology and Evolution*, *4*, 1199–1208. Available from https://doi.org/10.1002/ece3.953.

Huang, M. J., Rao, M. P. N., Salam, N., Xiao, M., Huang, H. Q., Li, W. J. (2017). *Allostreptomyces psammosilenae* gen. Nov., sp. nov., an endophytic actinobacterium isolated from the roots of *Psammosilene tunicoides* and emended description of the family *Streptomycetaceae* [Waksman and Henrici (1943)AL] emend. Rainey et al. 1997, Emend. K. *International Journal of Systematic and Evolutionary Microbiology*, *67*, 288–293. Available from https://doi.org/10.1099/ijsem.0.001617.

Indananda, C., Thamchaipenet, A., Matsumoto, A., Inahashi, Y., Duangmal, K., Takahashi, Y. (2011). *Actinoallomurus oryzae* sp. nov., an endophytic actinomycete isolated from roots of a Thai jasmine rice plant. *International Journal of Systematic and Evolutionary Microbiology*, *61*, 737–741. Available from https://doi.org/10.1099/ijs.0.022509-0.

Iniguez, A. L., Dong, Y., Carter, H. D., Ahmer, B. M. M., Stone, J. M., Triplett, E. W. (2005). Regulation of enteric endophytic bacterial colonization by plant defenses. *Molecular Plant-Microbe Interactions*, *18*, 169–178. Available from https://doi.org/10.1094/MPMI-18-0169.

Jasim, B., Anish, M. C., Shimil, V., Jyothis, M, & Radhakrishnan, E. K. (2015). Studies on plant growth promoting properties of fruit-associated bacteria from *Elettaria cardamomum* and molecular analysis of ACC deaminase gene. *Applied Biochemistry and Biotechnology*, *177*, 175–189. Available from https://doi.org/10.1007/s12010-015-1736-6.

Jasim, B., Benny, R., Sabu, R., Mathew, J, & Radhakrishnan, E. K. (2016). Metabolite and mechanistic basis of antifungal property exhibited by endophytic *Bacillus amyloliquefaciens* BmB 1. *Applied Biochemistry and Biotechnology*, *179*, 830–845. Available from https://doi.org/10.1007/s12010-016-2034-7.

Jasim, B., Daya, P. S., Sreelakshmi, K. S., Sachidanandan, P., Aswani, R., Jyothis, M., Radhakrishnan, E. K. (2017). Bacopaside N1 biosynthetic potential of endophytic *Aspergillus* sp. BmF 16 isolated from *Bacopa monnieri*. *3 Biotech*, *7*, 210. Available from https://doi.org/10.1007/s13205-017-0788-4.

Jasim, B., John Jimtha, C., Jyothis, M., & Radhakrishnan, E. K. (2013). Plant growth promoting potential of endophytic bacteria isolated from *Piper nigrum*. *Plant Growth Regulation*, *71*, 1–11. Available from https://doi.org/10.1007/s10725-013-9802-y.

Jasim, B., Joseph, A. A., John, C. J., Mathew, J., & Radhakrishnan, E. K. (2014). Isolation and characterization of plant growth promoting endophytic bacteria from the rhizome of *Zingiber officinale*. *3 Biotech*, *4*, 197–204. Available from https://doi.org/10.1007/s13205-013-0143-3.

Jasim, B., Mathew, J., & Radhakrishnan, E. K. (2016). Identification of a Novel endophytic *Bacillus* sp. from *Capsicum annuum* with highly efficient and broad spectrum plant probiotic effect. *Journal of Applied Microbiology*, *121*. Available from https://doi.org/10.1111/jam.13214.

Jasim, B., Sreelakshmi, K. S., Mathew, J., & Radhakrishnan, E. K. (2016). Surfactin, iturin, and fengycin biosynthesis by endophytic *Bacillus* sp. from *Bacopa monnieri*. *Microbial Ecology*, *72*, 106–119. Available from https://doi.org/10.1007/s00248-016-0753-5.

Jia, F., Liu, C., Wang, X., Liu, Q., Zhang, J., Gao, R., Xiang, W. (2013). *Wangella harbinensis* gen. nov., sp. nov., a new member of the family *Micromonosporaceae*. *Antonie Van Leeuwenhoek*, *103*, 399–408. Available from https://doi.org/10.1007/s10482-012-9820-1.

Jiang, Z. K., Pan, Z., Li, F. N., Li, X.J., Liu, S. W., Tuo, L., Jiang, M. G., Sun, C. H. (2017). *Marmoricola endophyticus* sp. Nov., an endophytic actinobacterium isolated from *Thespesia populnea*. *International Journal of Systematic and Evolutionary Microbiology*, *67*, 4379–4384. Available from https://doi.org/10.1099/ijsem.0.002297.

Jiang, Z. K., Tuo, L., Huang, D. L., Osterman, I. A., Tyurin, A. P., Liu, S. W., Lukyanov, D. A., Sergiev, P. V., Dontsova, O. A., Korshun, V. A., Li, F. N., Sun, C. H. (2018). Diversity, novelty, and antimicrobial activity of endophytic actinobacteria from mangrove plants in Beilun Estuary National Nature Reserve of Guangxi, China. *Frontiers in Microbiology*, *9*, 1–11. Available from https://doi.org/10.3389/fmicb.2018.00868.

Jin, H., Yan, Z., Liu, Q., Yang, X., Chen, J., Qin, B. (2013). Diversity and dynamics of fungal endophytes in leaves, stems and roots of *Stellera chamaejasme* L. in northwestern China. *Antonie Van Leeuwenhoek*, *104*, 949–963. Available from https://doi.org/10.1007/s10482-013-0014-2.

John, C. J., Kumar, S., & Ge, M. (2020). Probiotic prospects of PGPR for green and sustainable agriculture. *Archives of Phytopathology and Plant Protection*, 1–16. Available from https://doi.org/10.1080/03235408.2020.1805901.

Joseph, C. M., & Phillips, D. A. (2003). Metabolites from soil bacteria affect plant water relations. *Plant Physiology and Biochemistry: PPB/Societe Francaise de Physiologie Vegetale*, *41*, 189–192. Available from https://doi.org/10.1016/S0981-9428(02)00021-9.

Juhas, M., Van Der Meer, J. R., Gaillard, M., Harding, R. M., Hood, D. W., Crook, D. W. (2009). Genomic islands: Tools of bacterial horizontal gene transfer and evolution. *FEMS Microbiology Reviews*, *33*, 376–393. Available from https://doi.org/10.1111/j.1574-6976.2008.00136.x.

Kaewkla, O., & Franco, C. M. M. (2013). Rational approaches to improving the isolation of endophytic actinobacteria from Australian native trees. *Microbial Ecology*, *65*, 384–393. Available from https://doi.org/10.1007/s00248-012-0113-z.

Kaewkla, O., Thamchaipinet, A., & Franco, C. M. M. (2017). *Micromonospora terminaliae* sp. Nov., an endophytic actinobacterium isolated from the surface-sterilized stem of the medicinal plant *Terminalia mucronata*. *International Journal of Systematic and Evolutionary Microbiology*, *67*, 225–230. Available from https://doi.org/10.1099/ijsem.0.001600.

Kandel, S. L., Joubert, P. M., & Doty, S. L. (2017). Bacterial endophyte colonization and distribution within plants. *Microorganisms*, *5*, 77. Available from https://doi.org/10.3390/microorganisms5040077.

Kharwar, R. N., Maurya, A. L., Verma, V. C., Kumar, A., Gond, S. K., Mishra, A. (2012). Diversity and antimicrobial activity of endophytic fungal community isolated from medicinal plant *Cinnamomum camphora. Proceedings of the National Academy of Sciences, India, Section B: Biological Sciences*, *82*, 557–565. Available from https://doi.org/10.1007/s40011-012-0063-8.

Kharwar, R. N., Verma, S. K., Mishra, A., Gond, S. K., Sharma, V. K., Afreen, T., Kumar, A. (2011). Assessment of diversity, distribution and antibacterial activity of endophytic fungi isolated from a medicinal plant Adenocalymma alliaceum Miers. *Symbiosis*, *55*, 39–46. Available from https://doi.org/10.1007/s13199-011-0142-2.

Kleczewski, N. M., Bauer, J. T., Bever, J. D., Clay, K., Reynolds, H. L. (2012). A survey of endophytic fungi of switchgrass (*Panicum virgatum*) in the Midwest, and their putative roles in plant growth. *Fungal Ecology*, *5*, 521–529. Available from https://doi.org/10.1016/j.funeco.2011.12.006.

Koukol, O., Kolařík, M., Kolářová, Z., & Baldrian, P. (2012). Diversity of foliar endophytes in wind- fallen *Picea abies* trees. *Fungal Divers*, *54*, 69–77. Available from https://doi.org/10.1007/s13225-011-0112-2.

Kuffner, M., De Maria, S., Puschenreiter, M., Fallmann, K., Wieshammer, G., Gorfer, M., Strauss, J., Rivelli, A. R., Sessitsch, A. (2010). Culturable bacteria from Zn- and Cd-accumulating Salix caprea with differential effects on plant growth and heavy metal availability. *Journal of Applied Microbiology*, *108*, 1471–1484. Available from https://doi.org/10.1111/j.1365-2672.2010.04670.x.

Kumar, A., Singh, R., Yadav, A., et al. (2016). Isolation and characterization of bacterial endophytes of *Curcuma longa* L. *3 Biotech*, *6*, 60. Available from https://doi.org/10.1007/s13205-016-0393-y.

Kumar, S., & Kaushik, N. (2013). Endophytic fungi isolated from oil-seed crop *Jatropha curcas* produces oil and exhibit antifungal activity. *PLoS One*, *8*, 1–8. Available from https://doi.org/10.1371/journal.pone.0056202.

Kumar, S., Kaushik, N., Edrada-Ebel, R., et al. (2011). Isolation, characterization, and bioactivity of endophytic fungi of Tylophora indica. *World Journal of Microbiology and Biotechnology*, *27*, 571–577. Available from https://doi.org/10.1007/s11274-010-0492-6.

Kurose, D., Furuya, N., Tsuchiya, K., et al. (2012). Endophytic fungi associated with *Fallopia japonica* (Polygonaceae) in Japan and their interactions with *Puccinia polygoni-amphibii* var. tovariae, a candidate for classical biological control. *Fungal Biology*, *116*, 785–791. Available from https://doi.org/10.1016/j.funbio.2012.04.011.

Kusari, P., Kusari, S., Spiteller, M., & Kayser, O. (2013). Endophytic fungi harbored in *Cannabis sativa* L.: Diversity and potential as biocontrol agents against host plant-specific phytopathogens. *Fungal Divers*, *60*, 137–151. Available from https://doi.org/10.1007/s13225-012-0216-3.

Li, J. H., Wang, E. T., Chen, W. F., & Chen, W. X. (2008). Genetic diversity and potential for promotion of plant growth detected in nodule endophytic bacteria of soybean grown in Heilongjiang province of China. *Soil Biology & Biochemistry*, *40*, 238–246. Available from https://doi.org/10.1016/j.soilbio.2007.08.014.

Kumar, A., Droby, S., Singh, V. K., Singh, S. K., White, J. F. (2020). Entry, colonization, and distribution of endophytic microorganisms in plants.

Li, L., Tang, Y. L., Wei, B., Xie, Q. Y., Deng, Z., Hong, K. (2013). *Micromonospora sonneratiae* sp. nov., isolated from a root of *Sonneratia apetala. International Journal of*

Systematic and Evolutionary Microbiology, *63*, 2383–2388. Available from https://doi.org/10.1099/ijs.0.043570-0.

Liu, H., Carvalhais, L. C., Crawford, M., Singh, E., Dennis, P. G., Pieterse, C. M. J., Schenk, P. M. (2017). Inner plant values: Diversity, colonization and benefits from endophytic bacteria. *Frontiers in Microbiology*, *8*, 1–17. Available from https://doi.org/10.3389/fmicb.2017.02552.

Liu, Q., Liu, C., Yu, J., Yan, J., & Qi, X (2012). Analysis of the Ketosynthase genes in *Streptomyces* and its implications for preventing reinvestigation of polyketides with bioactivities. *The Journal of Agricultural Science*, *4*, 262–270. Available from https://doi.org/10.5539/jas.v4n7p262.

Loaces, I., Ferrando, L., & Scavino, A. F. (2011). Dynamics, diversity and function of endophytic siderophore-producing bacteria in rice. *Microbial Ecology*, *61*, 606–618. Available from https://doi.org/10.1007/s00248-010-9780-9.

Lopez, B. R., Bashan, Y., & Bacilio, M. (2011). Endophytic bacteria of *Mammillaria fraileana*, an endemic rock-colonizing cactus of the southern Sonoran Desert. *Archives of Microbiology*, *193*, 527–541. Available from https://doi.org/10.1007/s00203-011-0695-8.

Li, L., Ma, J., Mohamad, O. A., et al. (2015) *Phytoactinopolyspora endophytica gen. nov., sp. nov., a halotolerant filamentous actinomycete isolated from the roots of* Glycyrrhiza *uralensis F. 2671–2677.* https://doi.org/10.1099/ijs.0.000322.

Lour, T., Qin, S., Bian, G., Tamura, T., Zhang, Y., Zhang, W., Cao, C., Jiang, J. (2013). An endophytic actinomycete isolated from the surface-sterilized stems of a coastal halophyte. pp. 2770–2775. Https://doi.org/10.1099/ijs.0.047456-0.

López-Gómez, B. F., Lara-Herrera, A., Bravo-Lozano, A. G., Lozano-Gutiérrez, J., Avelar-Mejía, J. J., Luna-Flores, M., Llamas-Llamas, J. J. (2012). Improvement of plant growth and yield in pepper by vermicompost application, in greenhouse conditions. *Acta Horticulturae*, *947*, 313–318. Available from https://doi.org/10.17660/ActaHortic.2012.947.40.

Madhaiyan, M., Hu, C. J., Kim, S. J., Weon, H. Y., Kwon, S. W., Ji, L. (2013). *Jatrophihabitans endophyticus* gen. nov., sp. nov., an endophytic actinobacterium isolated from a surface-sterilized stem of *Jatropha curcas* L. *International Journal of Systematic and Evolutionary Microbiology*, *63*, 1241–1248. Available from https://doi.org/10.1099/ijs.0.039685-0.

Maheswari, S., & Rajagopal, K. (2013). Biodiversity of endophytic fungi in Kigelia pinnata during two different seasons. *Current Science*, *104*, 515–518.

Mathesius, U., Mulders, S., Gao, M., Teplitski, M., Caetano-Anolles, G., Rolfe, B. G., Bauer, W. D. (2003). Extensive and specific responses of a eukaryote to bacterial quorum-sensing signals. *Proceedings of the National Academy of Sciences of the United States of America*, *100*, 1444–1449. Available from https://doi.org/10.1073/pnas.262672599.

Mei, C., & Flinn, B. S. (2010). The use of beneficial microbial endophytes for plant biomass and stress tolerance improvement. *Recent Patents on Biotechnology*, *4*, 81–95.

Mendoza-Mendoza, A., Zaid, R., Lawry, R., Hermosa, R., Monte, E., Horwitz, B. A., Mukherjee, P. K. (2018). Molecular dialogues between *Trichoderma* and roots: Role of the fungal secretome. *Fungal Biology Reviews*, *32*, 62–85. Available from https://doi.org/10.1016/j.fbr.2017.12.001.

Meneses, C., Gonçalves, T., Alquéres, S., Rouws, L., Serrato, R., Vidal, M., Baldani, J. I. (2017). *Gluconacetobacter diazotrophicus* exopolysaccharide protects bacterial cells against oxidative stress in vitro and during rice plant colonization. *Plant and Soil*, *416*, 133−147. Available from https://doi.org/10.1007/s11104-017-3201-5.

Miliute, I., Buzaite, O., Baniulis, D., & Stanys, V. (2015). Bakterinių endofitų reikšmė žemės ūkio augalų atsparumui stresui: Apžvalga. *Zemdirbyste*, *102*, 465−478. Available from https://doi.org/10.13080/z-a.2015.102.060.

Mishra, A., Gond, S. K., Kumar, A., Sharma, V. K., Verma, S. K., Kharwar, R. N., Sieber, T. N. (2012). Season and tissue type affect fungal endophyte communities of the Indian medicinal plant *Tinospora cordifolia* more strongly than geographic location. *Microbial Ecology*, *64*, 388−398. Available from https://doi.org/10.1007/s00248-012-0029-7.

Nimnoi, P., Pongsilp, N., & Lumyong, S. (2010). Endophytic actinomycetes isolated from *Aquilaria crassna* Pierre ex Lec and screening of plant growth promoters production. *World Journal of Microbiology and Biotechnology*, *26*, 193−203. Available from https://doi.org/10.1007/s11274-009-0159-3.

Orlandelli, R. C., Alberto, R. N., Rubin Filho, C. J., & Pamphile, J. A. (2012). Diversity of endophytic fungal community associated with *Piper hispidum* (Piperaceae) leaves. *Genetics and Molecular Research: GMR*, *11*, 1575−1585. Available from https://doi.org/10.4238/2012.May.22.7.

Palaniappan, P., Chauhan, P. S., Saravanan, V. S., Anandham, R., & Sa, T. (2010). Isolation and characterization of plant growth promoting endophytic bacterial isolates from root nodule of *Lespedeza* sp. *Biology and Fertility of Soils*, *46*, 807−816. Available from https://doi.org/10.1007/s00374-010-0485-5.

Pancher, M., Ceol, M., Corneo, P. E., Longa, C. M. O., Yousaf, S., Pertot, I., Campisano, A. (2012). Fungal endophytic communities in grapevines (*Vitis vinifera* L.) Respond to crop management. *Applied and Environmental Microbiology*, *78*, 4308−4317. Available from https://doi.org/10.1128/AEM.07655-11.

Park, M. S., Jung, S. R., Lee, M. S., Kim, K. O., Do, J. O., Lee, K. H., Kim, S. B., Bae, K. S. (2005). Isolation and characterization of bacteria associated with two sand dune plant species, Calystegia soldanella and *Elymus mollis*. *Journal of Microbiology (Seoul, Korea)*, *43*, 219−227.

Petrini, O. (1991). *Fungal endophytes of tree leaves* (pp. 179−197). New York: Springer.

Piromyou, P., Buranabanyat, B., Tantasawat, P., Tittabutr, P., Boonkerd, N., Teaumroong, N. (2011). Effect of plant growth promoting rhizobacteria (PGPR) inoculation on microbial community structure in rhizosphere of forage corn cultivated in Thailand. *European Journal of Soil Biology*, *47*, 44−54. Available from https://doi.org/10.1016/j.ejsobi.2010.11.004.

Plett, J. M., & Martin, F. M. (2018). Know your enemy, embrace your friend: Using omics to understand how plants respond differently to pathogenic and mutualistic microorganisms. *The Plant Journal: For Cell and Molecular Biology*, *93*, 729−746. Available from https://doi.org/10.1111/tpj.13802.

Prakash, O., Nimonkar, Y., Munot, H., Sharma, A., Vemuluri, V. R., Chavadar, M. S., Shouche, Y. S. (2014). Description of Micrococcus aloeverae sp . nov ., an endophytic actinobacterium isolated from Aloe vera 3427−3433. Https://doi.org/10.1099/ijs.0.063339-0.

Promputtha, I., Hyde, K. D., McKenzie, E. H. C., Peberdy, J. F., & Lumyong, S. (2010). Can leaf degrading enzymes provide evidence that endophytic fungi becoming saprobes? *Fungal Divers*, *41*, 89−99. Available from https://doi.org/10.1007/s13225-010-0024-6.

Puente, M. E., Li, C. Y., & Bashan, Y. (2009). Endophytic bacteria in cacti seeds can improve the development of cactus seedlings. *Environmental and Experimental Botany*, *66*, 402–408. Available from https://doi.org/10.1016/j.envexpbot.2009.04.007.

Qadri, M., Rajput, R., Abdin, M. Z., Vishwakarma, R. A., Riyaz-Ul-Hassan, S. (2014). Diversity, molecular phylogeny, and bioactive potential of fungal endophytes associated with the Himalayan blue pine (*Pinus wallichiana*). *Microbial Ecology*, *67*, 877–887. Available from https://doi.org/10.1007/s00248-014-0379-4.

Qi, F., Jing, T., & Zhan, Y. (2012). Characterization of endophytic fungi from *Acer ginnala* maxim. in an artificial plantation: Media effect and tissue-dependent variation. *PLoS One*, *7*, 1–6. Available from https://doi.org/10.1371/journal.pone.0046785.

Qin, S., Bian, G. K., Zhang, Y. J., Xing, K., Cao, C. L., Liu, C. H., Dai, C. C., Li, W. J., Jiang, J.H. (2013). *Modestobacter roseus* sp. nov., an endophytic actinomycete isolated from the coastal halophyte *Salicornia europaea* Linn., and emended description of the genus Modestobacter. *International Journal of Systematic and Evolutionary Microbiology*, *63*, 2197–2202. Available from https://doi.org/10.1099/ijs.0.044412-0.

Qin, S., Feng, W. W., Xing, K., Bai, J. L., Yuan, B., Liu, W. J., Jiang, J. H. (2015). Complete genome sequence of *Kibdelosporangium phytohabitans* KLBMP 1111T, a plant growth promoting endophytic actinomycete isolated from oil-seed plant *Jatropha curcas* L. *Journal of Biotechnology*, *216*, 129–130. Available from https://doi.org/10.1016/j.jbiotec.2015.10.017.

Qin, S., Xing, K., Jiang, J.-H., Xu, L. -H., Li, W. -J. (2011). Biodiversity, bioactive natural products and biotechnological potential of plant-associated endophytic actinobacteria. *Applied Microbiology and Biotechnology*, *89*, 457–473. Available from https://doi.org/10.1007/s00253-010-2923-6.

Qin, S., Yuan, B., Zhang, Y.-J., Bian, G. -K., Tamura, T., Sun, B. -Z., Li, W. -J., Jiang, J. -H. (2012). *Nocardioides panzhihuaensis* sp. nov., a novel endophytic actinomycete isolated from medicinal plant *Jatropha curcas* L. *Antonie Van Leeuwenhoek*, *102*, 353–360. Available from https://doi.org/10.1007/s10482-012-9745-8.

Rasool, U., & Hemalatha, S. (2017). Marine endophytic actinomycetes assisted synthesis of copper nanoparticles (CuNPs): Characterization and antibacterial efficacy against human pathogens. *Materials Letters*, *194*, 176–180. Available from https://doi.org/10.1016/j.matlet.2017.02.055.

Reinhold-Hurek, B., & Hurek, T. (2011). Living inside plants: Bacterial endophytes. *Current Opinion in Plant Biology*, *14*, 435–443. Available from https://doi.org/10.1016/j.pbi.2011.04.004.

Rhoden, S. A., Garcia, A., Rubin Filho, C. J., Azevedo, J. L., Pamphile, J. A. (2012). Phylogenetic diversity of endophytic leaf fungus isolates from the medicinal tree *Trichilia elegans* (Meliaceae). *Genetics and Molecular Research: GMR*, *11*, 2513–2522. Available from https://doi.org/10.4238/2012.June.15.8.

Ringelberg, D., Foley, K., & Reynolds, C. M. (2012). Bacterial endophyte communities of two wheatgrass varieties following propagation in different growing media. *Canadian Journal of Microbiology*, *58*, 67–80. Available from https://doi.org/10.1139/W11-122.

Rivera-Orduña, F. N., Suarez-Sanchez, R. A., Flores-Bustamante, Z. R., Gracida-Rodriguez, J. N., Flores-Cotera, L. B. (2011). Diversity of endophytic fungi of *Taxus globosa* (Mexican yew). *Fungal Divers*, *47*, 65–74. Available from https://doi.org/10.1007/s13225-010-0045-1.

Rosa, L. H., Almeida Vieira, M. D. L., Santiago, I. F., & Rosa, C. A. (2010). Endophytic fungi community associated with the dicotyledonous plant *Colobanthus quitensis*

(Kunth) Bartl. (Caryophyllaceae) in Antarctica. *FEMS Microbiology Ecology*, *73*, 178–189. Available from https://doi.org/10.1111/j.1574-6941.2010.00872.x.

Rosa, L. H., Tabanca, N., Techen, N., Wedge, D. E., Pan, Z., Bernier, U. R., Becnel, J. J., Agramonte, N. M., Walker, L. A., Moraes, R. M. (2012). Diversity and biological activities of endophytic fungi associated with micropropagated medicinal plant *Echinacea purpurea* (L.) Moench. *American Journal of Plant Sciences*, *03*, 1105–1114. Available from https://doi.org/10.4236/ajps.2012.38133.

Rosenblueth, M., & Martínez-Romero, E. (2006). Bacterial endophytes and their interactions with hosts. *Molecular Plant-Microbe Interactions: MPMI*, *19*, 827–837. Available from https://doi.org/10.1094/MPMI-19-0827.

Ryan, R. P., Germaine, K., Franks, A., Ryan, D. J., Dowling, D. N. (2008). Bacterial endophytes: Recent developments and applications. *FEMS Microbiology Letters*, *278*, 1–9. Available from https://doi.org/10.1111/j.1574-6968.2007.00918.x.

Salam, N., Khieu, T. N., Liu, M. J., Vu, T. T., Chu-Ky, S., Quach, N. T., Phi, Q. T., Narsing Rao, M. P., Fontana, A., Sarter, S., Li, W. J. (2017). Endophytic actinobacteria associated with *Dracaena cochinchinensis* Lour.: Isolation, diversity, and their cytotoxic activities. *BioMed Research International*, *2017*. Available from https://doi.org/10.1155/2017/1308563.

Sanjay, A., Purvi, N. P., Meghna, J. V., & Rao, G. G. (2014). Isolation and characterization of endophytic bacteria colonizing halophyte and other salt tolerant plant species from coastal Gujarat. *African Journal of Microbiology Research*, *8*, 1779–1788. Available from https://doi.org/10.5897/ajmr2013.5557.

Santoyo, G., Moreno-Hagelsieb, G., del Carmen Orozco-Mosqueda, M., & Glick, B. R. (2016). Plant growth-promoting bacterial endophytes. *Microbiological Research*, *183*, 92–99. Available from https://doi.org/10.1016/j.micres.2015.11.008.

Saravanan, V. S., Madhaiyan, M., Osborne, J., Thangaraju, M., & Sa, T. M. (2008). Ecological occurrence of *Gluconacetobacter diazotrophicus* and nitrogen-fixing Acetobacteraceae members: Their possible role in plant growth promotion. *Microbial Ecology*, *55*, 130–140.

Schikora, A., Schenk, S. T., & Hartmann, A. (2016). Beneficial effects of bacteria-plant communication based on quorum sensing molecules of the *N*-acyl homoserine lactone group. *Plant Molecular Biology*, *90*, 605–612. Available from https://doi.org/10.1007/s11103-016-0457-8.

Sessitsch, A., Hardoim, P., Döring, J., Weilharter, A., Krause, A., Woyke, T., Mitter, B., Hauberg-Lotte, L., Friedrich, F., Rahalkar, M., Hurek, T., Sarkar, A., Bodrossy, L., Van Overbeek, L., Brar, D., Van Elsas, J. D., Reinhold-Hurek, B. (2012). Functional characteristics of an endophyte community colonizing rice roots as revealed by metagenomic analysis. *Molecular Plant-Microbe Interactions*, *25*, 28–36. Available from https://doi.org/10.1094/MPMI-08-11-0204.

Sevilla, M., Burris, R. H., Gunapala, N., & Kennedy, C. (2001). Comparison of benefit to sugarcane plant growth and 15N2 incorporation following inoculation of sterile plants with acetobacter diazotrophicus wild-type and Nif- mutant strains. *Molecular Plant-Microbe Interactions*, *14*, 358–366. Available from https://doi.org/10.1094/MPMI.2001.14.3.358.

Sgroy, V., Cassán, F., Masciarelli, O., Del Papa, M. F., Lagares, A., Luna, V. (2009). Isolation and characterization of endophytic plant growth-promoting (PGPB) or stress homeostasis-regulating (PSHB) bacteria associated to the halophyte *Prosopis strombulifera*. *Applied Microbiology and Biotechnology*, *85*, 371–381. Available from https://doi.org/10.1007/s00253-009-2116-3.

Shen, Y., Liu, C., Wang, X., Zhao, J., Jia, F., Zhang, Y., Wang, L., Yang, D., Xiang, W. (2013). *Actinoplanes hulinensis* sp. nov., a novel actinomycete isolated from soybean root (*Glycine max* (L.) Merr). *Antonie Van Leeuwenhoek*, *103*, 293–298. Available from https://doi.org/10.1007/s10482-012-9809-9.

Shi, Y., Yang, H., Zhang, T., Sun, J, Lou, K. J. K. (2014). Illumina-based analysis of endophytic bacterial diversity and space-time dynamics in sugar beet on the north slope of Tianshan mountain. *Applied Microbiology and Biotechnology*, *98*, 6375–6385. Available from https://doi.org/10.1007/s00253-014-5720-9.

Shin, M. N., Shim, J., You, Y., Myung, H., Bang, K. S., Cho, M., Kamala-Kannan, S., Oh, B. T. (2012). Characterization of lead resistant endophytic *Bacillus* sp. MN3–4 and its potential for promoting lead accumulation in metal hyperaccumulator *Alnus firma*. *Journal of Hazardous Materials*, *199–200*, 314–320. Available from https://doi.org/10.1016/j.jhazmat.2011.11.010.

Singh, V. K., Singh, A. K., & Kumar, A. (2017). Disease management of tomato through PGPB: Current trends and future perspective. *3 Biotech*, *7*, 1–10. Available from https://doi.org/10.1007/s13205-017-0896-1.

Singh, V. K., Singh, A. K., Singh, P. P., & Kumar, A. (2018). Interaction of plant growth promoting bacteria with tomato under abiotic stress: A review. *Agriculture, Ecosystems & Environment*, *267*, 129–140. Available from https://doi.org/10.1016/j.agee.2018.08.020.

Spaepen, S., Versées, W., Gocke, D., Pohl, M., Steyaert, J., Vanderleyden, J. (2007). Characterization of phenylpyruvate decarboxylase, involved in auxin production of *Azospirillum brasilense*. *Journal of Bacteriology*, *189*, 7626–7633. Available from https://doi.org/10.1128/JB.00830-07.

Straub, T., Zabel, A., Gilfillan, G. D., Feller, C., Becker, P. B. (2013). Different chromatin interfaces of the Drosophila dosage compensation complex revealed by high-shear ChIP-seq. *Genome Research*, *23*, 473–485. Available from https://doi.org/10.1101/gr.146407.112.

Sun, L. N., Zhang, Y. F., He, L. Y., Chen, Z.J., Wang, Q. Y., Qian, M., Sheng, X. F. (2010). Genetic diversity and characterization of heavy metal- resistant-endophytic bacteria from two copper-tolerant plant species on copper mine wasteland. *Bioresource Technology*, *101*, 501–509. Available from https://doi.org/10.1016/j.biortech.2009.08.011.

Sun, X., Guo, L.-D., & Hyde, K. D. (2011). Community composition of endophytic fungi in Acer truncatum and their role in decomposition. *Fungal Divers*, *47*, 85–95. Available from https://doi.org/10.1007/s13225-010-0086-5.

Taghavi, S., Garafola, C., Monchy, S., Newman, L., Hoffman, A., Weyens, N., Barac, T., Vangronsveld, J., van der Lelie, D. (2009). Genome survey and characterization of endophytic bacteria exhibiting a beneficial effect on growth and development of poplar trees. *Applied and Environmental Microbiology*, *75*, 748–757. Available from https://doi.org/10.1128/AEM.02239-08.

Taghavi, S., van der Lelie, D., Hoffman, A., Zhang, Y. B., Walla, M. D., Vangronsveld, J., Newman, L., Monchy, S. (2010). Genome sequence of the plant growth promoting endophytic bacterium *Enterobacter* sp. 638. *PLoS Genetics*, *6*, 19. Available from https://doi.org/10.1371/journal.pgen.1000943.

Tan, X.-M., Chen, X.-M., Wang, C.-L., Jin, X. -H., Cui, J. -L., Chen, J., Guo, S. -X., Zhao, L. -F. (2012). Isolation and identification of endophytic fungi in roots of nine holcoglossum plants (*Orchidaceae*) collected from Yunnan, Guangxi, and Hainan Provinces of China. *Current Microbiology*, *64*, 140–147. Available from https://doi.org/10.1007/s00284-011-0045-8.

Taulé, C., Mareque, C., Barlocco, C., Hackembruch, F., Reis, V. M., Sicardi, M., Battistoni, F. (2012). The contribution of nitrogen fixation to sugarcane (*Saccharum officinarum* L.), and the identification and characterization of part of the associated diazotrophic bacterial community. *Plant and Soil*, *356*, 35–49. Available from https://doi.org/10.1007/s11104-011-1023-4.

Tejesvi, M. V., Kajula, M., Mattila, S., & Pirttilä, A. M. (2011). Bioactivity and genetic diversity of endophytic fungi in *Rhododendron tomentosum* Harmaja. *Fungal Divers*, *47*, 97–107. Available from https://doi.org/10.1007/s13225-010-0087-4.

Thamchaipenet, A., Indananda, C., Bunyoo, C., Duangmal, K., Matsumoto, A., Takahashi, Y. (2010). *Actinoallomurus acaciae* sp. nov., an endophytic actinomycete isolated from *Acacia auriculiformis* A. Cunn. ex Benth. *International Journal of Systematic and Evolutionary Microbiology*, *60*, 554–559. Available from https://doi.org/10.1099/ijs.0.012237-0.

Thawai, C. (2015). *Micromonospora costi* sp. nov., isolated from a leaf of *Costus speciosus*. *International Journal of Systematic and Evolutionary Microbiology*, *65*, 1456–1461. Available from https://doi.org/10.1099/ijs.0.000120.

Thongsandee, W., Matsuda, Y., & Ito, S. (2012). Temporal variations in endophytic fungal assemblages of *Ginkgo biloba* L. *Journal of Forestry Research*, *17*, 213–218. Available from https://doi.org/10.1007/s10310-011-0292-3.

Tichá, M., Illésová, P., Hrbáčková, M., Basheer, J., Novák, D., Hlaváčková, K., Šamajová, O., Niehaus, K., Ovečka, M., Šamaj, J. (2020). Tissue culture, genetic transformation, interaction with beneficial microbes, and modern bio-imaging techniques in alfalfa research. *Critical Reviews in Biotechnology*, *40*, 1265–1280. Available from https://doi.org/10.1080/07388551.2020.1814689.

Timmusk, S., Paalme, V., Pavlicek, T., Bergquist, J., Vangala, A., Danilas, T., Nevo, E. (2011). Bacterial distribution in the rhizosphere of wild barley under contrasting microclimates. *PLoS One*, *6*, 1–7. Available from https://doi.org/10.1371/journal.pone.0017968.

Trdá, L., Fernandez, O., Boutrot, F., Héloir, M. C., Kelloniemi, J., Daire, X., Adrian, M., Clément, C., Zipfel, C., Dorey, S., Poinssot, B. (2014). The grapevine flagellin receptor VvFLS2 differentially recognizes flagellin-derived epitopes from the endophytic growth-promoting bacterium *Burkholderia phytofirmans* and plant pathogenic bacteria. *The New Phytologist*, *201*, 1371–1384. Available from https://doi.org/10.1111/nph.12592.

Tseng, T. T., Tyler, B. M., & Setubal, J. C. (2009). Protein secretion systems in bacterial-host associations, and their description in the Gene Ontology. *BMC Microbiology*, *9*, 1–9. Available from https://doi.org/10.1186/1471-2180-9-S1-S2.

Vieira, M. L. A., Hughes, A. F. S., Gil, V. B., Vaz, A. B. M., Alves, T. M. A., Zani, C. L., Rosa, C. A., Rosa, L. H. (2012). Diversity and antimicrobial activities of the fungal endophyte community associated with the traditional Brazilian medicinal plant Solanum cernuum vell. (Solanaceae). *Canadian Journal of Microbiology*, *58*, 54–66. Available from https://doi.org/10.1139/W11-105.

Visioli, G., D'Egidio, S., Vamerali, T., Mattarozzi, M., & Sanangelantoni, A. M. (2014). Culturable endophytic bacteria enhance Ni translocation in the hyperaccumulator *Noccaea caerulescens*. *Chemosphere*, *117*, 538–544. Available from https://doi.org/10.1016/j.chemosphere.2014.09.014.

Wang, H. F., Zhang, Y. G., Chen, J. Y., Hozzein, W. N., Liu, W. -H., Li, L., Chen, J. -Y., Guo, J. -W., Zhang, Y. -M., Li, W. -J. (2015). *Frigoribacterium endophyticum* sp. nov., an endophytic actinobacterium isolated from the root of *Anabasis elatior* (C. A. Mey.) Schischk. *International Journal of Systematic and Evolutionary Microbiology*, *65*, 1207–1212. Available from https://doi.org/10.1099/ijs.0.000081.

Wang, X., Zhao, J., Liu, C., Wang, J., Shen, Y., Jia, F., Wang, L., Zhang, J., Yu, C., Xiang, W. (2013). Isolated from eggplant root (Solanum melongena L .) 2418–2423. Https://doi.org/10.1099/ijs.0.045617-0.

Wang, H.-F., Zhang, Y.-G., Cheng, J., Guo, J. W., Li, L., Hozzein, W. N., Zhang, Y. M., Wadaan, M. A. M., Li, W. J. (2015). *Labedella endophytica* sp. nov., a novel endophytic actinobacterium isolated from stem of *Anabasis elatior* (C. A. Mey.) Schischk. *Antonie Van Leeuwenhoek*, *107*, 95–102. Available from https://doi.org/10.1007/s10482-014-0307-0.

Weilharter, A., Mitter, B., Shin, M. V., Chain, P. S. G., Nowak, J., Sessitsch, A. (2011). Complete genome sequence of the plant growth-promoting endophyte *Burkholderia phytofirmans* strain PsJN. *Journal of Bacteriology*, *193*, 3383–3384.

Wiewióra, B., Żurek, G., & Pańka, D. (2015). Is the vertical transmission of *Neotyphodium lolii* in perennial ryegrass the only possible way to the spread of endophytes? *PLoS One*, *10*, 1–11. Available from https://doi.org/10.1371/journal.pone.0117231.

Wu, H. (2012). Culture-dependent diversity of actinobacteria associated with seagrass (*Thalassia hemprichii*). *African Journal of Microbiology Research*, *6*, 87–94. Available from https://doi.org/10.5897/ajmr11.981.

Wu, H., Yang, H.-Y., You, X.-L., & Li, Y.-H. (2013). Diversity of endophytic fungi from roots of *Panax ginseng* and their saponin yield capacities. *Springerplus*, *2*, 107. Available from https://doi.org/10.1186/2193-1801-2-107.

Xie, Q. Y., Wang, C., Wang, R., Lin, H. P., Goodfellow, M., Hong, K. (2011). *Jishengella endophytica* gen. nov., sp. nov., a new member of the family *Micromonosporaceae*. *International Journal of Systematic and Evolutionary Microbiology*, *61*, 1153–1159. Available from https://doi.org/10.1099/ijs.0.025288-0.

Xing, K., Liu, W., Zhang, Y.-J., Bian, G. -K., Zhang, W. -D., Tamura, T., Lee, J. -S., Qin, S., Jiang, J. -H. (2013). *Amycolatopsis jiangsuensis* sp. nov., a novel endophytic actinomycete isolated from a coastal plant in Jiangsu, China. *Antonie Van Leeuwenhoek*, *103*, 433–439. Available from https://doi.org/10.1007/s10482-012-9823-y.

Xing, K., Qin, S., Zhang, W.-D., Cao, C. -L., Ruan, J. -S., Huang, Y., Jiang, J. -H. (2014). *Glycomyces phytohabitans* sp. nov., a novel endophytic actinomycete isolated from the coastal halophyte in Jiangsu, East China. *The Journal of Antibiotics*, *67*, 559–563. Available from https://doi.org/10.1038/ja.2014.40.

Xing, Y.-M., Chen, J., Cui, J.-L., Chen, X. -M., Guo, S. -X. (2011). Antimicrobial activity and biodiversity of endophytic fungi in *Dendrobium devonianum* and *Dendrobium thyrsiflorum* from Vietman. *Current Microbiology*, *62*, 1218–1224. Available from https://doi.org/10.1007/s00284-010-9848-2.

Xiong, Z., Wang, J., Hao, Y., & Wang, Y. (2013). Recent advances in the discovery and development of marine microbial natural products. *Marine Drugs*, *11*, 700–717. Available from https://doi.org/10.3390/md11030700.

Xiong, Z. J., Zhang, J. L., Zhang, D. F., Zhou, Z. L., Liu, M. J., Zhu, W. Y., Zhao, L. X., Xu, L. H., Li, W. J. (2013). *Rothia endophytica* sp. nov., an actinobacterium isolated from *Dysophylla stellata* (Lour.) benth. *International Journal of Systematic and

Evolutionary Microbiology, *63*, 3964–3969. Available from https://doi.org/10.1099/ijs.0.052522-0.

Yan, L., & Khan, R. A. A. (2021). Biological control of bacterial wilt in tomato through the metabolites produced by the biocontrol fungus, *Trichoderma harzianum*. *Egyptian Journal of Biological Pest Control*, *31*. Available from https://doi.org/10.1186/s41938-020-00351-9.

You, J., & Chan, Z. (2015). Ros regulation during abiotic stress responses in crop plants. *Frontiers in Plant Science*, *6*, 1–15. Available from https://doi.org/10.3389/fpls.2015.01092.

Zamioudis, C., & Pieterse, C. M. J. (2012). Modulation of host immunity by beneficial microbes. *Molecular Plant-Microbe Interactions*, *25*, 139–150. Available from https://doi.org/10.1094/MPMI-06-11-0179.

Zhang, W., Xu, L., Yang, L., Huang, Y., Li, S., Shen, Y. (2014). Phomopsidone A, a novel depsidone metabolite from the mangrove endophytic fungus *Phomopsis* sp. A123. *Fitoterapia*, *96*, 146–151. Available from https://doi.org/10.1016/j.fitote.2014.05.001.

Zhang, X., Lin, L., Zhu, Z., Yang, X., Wang, Y., An, Q. (2013). Colonization and modulation of host growth and metal uptake by endophytic bacteria of *Sedum alfredii*. *International Journal of Phytoremediation*, *15*, 51–64. Available from https://doi.org/10.1080/15226514.2012.670315.

Zhang, Y. G., Wang, H. F., Alkhalifah, D. H. M., Xiao, M., Zhou, X. K., Liu, Y. H., Hozzein, W. N., Li, W. J. (2018). *Glycomyces anabasis* sp. nov., a novel endophytic actinobacterium isolated from roots of *Anabasis aphylla* L. *International Journal of Systematic and Evolutionary Microbiology*, *68*, 1285–1290. Available from https://doi.org/10.1099/ijsem.0.002668.

Zinniel, D. K., Lambrecht, P., Harris, N. B., Feng, Z., Kuczmarski, D., Higley, P., Ishimaru, C. A., Arunakumari, A., Barletta, R. G., Vidaver, A. K. (2002). Isolation and characterization of endophytic colonizing bacteria from agronomic crops and prairie plants. *Applied and Environmental Microbiology*, *68*, 2198–2208.

Zipfel, C., & Oldroyd, G. E. D. (2017). Plant signalling in symbiosis and immunity. *Nature*, *543*, 328–336. Available from https://doi.org/10.1038/nature22009.

Zhu, W., Zhang, J., Qin, Y., Xiong, Z., Zhang, D., Klenk, H., Zhao, L., Xu, L., Li, W. (2013). Bacterium isolated from Camptotheca acuminata 3269–3273.

Zhu, W., Zhao, L., Zhao, G., Duan, X., Qin, S., Li, J., Xu, L., Li, W. (2012). Actinomycete isolated from Camptotheca acuminata Decne ., reclassification of Actinaurispora siamensis as Plantactinospora siamensis comb . nov . and emended descriptions of the genus Plantactinospora and Plantactinospora mayteni Printed in Great Britain 2435–2442. Available from: https://doi.org/10.1099/ijs.0.036459-0

CHAPTER

Recent trends in characterization of endophytic microorganisms

2

Ayesha T. Tahir[1], Jun Kang[2], Musfirah Bint-e-Mansoor[1], Javeria Ayub[1,4], Zakira Naureen[3] and Fauzia Yusuf Hafeez[1]

[1]Department of Biosciences, COMSATS University Islamabad, Islamabad, Pakistan
[2]School of Life Sciences, Tianjin University, Tianjin, P.R. China
[3]Department of Biological Sciences and Chemistry, College of Arts and Science, University of Nizwa, Nizwa, Oman
[4]Department of Microbiology, Balochistan University of Information Technology, Engineering and Management Sciences Quetta, Quetta, Pakistan

2.1 Introduction

Plants dwell in dynamic relationship with a variety of microbial communities, out of which some reside inside the plant as endosymbionts without causing any disease or adverse effects and are named endophytes. These endosymbionts were first described by the German botanist Friedrich in 1809 (Link, 1809). The term "endophytes" has been derived from two Greek words "endon" and "phyton," which means within plant. Endophytes are ubiquitous and can be isolated from wide variety of plants, including pteridophytes, bryophytes, angiosperms, and gymnosperms. The plant microecosystem comprises mainly of bacteria and fungi (Manias, Verma, & Soni, 2020). Among these, endophytic bacteria/fungi live asymptomatically in the plants and might develop a symbiotic association with them. However, endophytes can sometimes change their regime and turn into a pathogen, ending up causing diseases in the host (Martinez-Klimova, Rodríguez-Peña, & Sánchez, 2017).

Both plants and endophytes share mutual benefits. Plant provides endophytes with food and shelter and, in turn, endophytes help to induce biotic and abiotic stress tolerance, mediate nutrient uptake, and enhance growth thereby increasing the productivity. In addition, the endophytes protect the plants from herbivores, pests, and pathogens by direct antagonism, quorum sensing, and induction of systemic resistance (Hardoim et al., 2015). They can also enhance the levels of plant growth−promoting (PGP) hormones (cytokines, auxin, and gibberellins) and facilitate nutrient absorption by solubilizing them from complex to simple form (Santoyo, Moreno-Hagelsieb, del Carmen Orozco-Mosqueda, & Glick, 2016).

Besides being beneficial to plants in a variety of ways, the endophytes are important from industrial point of view as a source of important enzymes, novel

Biocontrol Mechanisms of Endophytic Microorganisms. DOI: https://doi.org/10.1016/B978-0-323-88478-5.00012-2

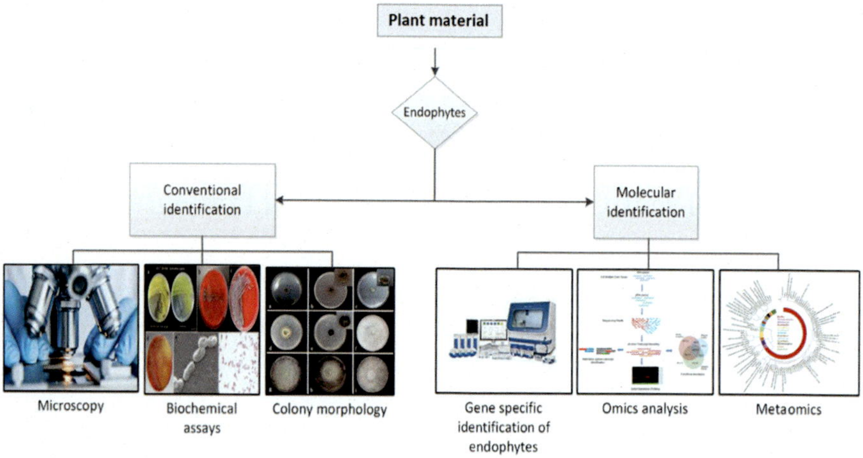

FIGURE 2.1

Schematic diagram showing conventional and modern molecular techniques involved in the identification of endophytic microorganisms.

metabolites, and pharmaceutically important compounds (Kaul, Gupta, Ahmed, & Dhar, 2012; Mousa & Raizada, 2013; Premjanu & Jayanthy, 2012; Strobel & Daisy, 2003). In addition, these plant-endophyte symbiotic associations are also beneficial for phytoremediation by removing pollutants from water and soil.

A large variety of endophytic microorganisms have been isolated from various plant tissues and according to National Center for Biotechnology Information (NCBI) nucleotide database, nearly 70% of the isolated endophytes are fungi, while the remaining 30% are bacteria. The majority of uncultivable endophytes are quite difficult to grow in vitro. Therefore isolation and cultivation processes differ between endophytic species, and the correct choice of method can greatly influence the outcome. Advances in molecular techniques have made the characterization of uncultivable species easier.

In this chapter, we attempt to describe the commonly used techniques and methods for the endophytes characterization followed by modern approaches such as genomics, proteomics, metabolomics, and metaomics that are used to find potent endophytic bacterial agents and to explore plant–endophyte interactions. A schematic diagram has been shown in Fig. 2.1 to describe the conventional alongwith modern techniques used for the identification of endophytic microorganisms.

2.2 Conventional characterization of endophytes

The common steps that are involved in the isolation of endophytes are shown in Fig. 2.2. Preliminary steps from sample collection to surface sterilization and

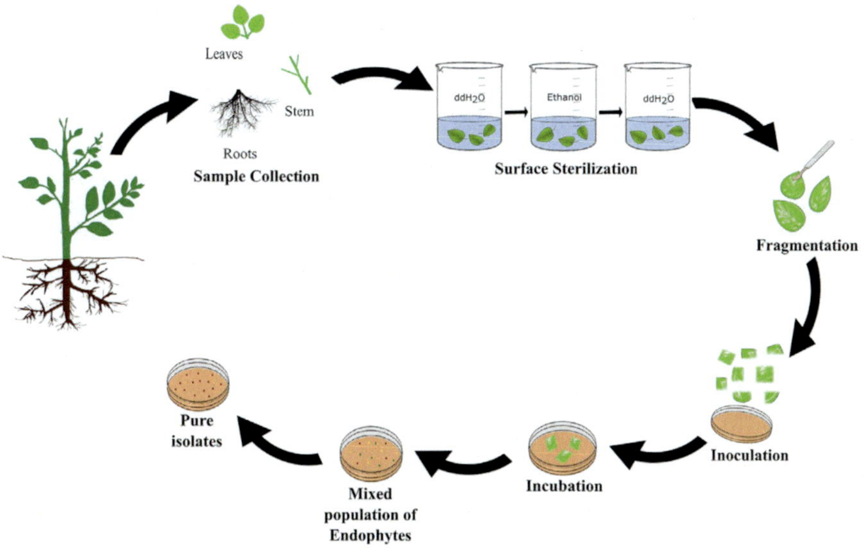

FIGURE 2.2

Steps involved in isolation of endophytes (Manias et al., 2020).

plant tissue fragmentation are similar for both bacteria and fungi isolation (Manias et al., 2020).

The growth medium selection is very crucial step in the isolation of endophytes from any plant tissue for instance root tissue. Most commonly used medium for the isolation of endophytic bacteria is nutrient agar (NA) which is an undefined general purpose medium as there is no component in the nutrient agar that suppresses the growth of fungi so it is usually supplemented with antifungal agent such as nystatin (30 μg/mL) (Nawed & Chandra, 2015).

For the isolation of endophytic agents first healthy unbroken plant tissue (leave, root nodule, stem, or seeds) is taken, surface sterilized with 2% sodium hypochlorite for 10 minutes followed by 95% ethanol washing, finally rinsed with water four to five times. The efficacy of disinfection process can be evaluated by surface sterility test. Thereafter, plant sample (1 g) is aseptically ground in saline phosphate buffer (KCl, NaCl, KH_2PO_4, and Na_2HPO_4, pH 7.4) and centrifuged for 1 minute at 4°C. The supernatant is serially diluted up to 10^{-5} fold and culture on growth media, after subculturing pure culture is obtained and their glycerol stock is preserved for further characterization (Aravind, Kumar, Eapen, & Ramana, 2009).

2.2.1 Morphological characterization

These isolates are morphologically characterized on the basis of colony phenotypic characteristics such as color, shape, margin, form (filamentous, circular, and irregular) elevation, surface, texture, and opacity (Pelczar Jr & Reid, 1965).

Further, Gram's staining as well as endospore staining is also used to characterize them. Endophytic fungi are characterized preliminary on the basis of their hyphae structure.

2.2.2 Biochemical and physiological characterization

Different biochemical tests such as oxidase, catalase, Voges–Proskauer test, methyl red, hydrogen cyanide (HCN) production, ammonia production, starch hydrolysis test, casein hydrolysis test, fermentation assay, utilization of citrate and NaCl tolerance can be assessed for endophytic microorganisms (Bergey, Buchanan, Gibbons, & American Society for Microbiology, 1974).

2.2.3 Plant growth promoting and biocontrol activities

Plant root–inhabiting bacteria possess a myriad of activities that help the plant to sequester nutrients and grow better. These bacteria can fix nitrogen, solubilize micro-/macronutrients and produce phytohormones such as Indole acetic acid (IAA) and Gibberellic acid (GA). In addition to that they produce organic acids and ammonia. Some of the bacteria produce hydrolytic enzymes, such as hydrogen cyanide (HCN), and antibiotics that help to eradicate the pathogenic microorganisms in the vicinity. Conventional methodologies used in characterizing newly isolated bacterial strains mainly include microbiological assays using various substrates. A list of commonly used microbiological assays for detecting growth-promoting capabilities of plant root–associated bacteria and endophytes is shown in Table 2.1.

2.3 Characterization of endophytes using modern techniques

Omic tools, including genomics, proteomics, transcriptomics, and metabolomics, are increasingly used for endophyte characterization. These omics strategies allow the analysis of microbiota such as fungal and bacterial strains, their interaction with plants, as well as analysis of whole genome, proteome, transcriptome, and metabolome to enhance interaction of both in a positive manner (Fig. 2.3).

2.3.1 Genomics/metagenomics

Endophyte characterization based on morphology is not comprehensive due to the complex interaction between host and endophytes. Genomic analysis is important not only to unravel the intricate web of interactions but also to manipulate genetic information to enhance positive association between plant and its endophytes (del Carmen Orozco-Mosqueda & Santoyo, 2020).

Table 2.1 Conventional methods of characterization for plant growth promotion and biocontrol activities of plant root–associated bacteria.

Activities	Microbiological assays/reagents	Media	References
Nitrogen fixation	ARA	Nitrogen-free malate medium	Hardy, Burns, and Holsten (1973)
Ammonia production	Nesseler's reagent	Peptone liquid medium	Demutskaya and Kalinichenko (2010)
P-solubilization	Agar medium containing insoluble tricalcium phosphate	PKV's agar	Nautiyal (1999)
Zn solubilization	Agar medium containing insoluble zinc 0.1% zinc oxide and zinc sulfate	Various media amended with insoluble zinc	Costerousse, Schönholzer-Mauclaire, Frossard, and Thonar (2018)
Organic acid production	PKV's medium and alizarin red S pH indicator	PKV's agar	Sánchez-de Prager and Cisneros-Rojas (2017)
Siderophore production	CAS	CAS agarStandard succinate medium	Schwyn and Neilands (1987), Naureen, Price, Hafeez, and Roberts (2009)
IAA production	Salkowski colorimetric technique	Agar medium supplemented with tryptophan such as Malate medium supplemented with tryptophan and Salkowski reagent for quantification	Gordon and Weber (1951), Okon, Albrecht, and Burris (1977)
GA production	Folin-Wu reagent	Trypticase soy broth supplemented with glucose	Graham and Henderson (1961)
ACC deaminase activity	Medium with ACC as sole carbon source	ACC medium	Penrose and Glick (2003)
Volatile antibiotic production	Dual culture assay	Potato dextrose agar	Naureen et al. (2009)
Diffusible antibiotic production	Dual culture assay	Potato dextrose agar	Naureen et al. (2009)

(Continued)

Table 2.1 Conventional methods of characterization for plant growth promotion and biocontrol activities of plant root–associated bacteria. *Continued*

Activities	Microbiological assays/reagents	Media	References
HCN production	Picric acid dipped filter paper method	NA amended with glycine	Schippers, Bakker, Bakker, and Van Peer (1990)
Surfactant production	Drop collapse assayBlue agar assay	Blue agar; tryptic soy agar amended with methylene blue and CTAB	Siegmund and Wagner (1991)
Antagonistic activity	Dual culture assay	Potato dextrose agar	Naureen et al. (2009)
Proteases	Skimmed milk agar	Agar media containing skimmed milk only	Abo-Aba, Soliman, and Nivien (2006)
Chitinases	Colloidal chitin method	Agar medium amended with colloidal chitin	O'Brien and Colwell (1987)
Cellulases	CMC	Agar medium amended with CMC	Kasana, Salwan, Dhar, Dutt, and Gulati (2008)
Amylases	Potassium iodide	Starch agar	Malleswari and Bagyanarayana (2013)

ACC, 1-Amino-cyclopropane-1-carboxylate*; ARA,* acetylene reduction assay*; CAS,* chrome azurol S*; CMC,* carboxymethyl cellulose*; PKV,* Pikovskaya*, NA,* nutrient agar*; CTAB,* cetyl trimethylammonium bromide.

Bacterial endophytes are important for sustainable agriculture by promoting plant growth as well as protecting them from phytopathogens (Khatoon et al., 2020). Roots are the main habitat for the bacterial endophytes, with significantly higher diversity and abundance than any other tissue of the plant. Further, root endophytes might act as second line of defense (as first line is rhizobacteria) against phytopathogens in plants (Carrión et al., 2019). Since the advent of next-generation sequencing (NGS) techniques, genomes of various endophytic bacteria have been sequenced and summarized in various studies, such as *Bacillus subtilis* (Deng et al., 2011), *Herbaspirillum seropedicae* strain SmR1 (Pedrosa et al., 2011), *Stenotrophomonas maltophilia* RR-10 (Zhu et al., 2012), *Enterobacter cloacae* subsp. Cloacae strain ENHKU01 (Liu et al., 2012), *Enterobacter radicincitans* DSM16656(T) (Witzel, Gwinn-Giglio, Nadendla, Shefchek, & Ruppel, 2012), *Rhizobium* sp. strain IRBG74 (Crook, Mitra, Ané, Sadowsky, & Gyaneshwar, 2013), *Bacillus thuringiensis* KB1 (Jeong, Jo, Hong, & Park, 2016), *Staphylococcus epidermidis* strains (Chaudhry & Patil, 2016), *Bacillus* sp. Strain E25 (Pérez-Equihua & Santoyo, 2021), and many more completed and are under process.

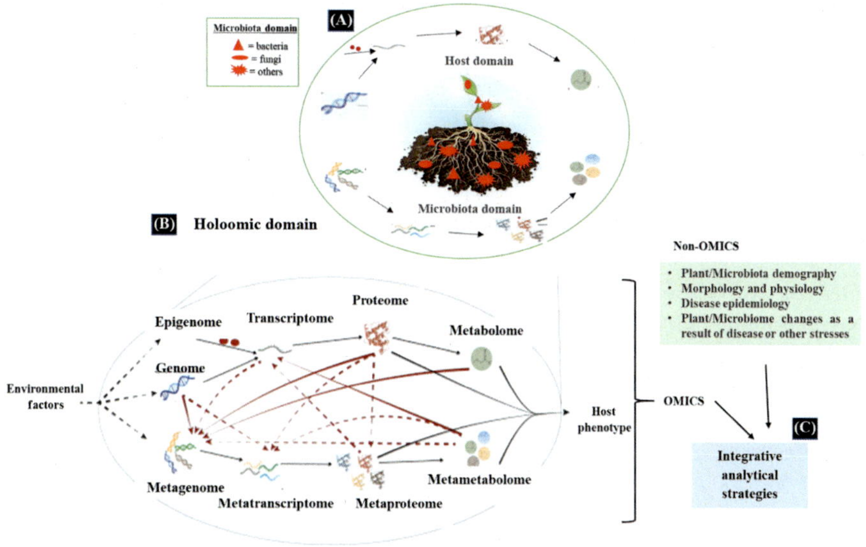

FIGURE 2.3

Plant (host)–endophyte (microbiota) interaction within holoomic approach is illustrated. (A) Host plant and associated microbiota with multiomics are presented. (B) Holoomic interactions are presented, with already reported and possible ways of interactions among omics domains. (C) Omics and nonomics data integration: potential domain to implement for better understanding of host–microbe interactions.

Adapted from Nyholm, L., Koziol, A., Marcos, S., Botnen, A. B., Aizpurua, O., Gopalakrishnan, S., ... & Alberdi, A. (2020). Holo-omics: Integrated host-microbiota multi-omics for basic and applied biological research. iScience, 23, 101414.

Genomic analysis facilitates for the interpretation of beneficial traits via in silico analysis, further greenhouse inoculation experiments can help to confirm these interpretations. Such as, single-molecule real-time sequencing was used for *Sphingomonas* sp. LK11 genome analysis. Functional analysis of LK11 showed that it encodes for the nutrient acquisition genes that are beneficial for plant growth as it was tested positive on soybean. This study concluded *Sphingomonas* sp. LK11 as ecofriendly agent for cleaning contaminated areas as well as plant growth promotion during environmental stresses (Asaf, Khan, Khan, Al-Harrasi, & Lee, 2018). Recently, complete genome analysis of *B. thuringiensis* CR71 from Mexican husk tomato plants unraveled multiple PGP actions of this strain, including phytohormone production as well as biofilm formation. Inoculation of this strain to cucumber showed 34.97% increase in plant production (Flores et al., 2020).

Bacterial genomes, due to their simpler compositions, were successfully sequenced in the mid-1990s. On the other hand, fungus being a eukaryote having multiple chromosomes, larger genome sizes, and complex genomic systems lagged behind in this race (Sharma, 2016). Endophytic fungal strains that have been sequenced include *Piriformospora indica* (Zuccaro et al., 2011), *Ascocoryne sarcoides* (Gianoulis et al., 2012), *Shiraia* sp. slf14 (Yang, Wang, Zhang, Yan, & Zhu, 2014), *Harpophora oryzae* (Xu et al., 2014), *Rhodotorula graminis* WP1 (Firrincieli et al., 2015), *Phialocephala scopiformis* DAOMC 229536 (Walker et al., 2016), and *Xylona heveae* (Gazis et al., 2016).

Analysis of genome sequences allowed genomic manipulations for functional validation of genes as well as for the development of strains with enhanced characteristics of desired traits. *Trichoderma atroviride* and *Trichoderma virens*, the two important biocontrol species, were sequenced and an analysis revealed the number of important genes that have role in interaction with plants and mycoparasitism, secondary metabolite production, polyketide synthases, and nonribosomal peptide synthetases (Kubicek et al., 2011). Owing to their importance, recently *T. virens* mutants were generated by gamma rays–induced mutagenesis. The mutant M7 showed several changes in morphology, production in metabolite production, as well as lacked functions that were involved in plant interaction and was no more able to use chitin and cellulose as carbon source. It was observed that genes involved in secondary metabolism and functions like hydrophobicity and transportation were downregulated. Furthermore, upon whole-genome sequencing of M7, 250 kb of genomic region was deleted and this region was predicted to have 71 Open Reading Frames (ORFs). This valuable work gave new insights of genetic morphogenesis, mycoparasitism, and other functionalities which can unravel beneficial traits for interaction between *Trichoderma* and host plants (Pachauri, Sherkhane, Kumar, & Mukherjee, 2020).

Unraveling beneficial properties and metabolic mechanisms are very important to completely understand and manipulate the association of endophytes with their host plant but, unfortunately, our understanding of this complex phenomenon is hindered by the fact that many endophytes are unable to be cultured in an artificial medium. In this regard, metagenomics is a helpful significant omics tool that is cultivation independent and involves genomic analysis of microbial communities originating from diverse ecologies. Hence, produce data a lot more than just simple genomic information from an individual member (Subudhi, Sahoo, Dey, Das, & Sahoo, 2018). High potential of entophytic community for the enhancement of plant growth, biocontrol, and bioremediation properties has been reported by metagenomics approach in roots of rice (Sessitsch et al., 2012).

One of the recent researches elucidated bacterial endophytes of *Panax ginseng* roots at different plant ages between 2 and 6 years, by shotgun sequencing method. Metagenomic analysis revealed that Proteobacteria and Actinobacteria were predominantly present in all plants of age 2–6 years, while α-proteobacteria was abundant in roots of 3-, 4-, and 5-year-old plant roots. Moreover, endophytic bacterial functions were predicted by SEED subsystem analysis (http://pubseed.theseed.org/) which showed that PGP activities were majorly found in plant having 3 years of the age which was also supported by gene frequency analysis of plant growth promotion-related genes. All these results showed that endophytic community in the roots of *P. ginseng* affect plant growth (Hong, Kim, Lee, Bang, & Jo, 2019).

To study diverse microbial communities, comparative metagenomics play its role. It has been successful to disclose not only morphological but also genetic diversity among endophytes, which may inhabit the same or different plants. Many studies have used comparative metagenomics approach to study diversity. Noronha, Júnior, Gilbert, and de Oliveira (2017) used the same approach to study taxonomic and functional patterns across different microbiome. Total 14 varieties of upland rice from different places of China were collected to find diversity of endophytes in their seeds. There were differences in abundance of Proteobacteria, dominant phyla in all samples, but not significant diversity in seed endophytic community (Wang et al., 2021).

Metagenomic studies are being made easier by using NGS that helps in rapid characterization of microbiota (Hong et al., 2019). After sequencing comes the postsequencing analysis that includes assembly, annotation, ORF analysis, taxonomic profiling, and metabolic reconstruction that are major challenging steps for comprehensive interpretations and are being resolved quickly by various upgraded bioinformatics tools.

2.3.2 Transcriptomics/metatranscriptomics

Genomics and metagenomics approaches highlight the presence of genes, whereas information regarding their expression remained elusive. Gene expression is strongly influenced by the environmental conditions; hence, pivotal to understand endophytes and their complex relationship with host plant. To understand phenomenon of differentially expressed genes, array of mRNA referred to as transcriptome is dissected using various tools. Just like comparative analysis of genome that, of transcriptome can be carried out to check expression variations of genes using host plant in the presence or absence of endophyte. This approach helped to unravel factors responsible for endophytic mechanism, the production of phytohormones, as well as the enhancement of plant growth and development (Ambrose & Belanger, 2012). *Bacillus mycoides*, a biocontrol agent as well as Plant Growth Promoting Rhizobacteria (PGPR) were isolated and purified from potatoes *endosphere* and soil of surrounding area. GFP (green fluorescent protein)-tagged *B. mycoides* EC18 colonizes plant successfully and strongly as compared to isolate SB8. As both the strains showed phenotypic differences, thus their transcriptome was analyzed which showed that endophytic strain EC18 responded more pronounced than SB8, the soil derived strain, and both showed high number of differentially expressed genes. Some genes involving membrane proteins and transcription factors encoding genes were significantly found to be upregulated in EC18 strain while downregulated in SB8. This study observed several strategies through transcriptome analysis that endophytes and soil isolate of *B. mycoides*, opt for survival (Yi, de Jong, Frenzel, & Kuipers, 2017).

Moreover, transcriptome analysis also helped to reveal the importance of endophytes in abiotic stresses and aids to unravel hidden mechanisms of stress mitigation. Tanscriptomic analysis of *Aspergillus montevidensis* ZYD4, an endophytic fungus, in the presence and absence of salt helped to understand molecular basis involved in osmoadaptive mechanisms of halophilic endophytic fungi. This study reported that genes involved in storage of sugars, accumulation of glycerol, organic acids, asexual sporulation, and pigment production showed differential expression under salt stress (Liu et al., 2017). Another study reported transcriptome analysis and differential gene expression in fescue having different endophytic strains under water stress (Dinkins, Nagabhyru, Young, West, & Schardl, 2019).

Metatranscriptomic analysis refers to the isolation of RNA or transcriptome of whole endophytic community which results into linkage between genome and its differential expression due to the ecological or environmental conditions. Complementary DNA (cDNA) formation was a challenging step in prokaryotes as understanding of polyadenylation lagged behind compared to its establishment in eukaryotes. The advancements in molecular biology enabled the use of mRNA and rRNA to develop double RNA method in which these RNA molecules are used to analyze different microbial communities and their expression in a specific environment. A study of metatranscriptomic analysis on soybean plant unraveled the number of small RNA sequences that were not related to its genome. Comparative transcriptomic analysis showed that these small RNA sequences are due to the presence of endophytic as well as pathogenic microbes in soybean plant tissues (Molina et al., 2012). Another study reported the use of combinatorial technique of Illumina RNA-seq of total RNA pyro-sequenced small subunit ribosomal ribonucleic acid (SSU-rRNA) genes to reveal the expression of different enzymes and microbial population that were responsible to degrade exoskeletons of peat arthropods (Ivanova, Wegner, Kim, Liesack, & Dedysh, 2016).

In addition to whole-transcriptome analysis, posttranscriptional modifications can be studied by isolating polyadenylated mRNA from sample using alternative extraction protocols (Yadav, Bragalini, Fraissinet-Tachet, Marmeisse, & Luis, 2016). Another approach is the "dual transcriptomics" in which the transcriptomes of host as well as its endophytes are analyzed for better understanding of their complex interaction, signaling, and symbiosis (Gómez-Godínez, Fernandez-Valverde, Romero, & Martínez-Romero, 2019). In addition to mRNA, noncoding RNAs such as miRNA have also been identified as key player in host−endophyte recognition and interaction (Pentimone et al., 2018).

Though transcriptome analysis gives a huge amount of information and a great insight of microbial diversity, community composition, and interaction between host and endophytes, yet it has certain limitations. Extraction of mRNA from environmental sample is not as easy as extraction of DNA due to its low concentration and the presence of other RNA molecules like transfer RNA and non-coding RNA. Scientists may have to add amplification steps to increase RNA concentrations. Moreover, protocols should be standardized to separate mRNA from non-mRNA which may impede transcription patterns (Abedi, Fatehi, Moradzadeh, & Gheisari, 2019).

2.3.3 Proteomics/metaproteomics

Postgenomic analysis gained much attention as the genomics and metagenomics approaches failed to reveal functional information and attributes about microbial communities. Proteomics/metaproteomics is one of the postgenomic tools where proteome (total protein content) of an organism or functional expression of whole community can be analyzed. Hence, provide information additional to those given by genomics and transcriptomics.

Microbial proteomics is the focus of researchers, and many protocols and technical advances have been achieved. Mainly studies focused to understand host–pathogen and plant–inducer interactions, where profiling of defense- and pathogenicity-related proteins have been performed using various pathogen–plant combinations (Ashwin et al., 2017). Fungal proteomics is subfield of microbial proteomics and helped to unravel the various fungal forms, including biotrophy, necrotrophy, hemibiotrophy, endophytism, and latency. These forms are obtained by studying differential gene expression and formation of those proteins which help fungi to maintain a specific form as well as switching between parasitism and mutualism. Such as *Colletotrichum* genus consisting of 29–700 species of fungi, which mostly supposed to act as phytopathogen in the past but can also form mutualistic endophytic relationship with the host. Proteome profiling could be beneficial to understand those proteins involved in transition of this form (Pillai, 2017). Yuan et al. reported postgenomics analysis at transcriptional and translational level to elucidate genes involved in the interaction of endophyte *Gilmaniella* sp. AL12 using *Atractylodes lancea* host plant (Yuan et al., 2019). Similarly, proteomics approach can also be used to unravel mechanisms involved for phytoremediation in both plants and microorganisms. Recently, Liquid Chromatography with tandem mass spectrometry (LC–MS-MS) was used to find that *Methylobacterium* sp. help to improve plant health under phytotoxic arsenic concentrations by changes in proteins related to energy metabolism and redox homeostasis (Alcántara-Martínez, Figueroa-Martínez, Rivera-Cabrera, Gutiérrez-Sánchez, & Volke-Sepúlveda, 2018).

To get a broader overview of microbial community functional profiles, inhabiting the plant tissue, metaproteomics plays a key role. Metaproteomics deals with direct identification and analysis of whole microbiota from an environmental sample. Metaproteomics can primarily be done by direct lysis method in which whole proteome is extracted from endosphere subjected to different environments and comparative analysis of their fingerprinting using gel-based or gel-free methods is performed (Maron, Ranjard, Mougel, & Lemanceau, 2007). The indirect method of lysis can also be implemented, where the proteome extraction of isolated endophytes under given stress treatment is done. Moreover, protein analysis of host plant in the presence or absence of endophytes enables to understand the role of proteins involved in endophytism (Li et al., 2012). Soil metaproteomic analyses have been conducted to elucidate functions and structures of microbial communities and the way they contribute to their respective ecosystem (Starke, Jehmlich, & Bastida, 2019). Interestingly, phylogenetic correlation of genomic and proteomic outcomes is quite comparable, hence encouraging to use proteomics for capturing not only functional information but also to figure out phylogenetic relations (Bastida et al., 2017; Starke et al., 2017).

Though the benefits of proteomics and metaproteomics applications are remarkable, we cannot ignore their limitations particularly those of phytomicrobial proteomics. Nonavailability of whole-genome sequence in plant species leads to inaccurate protein identification during in silico data analysis (Tahir et al., 2020). Moreover, protein extraction protocols are yet to standardize for various peculiar tissues. On the other hand, in spite of genome availability protein coverage is also one of the issues in proteomics. Low-abundant proteins are another impediment that can be overhauled by various enrichment procedures and subsequent modifications in analysis tools (Righetti & Boschetti, 2016).

2.3.4 Metabolomics/meta metabolomics

Metabolomics is essential for a complete understanding of complex mechanisms involving plant microbe's interaction, either at community or at individual level. Metabolomics deals with metabolome, set of metabolites, which any organism produces under given condition. Hence, it can measure organism's inherent functional status and can better represent phenotype than genomics measurements. Metabolites are organic, low molecular weight biomolecules and are present in cell, tissue, or body fluids. Plant metabolites are complex molecules and have a great diversity ranging between 200,000 and 1,000,000 but only few are characterized (Cragg & Newman, 2013). For efficient separation, isolation, detection, systemic characterization, and quantification of metabolites, targeted or nontargeted metabolomic methods are based mainly on mass spectrometry, nuclear magnetic resonance spectroscopy, or vibrational spectroscopy (Tahir, Fatmi, Nosheen, Imtiaz, & Khan, 2019). Metabolomics have significant contribution in field of microbial metabolomics, particularly related to environmental and agricultural samples. It can help in the development of biomarkers for disease detection, risk assessment against toxicants, and biotic as well as abiotic stresses.

Secondary metabolites produced by fungal endophytes mainly belong to polyketide, nonribosomal peptides, alkaloids, and terpenes groups (Liu & Liu, 2018). Many of these secondary metabolites are of great medicinal importance. A study by Wei et al. (2020) revealed that endophytic fungal strain Pr10 from medicinal plant *Brassica rapa* L. had significant antitumor activity as metabolomic extract was rich in amino acids and those sugar derivatives that had been previously proved to be effective against A549 cell line. Another study evidenced the importance of fungal secondary metabolites to be useful in the production of drugs. In this study, anticancer and antitrypanosome secondary metabolites from an endophytic fungus *Aspergillus flocculus* were identified using 2D-NMR (Tawfike et al., 2019). Similarly, another study showed that metabolome of endophytic fungus *Cercospora* sp. from the medicinal plant *Aerva javanica Juss. Ex.Schult* has biocontrol features exhibiting antioxidant and antimicrobial activities (Mookherjee, Mitra, Kutty, Mitra, & Maiti, 2020).

Metabolomics also played its role to unravel the PGP ability of bacterial endophytes. To understand the complexity of interaction between endophyte and its host, a group of researchers inoculated cucumber plants with isolated endophytic strains to decipher their role in plant growth. Metabolomic profiling showed that endophytic strains affected plant metabolomes, as inoculated plant metabolites involved in plant growth were 50% more concentrated than control plants without inoculum (Mahmood & Kataoka, 2020). In another study, metabolomic analysis using mass spectrometry was conducted to check the metabolomic changes involved in foliar inoculation system of *Medicago truncatula* by an endophyte *Arthrobacter agilis* UMCV2. The results suggested that the colonization of *A. agilis* UMCV2 is independent of the phenological age of leaf and modulate sugar/carbohydrate metabolism without disturbing other metabolic activities (Ramírez-Ordorica et al., 2020).

Though metabolomics can be used as powerful omics strategy to select beneficial traits to improve plant—endophyte interaction as well as to complement other omics-generated data, yet this emerging field must overcome several challenges which involve inability of data gained through metabolic profiling to generate pathways and determine regulatory element. For this purpose the use of integrated omics approach could be beneficial but is not easy due to the complexity in data integration (Huang, Chaudhary, & Garmire, 2017). On the other hand, modeling for the reconstruction of metabolic pathways, expert's evaluations having command on variant nature of metabolites, and thorough understanding of microbial mechanisms is needed. To overcome all these challenges, computational metabolomics tools have to be developed (Johnson, Ivanisevic, Benton, & Siuzdak, 2015) as well as automated metabolomic platforms are needed for the quantitative and qualitative identification of molecules.

Moreover, metametabolomics is the global study of small molecules and metabolites produced by a specific community under a particular environmental condition (O'Malley, 2013). Zhang and coworkers studied endophytic fungal communities and metabolites of a model plant *Ephedra sinica*. This study used global metabolomics approach and deciphered the differences in metabolic profiles of aerial and underground parts of the plants. Approximately 17 metabolites were differentially present in the stems and roots of *E. sinica* (Zhang, Xue, Miao, Cui, & Qin, 2020). Such studies can pave the way for better understanding of plant—microbiome interaction in different parts of the plants (roots, stems, leaves, and seeds) and can help in investigating metabolite profile, in particular, plant tissue.

2.3.5 Holo-OMICS: multi-OMICS integration from host and microbiota

The plant microbiome is complex and diverse comprising endophytic (residing within plant) and exophytic species (residing in surrounding environment)

(Compant, Samad, Faist, & Sessitsch, 2019; Fitzpatrick et al., 2020). The mixture of microbes interacts with each other (Wei et al., 2019) as well as with their host and environment, in complex trophic exchange networks. These microbiomes can shift drastically in composition, abundance, and activity over short timeframes, physical distances and are very sensitive to seasonal environmental factors (Griffiths et al., 2011; Prevost-Boure et al., 2014; Terrat et al., 2017). Together, these factors produce an interconnected biological system that is dynamic as well. Hence, is still a challenge to decipher the mechanisms that are responsible for these dynamics. Though huge data related to basal molecular mechanisms are available, yet additional linking of host and microbe data is required (Brunel et al., 2020) which could be beneficial to solve the puzzle of plant–microbiome interactions.

Aforementioned, multiomics approaches refer to various omics analysis from single sample. Data integration for comprehensive interpretations from these analyses referred to as integrated-omics. The incorporation of host and microbial multiomics data was recently coined as "holoomics" (Nyholm et al., 2020). In recent years, "holobiome" is the topic of investigation to capture endophytes of given plant species. Sum total of the component genomes in a eukaryotic organism; the genome of an individual member of a given taxon (the host genome) and the microbiome consortium (the genomes of the symbiotic microbiota) is termed holobiome (Guerrero, Margulis, & Berlanga, 2013) (Fig. 2.3). Despite huge challenges in holobiome analyses and even higher in holoomics, several studies are being reported. In sorghum, holoomics approach was used to better address the abiotic tolerance mechanisms (Xu et al., 2018).

Castrillo et al. (2017) used holoomic design in an integrated way to demonstrate that the plant root microbiome is responsible to link plant immune system and phosphate stress response.

In another study, holobiome approach was used to observe that root-specific transcription factor MYB72 regulates the excretion of an iron-mobilizing phenolic compound, coumarin scopoletin with selective antimicrobial activity (Stringlis et al., 2018).

For a successful implementation of holoomics approach, not only plant biologist and microbiologist should collaborate but statistical and computational biology experts are also needed. Multidisciplinary expertise is crucial to interpret and decode hidden connections within and between host and associated microbiota with appropriate statistical rigor. Xu, Pierroz, and Wipf (2021) comprehensively reviewed recent holomoics studies and also highlighted the approaches and tools to decipher plant–microbe interaction. Significant challenges include automated computational tools for omics data integration. Though various tools are available, they are either generalized or very target-specific and can be used for particular research scenario. Moreover, available tools are host- or microbiome-specific and are not able to simultaneously integrate data from both samples and to construct networks on the basis of correlations of two datasets.

2.4 Conclusion and perspective

A deep understanding of endophytes which mostly comprise bacteria and fungi, is important in realizing the use of endophytes as nutrient provider and potent biocontrol agent. They are also very important source of novel secondary metabolites, stress reliever of their host, produce industrially important enzymes, but many aspects of endophytic microorganism are unknown. Conventional as well as modern DNA-based approaches are used to unravel factors that are inevitable both for the development and maintenance of symbiotic association between plant and endophytic microbes.

Individual omics analysis is not able to provide complete information as there are multiple molecular layers of control that needs equal and parallel attention to reflect the crosstalk among host and associated microbiota. The understanding of this interaction is of immense importance because many functions of the endophytes depend on this continuous, complex, and dynamic relationship with plant as well as other microbiome members. Characterization of endophytes implementing a holoomic approach would allow researchers to reveal a range of biomolecular interactions responsible for shaping the phenotype of complex organisms, using a variety of molecular tools, and would ultimately provide great potential for application across many different fields of research (Xu et al., 2021). Moreover, models are being devised for the integration of omics and nonomics data to increase the algorithm's predictive ability (López de Maturana et al., 2019). Though most of such studies are focusing human microbe interaction, they can be implemented on that plant microbe in future.

Genome-wide association analysis (GWAS) is forward genetics approach that can also complement to dissect genetic loci through microbiome-based GWAS, involved in the recruitment of specific microbes (Beilsmith et al., 2019; Roman-Reyna, Pinili, & Borja, 2020). Many sophisticated approaches are successfully implementing to validate the hypothesis of aforesaid approaches. Currently, the genome engineering using CRISPR-Cas has emerged as an alternative to modify and promote positive interactions between microorganisms and plants to improve plant fitness (Huang et al., 2020; Rubin et al., 2020). This deeper knowledge with emerging technologies could be exploited more efficiently for agricultural improvement through exploring the role of potential endophytes in plant growth promotion, bioremediation, and biocontrol.

References

Abedi, M., Fatehi, R., Moradzadeh, K., & Gheisari, Y. (2019). Big data to knowledge: Common pitfalls in transcriptomics data analysis and representation. *RNA Biology*, *16* (11), 1531–1533.

Abo-Aba, S. E. M., Soliman, E. A. M., & Nivien, A. A. (2006). Enhanced production of extra cellular alkaline protease in *Bacillus circulans* through plasmid transfer. *Research Journal of Agriculture and Biological Sciences*, *16*, 526–530.

Alcántara-Martínez, N., Figueroa-Martínez, F., Rivera-Cabrera, F., Gutiérrez-Sánchez, G., & Volke-Sepúlveda, T. (2018). An endophytic strain of *Methylobacterium* sp. increases arsenate tolerance in *Acacia farnesiana* (L.) willd: A proteomic approach. *Science of The Total Environment*, *625*, 762–774.

Ambrose, K. V., & Belanger, F. C. (2012). SOLiD-SAGE of endophyte-infected red fescue reveals numerous effects on host transcriptome and an abundance of highly expressed fungal secreted proteins. *PLoS One*, *7*(12), e53214.

Aravind, R., Kumar, A., Eapen, S., & Ramana, K. (2009). Endophytic bacterial flora in root and stem tissues of black pepper (*Piper nigrum* L.) genotype: Isolation, identification and evaluation against *Phytophthora capsici*. *Letters in Applied Microbiology*, *48* (1), 58–64.

Asaf, S., Khan, A. L., Khan, M. A., Al-Harrasi, A., & Lee, I.-J. (2018). Complete genome sequencing and analysis of endophytic *Sphingomonas* sp. LK11 and its potential in plant growth. *3 Biotech*, *8*(9), 1–14.

Ashwin, N. M. R., Barnabas, L., Sundar, A. R., Malathi, P., Viswanathan, R., Masi, A., & Rakwal, R. (2017). Advances in proteomic technologies and their scope of application in understanding plant–pathogen interactions. *Journal of Plant Biochemistry and Biotechnology*, *26*(4), 371–386.

Bastida, F., Torres, I. F., Andrés-Abellán, M., Baldrian, P., López-Mondéjar, R., Větrovský, T., & Jehmlich, N. (2017). Differential sensitivity of total and active soil microbial communities to drought and forest management. *Global Change Biology*, *23* (10), 4185–4203.

Beilsmith, K., Thoen, M. P., Brachi, B., Gloss, A. D., Khan, M. H., & Bergelson, J. (2019). Genome-wide association studies on the phyllosphere microbiome: Embracing complexity in host–microbe interactions. *The Plant Journal*, *97*(1), 164–181.

Bergey, D. H., Buchanan, R. E., Gibbons, N. E., & American Society for Microbiology. (1974). *Bergey's manual of determinative bacteriology*. Baltimore, MD: Williams & Wilkins.

Brunel, C., Pouteau, R., Dawson, W., Pester, M., Ramirez, K. S., & van Kleunen, M. (2020). Towards unraveling macroecological patterns in rhizosphere microbiomes. *Trends in Plant Science*, *25*.

Carrión, V. J., Perez-Jaramillo, J., Cordovez, V., Tracanna, V., De Hollander, M., Ruiz-Buck, D., & Elsayed, S. S. (2019). Pathogen-induced activation of disease-suppressive functions in the endophytic root microbiome. *Science (New York, N.Y.)*, *366*(6465), 606–612.

Castrillo, G., Teixeira, P. J. P. L., Paredes, S. H., Law, T. F., de Lorenzo, L., Feltcher, M. E., & Dangl, J. L. (2017). Root microbiota drive direct integration of phosphate stress and immunity. *Nature*, *543*(7646), 513–518.

Chaudhry, V., & Patil, P. B. (2016). Genomic investigation reveals evolution and lifestyle adaptation of endophytic *Staphylococcus epidermidis*. *Scientific Reports*, *6*, 19263.

Compant, S., Samad, A., Faist, H., & Sessitsch, A. (2019). A review on the plant microbiome: Ecology, functions, and emerging trends in microbial application. *Journal of Advanced Research*, *19*, 29–37.

Costerousse, B., Schönholzer-Mauclaire, L., Frossard, E., & Thonar, C. (2018). Identification of heterotrophic zinc mobilization processes among bacterial strains isolated from wheat rhizosphere (*Triticum aestivum* L.). *Applied and Environmental Microbiology*, *84*(1).

Cragg, G. M., & Newman, D. J. (2013). Natural products: A continuing source of novel drug leads. *Biochimica et Biophysica Acta (BBA)-General Subjects*, *1830*(6), 3670–3695.

Crook, M. B., Mitra, S., Ané, J.-M., Sadowsky, M. J., & Gyaneshwar, P. (2013). Complete genome sequence of the Sesbania symbiont and rice growth-promoting endophyte *Rhizobium* sp. strain IRBG74. *Genome Announcements*, *1*(6).

del Carmen Orozco-Mosqueda, M., & Santoyo, G. (2020). Plant-microbial endophytes interactions: Scrutinizing their beneficial mechanisms from genomic explorations. *Current Plant Biology*, *25*, 100189.

Demutskaya, L., & Kalinichenko, I. (2010). Photometric determination of ammonium nitrogen with the Nessler reagent in drinking water after its chlorination. *Journal of Water Chemistry and Technology*, *32*(2), 90–94.

Deng, Y., Zhu, Y., Wang, P., Zhu, L., Zheng, J., Li, R., & Sun, M. (2011). Complete genome sequence of *Bacillus subtilis* BSn5, an endophytic bacterium of *Amorphophallus konjac* with antimicrobial activity for the plant pathogen *Erwinia carotovora* subsp. carotovora. *Journal of Bacteriology*, *193*.

Dinkins, R. D., Nagabhyru, P., Young, C. A., West, C. P., & Schardl, C. L. (2019). Transcriptome analysis and differential expression in tall fescue harboring different endophyte strains in response to water deficit. *The Plant Genome*, *12*(2), 180071.

Firrincieli, A., Otillar, R., Salamov, A., Schmutz, J., Khan, Z., Redman, R. S., & Doty, S. L. (2015). Genome sequence of the plant growth promoting endophytic yeast *Rhodotorula graminis* WP1. *Frontiers in Microbiology*, *6*, 978.

Fitzpatrick, C. R., Salas-González, I., Conway, J. M., Finkel, O. M., Gilbert, S., Russ, D., & Dangl, J. L. (2020). The plant microbiome: From ecology to reductionism and beyond. *Annual Review of Microbiology*, *74*, 81–100.

Flores, A., Diaz-Zamora, J. T., del Carmen Orozco-Mosqueda, M., Chávez, A., de Los Santos-Villalobos, S., Valencia-Cantero, E., & Santoyo, G. (2020). Bridging genomics and field research: Draft genome sequence of *Bacillus thuringiensis* CR71, an endophytic bacterium that promotes plant growth and fruit yield in *Cucumis sativus* L. *3 Biotech*, *10*(5), 1–7.

Gazis, R., Kuo, A., Riley, R., LaButti, K., Lipzen, A., Lin, J., & Henrissat, B. (2016). The genome of *Xylona heveae* provides a window into fungal endophytism. *Fungal Biology*, *120*(1), 26–42.

Gianoulis, T. A., Griffin, M. A., Spakowicz, D. J., Dunican, B. F., Sboner, A., Sismour, A. M., & Gerstein, M. B. (2012). Genomic analysis of the hydrocarbon-producing, cellulolytic, endophytic fungus *Ascocoryne sarcoides*. *PLoS Genetics*, *8*(3), e1002558.

Gómez-Godínez, L. J., Fernandez-Valverde, S. L., Romero, J. C. M., & Martínez-Romero, E. (2019). Metatranscriptomics and nitrogen fixation from the rhizoplane of maize plantlets inoculated with a group of PGPRs. *Systematic and Applied Microbiology*, *42*(4), 517–525.

Gordon, S. A., & Weber, R. P. (1951). Colorimetric estimation of indoleacetic acid. *Plant Physiology*, *26*(1), 192.

Graham, H. D., & Henderson, J. (1961). Reaction of gibberellic acid & gibberellins with Folin-Wu phosphomolybdic acid reagent & its use for quantitative assay. *Plant Physiology*, *36*(4), 405.

Griffiths, R. I., Thomson, B. C., James, P., Bell, T., Bailey, M., & Whiteley, A. S. (2011). The bacterial biogeography of British soils. *Environmental Microbiology*, *13*(6), 1642–1654.

Guerrero, R., Margulis, L., & Berlanga, M. (2013). Symbiogenesis: The holobiont as a unit of evolution. *International Microbiology: The Official Journal of the Spanish Society for Microbiology*, *16*(3), 133–143.

Hardoim, P. R., Van Overbeek, L. S., Berg, G., Pirttilä, A. M., Compant, S., Campisano, A., & Sessitsch, A. (2015). The hidden world within plants: Ecological and evolutionary considerations for defining functioning of microbial endophytes. *Microbiology and Molecular Biology Reviews*, *79*(3), 293–320.

Hardy, R., Burns, R. C., & Holsten, R. D. (1973). Applications of the acetylene-ethylene assay for measurement of nitrogen fixation. *Soil Biology and Biochemistry*, *5*(1), 47–81.

Hong, C. E., Kim, J. U., Lee, J. W., Bang, K. H., & Jo, I. H. (2019). Metagenomic analysis of bacterial endophyte community structure and functions in *Panax ginseng* at different ages. *3 Biotech*, *9*(8), 1–8.

Huang, P. W., Yang, Q., Zhu, Y. L., Zhou, J., Sun, K., Mei, Y. Z., & Dai, C. C. (2020). The construction of CRISPR-Cas9 system for endophytic *Phomopsis liquidambaris* and its PmkkA-deficient mutant revealing the effect on rice. *Fungal Genetics and Biology*, *136*, 103301.

Huang, S., Chaudhary, K., & Garmire, L. X. (2017). More is better: Recent progress in multi-omics data integration methods. *Frontiers in Genetics*, *8*, 84.

Ivanova, A. A., Wegner, C. E., Kim, Y., Liesack, W., & Dedysh, S. N. (2016). Identification of microbial populations driving biopolymer degradation in acidic peatlands by metatranscriptomic analysis. *Molecular Ecology*, *25*(19), 4818–4835.

Jeong, H., Jo, S. H., Hong, C. E., & Park, J. M. (2016). Genome sequence of the endophytic bacterium *Bacillus thuringiensis* strain KB1, a potential biocontrol agent against phytopathogens. *Genome Announcements*, *4*(2).

Johnson, C. H., Ivanisevic, J., Benton, H. P., & Siuzdak, G. (2015). Bioinformatics: The next frontier of metabolomics. *Analytical Chemistry*, *87*(1), 147–156.

Kasana, R. C., Salwan, R., Dhar, H., Dutt, S., & Gulati, A. (2008). A rapid and easy method for the detection of microbial cellulases on agar plates using Gram's iodine. *Current Microbiology*, *57*(5), 503–507.

Kaul, S., Gupta, S., Ahmed, M., & Dhar, M. K. (2012). Endophytic fungi from medicinal plants: A treasure hunt for bioactive metabolites. *Phytochemistry Reviews*, *11*(4), 487–505.

Khatoon, Z., Huang, S., Rafique, M., Fakhar, A., Kamran, M. A., & Santoyo, G. (2020). Unlocking the potential of plant growth-promoting rhizobacteria on soil health and the sustainability of agricultural systems. *Journal of Environmental Management*, *273*, 111118.

Kubicek, C. P., Herrera-Estrella, A., Seidl-Seiboth, V., Martinez, D. A., Druzhinina, I. S., Thon, M., & Mukherjee, P. K. (2011). Comparative genome sequence analysis underscores mycoparasitism as the ancestral life style of *Trichoderma*. *Genome Biology*, *12* (4), 1–15.

Li, H., Wang, J., Wang, J., Geng, G., Ju, H., & Creamer, R. (2012). Protein extraction methods for the two-dimensional gel electrophoresis analysis of the slow growing fungus *Undifilum oxytropis*. *African Journal of Microbiology Research*, *6*(4), 757–763.

Link, H.F. (1809). Observationes in ordines plantarum naturales. Dissertatio I. Mag Ges Naturf Freunde Berlin, 3, 3–42.

Liu, J., & Liu, G. (2018). *Analysis of secondary metabolites from plant endophytic fungi. Plant pathogenic fungi and oomycetes* (pp. 25–38). Springer.

Liu, K.-H., Ding, X.-W., Narsing Rao, M. P., Zhang, B., Zhang, Y.-G., Liu, F.-H., & Li, W.-J. (2017). Morphological and transcriptomic analysis reveals the osmoadaptive response of endophytic fungus *Aspergillus montevidensis* ZYD4 to high salt stress. *Frontiers in Microbiology*, *8*, 1789.

Liu, W.-Y., Chung, K. M.-K., Wong, C.-F., Jiang, J.-W., Hui, R. K.-H., & Leung, F. C.-C. (2012). Complete genome sequence of the endophytic *Enterobacter cloacae* subsp. cloacae strain ENHKU01. *Journal of Bacteriology*, *194*.

López de Maturana, E., Alonso, L., Alarcón, P., Martín-Antoniano, I. A., Pineda, S., Piorno, L., & Malats, N. (2019). Challenges in the integration of omics and non-omics data. *Genes*, *10*(3), 238.

Mahmood, A., & Kataoka, R. (2020). Metabolite profiling reveals a complex response of plants to application of plant growth-promoting endophytic bacteria. *Microbiological Research*, *234*, 126421.

Malleswari, D., & Bagyanarayana, G. (2013). Plant growth-promoting activities and molecular characterization of rhizobacterial strains isolated from medicinal and aromatic plants. *Journal of Pharmacy and Biological Sciences*, *6*, 30–37.

Manias, D., Verma, A., & Soni, D. K. (2020). *Isolation and characterization of endophytes: Biochemical and molecular approach. Microbial endophytes* (pp. 1–14). Elsevier.

Maron, P. A., Ranjard, L., Mougel, C., & Lemanceau, P. (2007). Metaproteomics: A new approach for studying functional microbial ecology. *Microbial Ecology*, *53*(3), 486–493.

Martinez-Klimova, E., Rodríguez-Peña, K., & Sánchez, S. (2017). Endophytes as sources of antibiotics. *Biochemical Pharmacology*, *134*, 1–17.

Molina, L. G., Cordenonsi da Fonseca, G., Morais, G. Ld, de Oliveira, L. F. V., Carvalho, J. Bd, Kulcheski, F. R., & Margis, R. (2012). Metatranscriptomic analysis of small RNAs present in soybean deep sequencing libraries. *Genetics and Molecular Biology*, *35*(1), 292–303.

Mookherjee, A., Mitra, M., Kutty, N. N., Mitra, A., & Maiti, M. K. (2020). Characterization of endo-metabolome exhibiting antimicrobial and antioxidant activities from endophytic fungus *Cercospora* sp. PM018. *South African Journal of Botany*, *134*, 264–272.

Mousa, W. K., & Raizada, M. N. (2013). The diversity of anti-microbial secondary metabolites produced by fungal endophytes: An interdisciplinary perspective. *Frontiers in microbiology*, *4*, 65.

Naureen, Z., Price, A. H., Hafeez, F. Y., & Roberts, M. R. (2009). Identification of rice blast disease-suppressing bacterial strains from the rhizosphere of rice grown in Pakistan. *Crop Protection*, *28*(12), 1052–1060.

Nautiyal, C. S. (1999). An efficient microbiological growth medium for screening phosphate solubilizing microorganisms. *FEMS Microbiology Letters*, *170*(1), 265–270.

Nawed, A., & Chandra, R. (2015). Endophytic bacteria: Optimizaton of isolation procedure from various medicinal plants and their preliminary characterization. *Asian Journal of Pharmaceutical and Clinical Research*, *8*(4), 233–238.

Noronha, M. F., Júnior, G. V. L., Gilbert, J. A., & de Oliveira, V. M. (2017). Taxonomic and functional patterns across soil microbial communities of global biomes. *Science of The Total Environment*, *609*, 1064–1074.

Nyholm, L., Koziol, A., Marcos, S., Botnen, A. B., Aizpurua, O., Gopalakrishnan, S., & Alberdi, A. (2020). Holo-omics: Integrated host-microbiota multi-omics for basic and applied biological research. *iScience*, *23*, 101414.

O'Malley, M.A. (2013). Metametabolomics. In Encyclopedia of *systems biology*, Springer, New York, NY, pp. 1296–1297.<https://doi.org/10.1007/978-1-4419-9863-7_903>.

O'Brien, M. A. R. K., & Colwell, R. R. (1987). A rapid test for chitinase activity that uses 4-methylumbelliferyl-*N*-acetyl-beta-D-glucosaminide. *Applied and Environmental Microbiology*, *53*(7), 1718–1720.

Okon, Y., Albrecht, S. L., & Burris, R. H. (1977). Methods for growing *Spirillum lipoferum* and for counting it in pure culture and in association with plants. *Applied and Environmental Microbiology*, *33*(1), 85–88.

Pachauri, S., Sherkhane, P. D., Kumar, V., & Mukherjee, P. K. (2020). Whole genome sequencing reveals major deletions in the genome of M7, a gamma ray-induced mutant of *Trichoderma virens* that is repressed in conidiation, secondary metabolism, and Mycoparasitism. *Frontiers in Microbiology*, *11*, 1030.

Pedrosa, F. O., Monteiro, R. A., Wassem, R., Cruz, L. M., Ayub, R. A., Colauto, N. B., & Hungria, M. (2011). Genome of *Herbaspirillum seropedicae* strain SmR1, a specialized diazotrophic endophyte of tropical grasses. *PLoS Genetics*, *7*(5), e1002064.

Pelczar, M., Jr, & Reid, R. (1965). *Microbiology* (2nd ed.). New York: McGraw-Hill.

Penrose, D. M., & Glick, B. R. (2003). Methods for isolating and characterizing ACC deaminase-containing plant growth-promoting rhizobacteria. *Physiologia Plantarum*, *118*(1), 10–15.

Pentimone, I., Lebrón, R., Hackenberg, M., Rosso, L. C., Colagiero, M., Nigro, F., & Ciancio, A. (2018). Identification of tomato miRNAs responsive to root colonization by endophytic *Pochonia chlamydosporia. Applied Microbiology and Biotechnology*, *102*(2), 907–919.

Pérez-Equihua, A., & Santoyo, G. (2021). Draft genome sequence of *Bacillus* sp. strain E25, a biocontrol and plant growth-promoting bacterial endophyte isolated from Mexican husk tomato roots (*Physalis ixocarpa* Brot. Ex Horm.). *Microbiology Resource Announcements*, *10*(1), e01112–e01120. Available from https://doi.org/10.1128/MRA.01112-20, PMID. Available from 33414296.

Pillai, T. G. (2017). Pathogen to endophytic transmission in fungi-a proteomic approach. *SOJ Microbiology and Infectious Diseases*, *5*, 1–5.

Premjanu, N., & Jayanthy, C. (2012). Endophytic fungi a repository of bioactive compounds-A review. *International Journal of Institutional Pharmacy and Life Sciences*, *2*(1), 135–162.

Prevost-Boure, N. C., Dequiedt, S., Thioulouse, J., Lelievre, M., Saby, N. P., Jolivet, C., & Ranjard, L. (2014). Similar processes but different environmental filters for soil bacterial and fungal community composition turnover on a broad spatial scale. *PLoS One*, *9*(11), e111667.

Ramírez-Ordorica, A., Valencia-Cantero, E., Flores-Cortez, I., Carrillo-Rayas, M. T., Elizarraraz-Anaya, M. I. C., Montero-Vargas, J., & Macías-Rodríguez, L. (2020). Metabolomic effects of the colonization of *Medicago truncatula* by the facultative endophyte *Arthrobacter agilis* UMCV2 in a foliar inoculation system. *Scientific Reports*, *10*(1), 1–11.

Righetti, P. G., & Boschetti, E. (2016). Global proteome analysis in plants by means of peptide libraries and applications. *Journal of Proteomics*, *143*, 3–14.

Roman-Reyna, V., Pinili, D., Borja, F. N., et al. (2020). Characterization of the leaf microbiome from whole-genome sequencing data of the 3000 rice genomes project. *Rice*, *13*, 72. Available from https://doi.org/10.1186/s12284-020-00432-1.

Rubin, B.E., Diamond, S., Cress, B.F., Crits-Christoph, A., He, C., Xu, M., … & Doudna, J.A. (2020). *Targeted genome editing of bacteria within microbial communities*. bioRxiv.

Sánchez-de Prager, M., & Cisneros-Rojas, C. A. (2017). Organic acids production by rhizosphere microorganisms isolated from a Typic Melanudands and its effects on the inorganic phosphates solubilization. *Acta Agronómica*, *66*(2), 241–247.

Santoyo, G., Moreno-Hagelsieb, G., del Carmen Orozco-Mosqueda, M., & Glick, B. R. (2016). Plant growth-promoting bacterial endophytes. *Microbiological Research*, *183*, 92–99.

Schippers, B., Bakker, A. W., Bakker, P. A. H. M., & Van Peer, R. (1990). Beneficial and deleterious effects of HCN-producing pseudomonads on rhizosphere interactions. *Plant and Soil*, *129*(1), 75–83.

Schwyn, B., & Neilands, J. (1987). Universal chemical assay for the detection and determination of siderophores. *Analytical Biochemistry*, *160*(1), 47–56.

Sessitsch, A., Hardoim, P., Döring, J., Weilharter, A., Krause, A., Woyke, T., & Rahalkar, M. (2012). Functional characteristics of an endophyte community colonizing rice roots as revealed by metagenomic analysis. *Molecular Plant-Microbe Interactions*, *25*(1), 28–36.

Sharma, K. K. (2016). Fungal genome sequencing: Basic biology to biotechnology. *Critical Reviews in Biotechnology*, *36*(4), 743–759.

Siegmund, I., & Wagner, F. (1991). New method for detecting rhamnolipids excreted by *Pseudomonas* species during growth on mineral agar. *Biotechnology Techniques*, *5*(4), 265–268.

Starke, R., Bastida, F., Abadía, J., García, C., Nicolás, E., & Jehmlich, N. (2017). Ecological and functional adaptations to water management in a semiarid agroecosystem: A soil metaproteomics approach. *Scientific Reports*, *7*(1), 1–16.

Starke, R., Jehmlich, N., & Bastida, F. (2019). Using proteins to study how microbes contribute to soil ecosystem services: The current state and future perspectives of soil metaproteomics. *Journal of Proteomics*, *198*, 50–58.

Stringlis, I. A., Yu, K., Feussner, K., De Jonge, R., Van Bentum, S., Van Verk, M. C., & Pieterse, C. M. (2018). MYB72-dependent coumarin exudation shapes root microbiome assembly to promote plant health. *Proceedings of the National Academy of Sciences*, *115*(22), E5213–E5222.

Strobel, G., & Daisy, B. (2003). Bioprospecting for microbial endophytes and their natural products. *Microbiology and Molecular Biology Reviews*, *67*(4), 491–502.

Subudhi, E., Sahoo, R., Dey, S., Das, A., & Sahoo, K. (2018). Unraveling plant-endophyte interactions: An omics insight. *Endophytes and Secondary Metabolites*, 1–19.

Tahir, A., Kang, J., Choulet, F., Ravel, C., Romeuf, I., Rasouli, F., & Branlard, G. (2020). Deciphering carbohydrate metabolism during wheat grain development via integrated transcriptome and proteome dynamics. *Molecular Biology Reports*, *47*(7), 5439–5449.

Tahir, A. T., Fatmi, Q., Nosheen, A., Imtiaz, M., & Khan, S. (2019). *Metabolomic approaches in plant research*, . *Essentials of bioinformatics* (Vol. III, pp. 109–140). Springer.

Tawfike, A. F., Romli, M., Clements, C., Abbott, G., Young, L., Schumacher, M., & Edrada-Ebel, R. (2019). Isolation of anticancer and anti-trypanosome secondary metabolites from the endophytic fungus *Aspergillus flocculus* via bioactivity guided isolation and MS based metabolomics. *Journal of Chromatography B*, *1106*, 71–83.

Terrat, S., Horrigue, W., Dequietd, S., Saby, N. P., Lelièvre, M., Nowak, V., & Ranjard, L. (2017). Mapping and predictive variations of soil bacterial richness across France. *PLoS One*, *12*(10), e0186766.

Walker, A. K., Frasz, S. L., Seifert, K. A., Miller, J. D., Mondo, S. J., LaButti, K., & Grigoriev, I. V. (2016). Full genome of *Phialocephala scopiformis* DAOMC 229536, a fungal endophyte of spruce producing the potent anti-insectan compound rugulosin. *Genome Announcements*, *4*(1).

Wang, Z., Zhu, Y., Jing, R., Wu, X., Li, N., Liu, H., ... Liu, Y. (2021). High-throughput sequencing-based analysis of the composition and diversity of endophytic bacterial community in seeds of upland rice. *Archives of Microbiology*, *203*(2), 609–620. Available from https://doi.org/10.1007/s00203-020-02058-9, Epub 2020 Sep 29. PMID. Available from 32995980.

Wei, J., Chen, F., Liu, Y., Abudoukerimu, A., Zheng, Q., Zhang, X., & Yimiti, D. (2020). Comparative metabolomics revealed the potential antitumor characteristics of four endophytic fungi of *Brassica rapa* L. *ACS Omega*, *5*(11), 5939–5950.

Wei, Z., Gu, Y., Friman, V. P., Kowalchuk, G. A., Xu, Y., Shen, Q., & Jousset, A. (2019). Initial soil microbiome composition and functioning predetermine future plant health. *Science Advances*, *5*(9), eaaw0759.

Witzel, K., Gwinn-Giglio, M., Nadendla, S., Shefchek, K., & Ruppel, S. (2012). Genome sequence of *Enterobacter radicincitans* DSM16656T, a plant growth-promoting endophyte. *Journal of Bacteriology*, *194*.

Xu, L., Naylor, D., Dong, Z., Simmons, T., Pierroz, G., Hixson, K. K., & Coleman-Derr, D. (2018). Drought delays development of the sorghum root microbiome and enriches for monoderm bacteria. *Proceedings of the National Academy of Sciences*, *115*(18), E4284–E4293.

Xu, L., Pierroz, G., Wipf, H. M. L., et al. (2021). Holo-omics for deciphering plant-microbiome interactions. *Microbiome*, *9*, 69. Available from https://doi.org/10.1186/s40168-021-01014-z.

Xu, X.-H., Su, Z.-Z., Wang, C., Kubicek, C. P., Feng, X.-X., Mao, L.-J., & Zhang, C.-L. (2014). The rice endophyte *Harpophora oryzae* genome reveals evolution from a pathogen to a mutualistic endophyte. *Scientific Reports*, *4*, 5783.

Yadav, R. K., Bragalini, C., Fraissinet-Tachet, L., Marmeisse, R., & Luis, P. (2016). *Metatranscriptomics of soil eukaryotic communities. Microbial environmental genomics (MEG)* (pp. 273–287). Springer.

Yang, H., Wang, Y., Zhang, Z., Yan, R., & Zhu, D. (2014). Whole-genome shotgun assembly and analysis of the genome of *Shiraia* sp. strain Slf14, a novel endophytic fungus producing huperzine A and hypocrellin A. *Genome Announcements*, *2*(1).

Yi, Y., de Jong, A., Frenzel, E., & Kuipers, O. P. (2017). Comparative transcriptomics of *Bacillus mycoides* strains in response to potato-root exudates reveals different genetic adaptation of endophytic and soil isolates. *Frontiers in Microbiology*, *8*, 1487.

Yuan, J., Zhang, W., Sun, K., Tang, M.-J., Chen, P.-X., Li, X., & Dai, C.-C. (2019). Comparative transcriptomics and proteomics of *Atractylodes lancea* in response to endophytic fungus *Gilmaniella* sp. AL12 reveals regulation in plant metabolism. *Frontiers in Microbiology*, *10*, 1208.

Zhang, Q., Xue, X. Z., Miao, S. M., Cui, J. L., & Qin, X. M. (2020). Differential relationship of fungal endophytic communities and metabolic profiling in the stems and roots of *Ephedra sinica* based on metagenomics and metabolomics. *Symbiosis*, *81*, 115–125.

Zhu, B., Liu, H., Tian, W.-X., Fan, X.-Y., Li, B., Zhou, X.-P., & Xie, G.-L. (2012). Genome sequence of *Stenotrophomonas maltophilia* RR-10, isolated as an endophyte from rice root. *Journal of Bacteriology*, *194*.

Zuccaro, A., Lahrmann, U., Güldener, U., Langen, G., Pfiffi, S., Biedenkopf, D., & Basiewicz, M. (2011). Endophytic life strategies decoded by genome and transcriptome analyses of the mutualistic root symbiont *Piriformospora indica*. *PLoS Pathogens*, *7* (10), e1002290.

CHAPTER

Biocontrol mechanism of endophytic microorganisms

3

Gayathri Segaran, Saranya Shankar and Mythili Sathiavelu
School of Biosciences and Technology, Vellore Institute of Technology, Vellore, India

3.1 Introduction

Biological control refers to the use of microorganisms or their metabolites to completely remove or eliminate the negative effects of phytopathogens on plants in terms of health and yield. Naturally available biocontrol agents (BCAs) have an antagonistic effect against pathogens, and thereby it reduces their harmful effect. The antagonistic microbe should be genetically stable and potent even at low concentrations for a wide range of microbial pathogens that affect numerous vegetables and fruits (Ulloa-Ogaz, Munoz-Castellanos, & Nevarez-Moorillon, 2015). For more than 70 years, biological methods are under investigation to control phytopathogens; however, synthetic pesticides dominate over the biocontrol products, which is estimated to be only 3.5% of the global pesticide market (Bolívar-Anillo, Garrido, & Collado, 2020). The expected outcome for using BCAs is to lower the population of pests/pathogens below a threshold of economic and ecological impact, while it also enables the host plant to return to a healthy state. As BCA source species are from the host's environment, it can specifically focus on their target in some cases and is less harmful to nontarget organisms. Fungi, bacteria, viruses, and few higher organisms such as insects and nematodes are the various sources of BCAs used in diverse forms (Rabiey et al., 2019). The advantage of using BCA with various mechanisms of action is that it improves BCAs efficiency and the target pathogen population delay in developing resistance toward the biocontrol products. Microorganisms with simple nutrition, diverse mechanisms of action, rapid growth, endurance toward particular fungicides, good adaptability, and a broader environment are the most preferred BCA product (Card, Johnson, Teasdale, & Caradus, 2016).

The population of the pathogen is decreased with different disease control strategies like the use of chemical, biological, cultural, and integrated methods. Upon the application of copper compounds, the plant tissues are secured from infections and it also lowers the secondary inoculum, but the effectiveness of this treatment is only on plant surfaces. Inside the cucumber plants, the movement of a systemic pathogen cannot be controlled by copper. In controlling plant disease, the excessive use of copper-based agrochemicals may lead to harmful effects on nontarget

Biocontrol Mechanisms of Endophytic Microorganisms. DOI: https://doi.org/10.1016/B978-0-323-88478-5.00015-8

organisms, environmental contamination, chemical residues in foods, and the rise of pathogen resistance (Akbaba & Ozaktan, 2018). In the agro-ecosystem targeted, BCAs can either be used before-planting as a preventive tool or post-planting as a palliative tool (Mercado-Blanco & Lugtenberg, 2014). Outgrow, outcompete, and blockage of pathogens entry points are done by certain BCAs whereas these agents can also have the ability to prevent propagules germination (Chowdhary & Sharma, 2020). In the dynamic phytosphere, the functional partners include endophytic microbes, and exploring their beneficial role in the agricultural sector will be a prospective research field (Chowdhary & Sharma, 2020).

3.2 Endophytes and its role

Endophytic microorganisms can increase plant fitness by enhancing the plant tolerance toward drought and heavy metals, thereby provokes plant growth and lowers phytopathogen settling. The isolation of fungal endophyte communities from various plants of Arctic, temperate, tropical, and subtropical ecosystems was recorded (Paul, Deng, Sang, Choi, & Yu, 2012). The source of beneficial endophytic microbes is rich in plants that are developing in humid tropical conditions (Savani, Bhattacharyya, & Baruah, 2020). Endophytes can be actinomycetes, bacteria, and fungi that colonize in the internal tissue of plants (Akbaba & Ozaktan, 2018). Endophytes are not directly influenced by the environment because when their host plant grows, endophytes expand inside their intercellular spaces and colonize the whole plant axis. This unique habitat was adapted by the endophytic niche to restraint pathogens (Bacon & Hinton, 2006). The transmission of endophytes is most probably as an animal, water, or wind-borne spores and in tress, it gets transmitted horizontally. Endophytes can be detached from seeds of few plants, but infection cannot be transmitted to the seedlings. The community and composition of endophytic fungi may change according to their size, age, and distance from the source (Bayman, 2007). Notably, azotrophic endophytes can fix nitrogen, which has a beneficial contribution to an ecological intensification of agriculture (Dheeman, Maheshwari, & Baliyan, 2017). Fungi were the most studied endophytes followed by bacteria and these bacterial endophytes belonging to 200 genera from 16 different phyla, of which most of the species belonging to phyla *Firmicutes, Actinobacteria*, and *Proteobacteria. Agrobacterium*, *Achromobacter*, *Acinetobacter*, *Brevibacterium*, *Bacillus*, *Pseudomonas*, and *Xanthomonas* are major Gram-positive and negative bacterial species that contribute to bacterial endophytes diversity (Sahoo, Sarangi, & Kerry, 2018).

3.3 Symbiotic relationship between host and endophyte

Plant ecosystems hardly rely on their microbial communities to boost health, and the fine balance between disease and mutualism leads to this intimate association

(Rabiey et al., 2019). Depends on life history, chemotypic adaptations, the morphology of host plant and fungus, the interactions between plants and fungal endophytes ranging from mutualistic to antagonistic (Compant, Saikkonen, Mitter, Campisano, & Mercado-Blanco, 2016). The adaptations of endophytes are unique for a specific chemical environment of the host plant and these microbes are designed to be an unrevealed source of natural products (Jasim, Joseph, John, Mathew & Radhakrishnan, 2014). The interaction between plant-associated fungi and plant pathogens are of different kinds. These kinds of interactions can be direct or indirect contact and have various effects on plant health. Amensalism, competition, commensalism, mutualism, neutralism, parasitism, predation, and protocooperation are the different interactions between the two populations. Both microscopic and macroscopic levels of analysis can be done to observe the interaction types. When the association of two or more organisms results in a beneficial outcome for both, then this interaction refers to mutualism. Protocooperation is the form of mutualism, but the interacting organism doesn't depend on each other for their survival. As most of the BCAs are hardly dependent on the host for their survival, they are considered to be facultative mutualists. Commensalism refers to the symbiotic association between two organisms, in which one is benefited and the other neither harmed nor benefited. The neutralistic association between two interacting organisms does not show any effect but when one organism exerts adverse effects on another during the interaction, then it refers to antagonism (Narayanasamy, 2013).

3.4 An overview of endophytes as a biocontrol agent

Endophytes can be identified, and the beneficial functions of their communities can be unraveled using novel advances in cultivation-independent techniques like next-generation sequencing technology, association analyses, and network inference modeling. The origin of microbes (native/introduced), intended duration of the control (permanent/temporal) and capability of reproduction after field release distinguish their biocontrol strategies into four groups. This includes (1) classical biological control, (2) inoculation biological control, (3) inundation biological control, and (4) conservation biological control. Classical biological control is a long-term control method applied to restrict insect pests and weeds, but not for phytopathogens. An exotic species with biological control potential is released into a new environment for perpetual protection. Inoculation biological control implies BCAs that can multiply after the introduction to the field and this protection strategy is for a long-lasting period but not permanent. In the case of inundation biological control, the biocontrol organisms are released in high numbers to attain the instant effect. The reproduction of BCAs is restricted in the environment and they couldn't provide long-term control over pathogens. Conservation biological control is achieved by modifying the environment to enhance the development of particular natural enemies that can target specific pathogens. This method does not need any separate control agents to release to the environment (Terhone, Kovalchuk, Zarsav, & Asiegbu, 2018).

The difficulties in the administration and establishment of microorganisms to plants can be overcome with endophytes because their life cycle is different from others and can be easily delivered and survived in the plant environment. The success and effectiveness of endophytic microorganisms as the biocontrol agent depends on several factors including growth phase and physiological state of the plant, the physical structure of the soil, environmental conditions, host specificity, colonization patterns, population dynamics, ability to move within host tissue and to induce systemic resistance. The ability of endophytic microbes to colonize an ecological niche that is similar to a few phytopathogens clearly shows their potential to be used as BCAs. In addition to these, climate variation, the intrinsic traits of the BCAs, the quality of formulated products, and the exertion of selection pressure also have an impact on BCA's effectiveness. Likewise, the evolution of the pathogen in response to selection pressures and their genetic diversity must be taken into considerations. Due to the genetic tractability and wide host range, *Piriformospora indica* was used as the model system for investigating the mechanisms involved in endophyte-root compatibility (Bolívar-Anillo et al., 2020; Card et al., 2016).

The bacterial BCAs are mostly derived from the plant's root zone, few were also isolated from the espermosphere, the phyllosphere, and the endosphere. The production of allelochemicals, induction of resistance pathways, and plants niche competition are the few biocontrol mechanisms of bacteria (Bolívar-Anillo et al., 2020). As host plants seem to be highly acceptable for bacterial endophytes, the genetic modifications of plants are not much needed for infection. The one-time application of bacterial strain to seeds or plants before or during the time of planting is sufficient for natural infection (Bacon & Hinton, 2006). The genus Epichloë (Ascomycota: Hypocreales: Clavicipitaceae) includes protective fungal endophytes for managing pasture pests. Insects infections are caused by the specific group of endophytes called fungal entomopathogens from the order Hypocreales (McKinnon et al., 2017). Chestnut blight can be controlled by using hypovirulent strains of *Cryphonectria parasitica* and dsRNA mycoviruses causes hypovirulence in the chestnut pathogen. When virus-carrying strains are inoculated on chestnut trees, through hyphal anastomoses virus spreads in the pathogen population and causes the formation of slower canker. In addition to this effect, viral infection reduces the virulence of plant pathogens (Terhone et al., 2018). The incidence of pathogenic fungi may be reduced by its similar nonpathogenic endophyte. Double-strand RNA mycoviruses in few pathogenic fungi act as endophytes and thereby reduces the damages caused by pathogenic strains (Lacava & Azevedo, 2014).

Numerous documents reported alkaloids, phenols, peptides, terpenoids, steroids, flavonoids, quinones, and their enzymes like chitinases, glucanases, laminarinases, and hydrolases as the major antimicrobial compounds liable for inhibiting pathogens (Ting, 2014). With the combination of antibiotics, competition, parasitism, or individual modes of action, microbes archives their antagonistic activity. In parasitism, antagonists produce numerous hydrolytic enzymes to degrade pathogenic fungi's cell wall (El-Katatny, Somitsch, Robra, El-Katatny, & Gübitz, 2000). The single BCAs can perform various modes of action to control the disease. The mechanism that instigates the host response toward infection and the mechanism of reducing pathogen growth are the two major modes of action that are performed by BCAs. The biocontrol organisms interact with their corresponding pathogens through direct (physical contact) or indirect contact (Terhone et al., 2018).

In dual culture antagonistic study, the fungal endophytes of *Capsicum annuum* L. were found to be efficient toward *Phytophthora capsici*, a serious fungal pathogen of chili pepper (Paul et al., 2012). Upon cutting process, *Phlebiopsis gigantean* was immediately applied on the *Heterobasidion annosum* s.l to prevent tree stumps infection, and the application of *Gliocladium* and *Trichoderma* species restraint the wound infections (Terhone et al., 2018). *Pseudomonas alcaligenes*, *Pseudomonas fluorescens*, *Pseudomonas putida*, *Bacillus pumilus*, *B. megaterium*, *B. subtilis*, *Clavibacter michiganensis*, *Alcaligenes* spp., *Kluyvera* spp., *Microbacterium* spp., are the different bacterial endophytes shown antagonism toward phytopathogens. When *Curtobacterium flaccumfaciens*, an endophyte of citrus plants inoculated to the model plant *C. roseus*, it showed inhibitory effect toward *Xylella fastidiosa* (causative agent of citrus variegated chlorosis) in both in vitro and in vivo conditions (Lacava & Azevedo, 2014). Under drought stress conditions, endophytic isolates, *Pseudomonas aeruginosa* strain NFTR and strain FTR, and *Enterobacter asburae* strain MRC12 shown antagonistic activity against fungal pathogens (Sandhya, Shrivastava, Ali, & Sai Shiva Krishna Prasad, 2017). Among 127 bacterial endophytes of winter wheat plants, two *Pseudomonas* sp. Strains (JD204 and JC186) were identified as members of *P. putida* based on phylogenetic analyses. Disease index and disease incidence of stripe rust was reduced when wheat cultivars were given a one-time treatment of strain JD204, additionally, it also improved yields on tested cultivars. The field trial has proven the capability of strain JD204 as an assuring yield-enhancing and biocontrol agent for eco-friendly wheat production (Pang et al., 2016). *Bacillus* isolates were found to be effective against *Verticillium dahlia*, this might be due to its chitinolytic activity and production of auxin and antibiotic (Alström & Van Vuurde, 2001).

3.5 Mycoparasitic interaction between biocontrol agent and plant pathogens

Mycoparasitism is the direct antagonistic mechanism between endophyte and pathogen, where the endophytic microbes feed on the plant pathogens (Witzell, Martín, & Blumenstein, 2014). Endophyte behinds its mycoparasitic activity by penetrating hyphae and coil around it. Disruption/deletion of the prey hyphae is the final step in this mechanism (Latz, Jensen, Collinge, & Jørgensen, 2018). In mycoparasitism or hyperparasitism, the biocontrol agent parasitizes on the targeted phytopathogen through direct contact. Haustoria is the specialized structures used by hyperparasite to utilize pathogen's cells content as the nutrients source and the proximity between two interacting species are much required in mycoparasitism. *Ampelomyces quisqualis* inhibits powdery mildew fungi by parasitizing its conidia and hyphae and *Coniothyrium minitans* focuses on destroying sclerotia of *Sclerotinia* spp. (Terhone et al., 2018). In the microscopic analysis, *Phragmites australis* penetrates the hyphae and degrades cytoplasm, thereby it was found to be active against eight soil-borne pathogens. In *in vitro* analysis, the mycoparasitic activity of fungi is studied only in the existence of a plant pathogen without the presence of the plant, this clearly states that the plant was not there to impede the interaction, even though the tested fungi is an endophytic strain. Parasitism is initiated only when the distance between the parasite and its prey is much closer inside the plant (Latz et al., 2018).

Examples of mycoparasitism by endophytic fungi include the work by Donayre. In *Coffea arabica*, mycoparasitic activity was found in *Trichoderma flagellatum* against tracheomycosis inducing pathogen *Fusarium* sp. The endophyte *Trichoderma harzanium* isolated from inflorescence tissues of *Aloe vera* L. secretes stigmasterol and ergosterol, these compounds displayed noticeable mycoparasitic activity toward *Sclerotinia rolfsii* and *Rhizoctonia solani* (Chowdhary & Sharma, 2020). The resistance against *Pythium ultimum* was induced in cucumber by a nonpathogenic *Fusarium oxysporum*, by the combination of mycoparasitism and antibiosis mechanisms (Paul et al., 2012). *F. proliferatum* is the cold-tolerant, mycoparasitic fungus that controls the growth of *Plasmopara viticola* (causative of grapevine downy mildew) by secreting its extracellular glucanolytic enzymes (Liarzi & Ezra, 2014). In *Hevea brasiliensis*, abnormal leaf fall was caused by the fungal pathogen *Phytophthora meadii*. *Alcaligenes* sp. (EIL-2), an endophytic bacterium of healthy rubber tree leaves was able to suppress *P. meadii* and *in vitro* dual cultures, hyphal growth of the pathogen was inhibited by endophyte's substances (Rabiey et al., 2019).

3.6 Antibiosis and secondary metabolite-mediated plant protection

Antibiotics have a different mode of action to inhibit the pathogens by affecting cellular membrane or ribosomes or other cellular constituents. At subinhibitory concentrations, antibiotics cause extreme physiological effects on target microorganisms and suppress the extracellular virulence factors production and adherence mechanisms in bacteria (Ulloa-Ogaz et al., 2015). Antimicrobial peptides (AMPs) could be a potent replacement for fungicides use on agricultural applications, as these molecules possess a wide range of antimicrobial activity with rapid action.

The existence of cations such as, Na^+, Ca^{2+}, K^+, and Mg^{2+} that are prevalent in the agricultural, biosphere, and food products have a negative impact on the antimicrobial activity of AMPs. AMP can be used for agricultural purposes, only when it is active in the presence of cations at biologically relevant concentrations. The selectivity and activity of peptides, the mechanism involved in antimicrobial activity, cellular properties of target microbes, and their cultural conditions have a major impact on the antimicrobial activity of peptides (Troskie, de Beer, Vosloo, Jacobs, & Rautenbach, 2014).

Antibiotics from natural origin inhibit the growth of various microbes in the sporulation and development stage. Antibiotic peptides are produced by *Bacillus* and *Paenibacillus* (Peeran, Prasad & Kamil, 2018; Senthilkumar, Swarnalakshmi, Govindasamy, Lee, & Annapurna, 2009). The transfer of encoded genes and cross-resistance is the restriction in the use of antibiotic-producing bacteria (Vijayabharathi, Sathya & Gopalakrishnan, 2016). The members from the *Streptomyces* genus are a rich source of antimicrobial compounds and endophytic *Streptomyces* inhibits fungal phytopathogens and suppresses plant diseases via antibiosis mechanisms in both *in vitro* and *in vivo* studies. The establishment of the microbial pathogen was restrained by the fast-growing *Trichoderma* through antibiosis or competition mechanism and this endophyte can colonize plant tissue (Lacava & Azevedo, 2014).

Cytochalasins H and J from *Diaporthe miriciae* are found to be potent against *Phomopsis* species in microdilution broth assays. Diaporthe species can produce various cytochalasins with antifungal potential to control fungal diseases and also maintains its antagonistic activity in plants. The antifungal activity of Cytochalasin H against microbes like *Sclerotinia sclerotiorum, Gaeumannomyces graminis* var. *tritici*, *Rhizoctonia cerealis, Bipolaris maydis, F. oxysporum, Bipolaris sorokiniana, Botrytis cinerea, Cladosporium cladosporioides* and *Cladosporium sphaerosphermum* was documented in previous reports. In target microbes, cytochalasins H and J generally provoke minor mycelial growth (hormesis). Hormesis is the growth stimulatory effect raised during the presence of subtoxic concentrations of a toxin and generally noted during *in vitro* fungicide testing (de Carvalho, Ferreira-D'Silva, Wedge, Cantrell, & Rosa, 2018). In another study 2,4-diacetylphloroglucinol, an antibiotic from *Pseudomonas fluorescens* strain (PDY7) was extremely efficient in decreasing the bacterial blight disease of rice in both glasshouse and field condition with inhibition of 58.83%–51.88% (Muthukumar, Udhayakumar, & Naveenkumar, 2017).

Phomopsis cassia, an endophytic fungus of *Cassia spectabilis*, produces two novel bioactive metabolites called "phomopsilactone," and "ethyl 2, 4-dihydroxy-5,6-dimethylbenzoate" exhibited good antifungal activity toward *Cladosporium sphaerospermum* and *Cladosporium cladosporioides* (Indira Devi & Momota, 2015). *S. sclerotiorum* (Lib) de Bary, a necrotrophic plant pathogenic fungus is the causative agent of sclerotinia stem rot affecting a wide range of crops in both horticultural and agricultural sectors. *S. sclerotiorum, Aspergillus flavus*, and *R. solani* growth were highly inhibited by *T. harzianum* TH 10-2-2 and *T. harzianum*

TH 5-1-2. The presence of antifungal compounds (39.4%) hexadecanoic acid, 2, 3-bis [(trimethylsilyl) oxy] propyl ester, linoleic, and palmitic acid (unsaturated fatty acids) in TH hex are detected by GC-MS analysis (Chowdhary & Sharma, 2020). Bacillaene, macrolactin, difficidin, siderophore, surfactin, and fengycin are the set of bioactive compounds from *Bacillus amyloliquefaciens* subsp. plantarum strain Fito_F321 with antimicrobial activity and has the potential to biocontrol of grapevine diseases. This natural protector of vineyard showed positive plant-microbial interactions (Pinto et al., 2018). Both volatile and nonvolatile metabolites produced by *P. putida* BP25, an endophyte associated with Black pepper proclaimed to possess an inhibitory effect on the mycelium of *Magnaporthe oryzae*. Further, monocot rice was endogenously colonized by *P. putida* BP25, the activation of defense mechanism and root growth was altered in a density-dependent manner (Ashajyothi et al., 2020).

S. rolfsii is a soil-borne, necrotrophic pathogen infecting chili and causes crop loss ranging from 16% to 80%. Among 120 fungal isolates of *Nothapodytes nimmoniana*, the endophyte *Alternaria* sp. was found to inhibit *S. rolfsii* with 46.62% inhibition in dual culture assay on PDA by producing tenuazonic acid and mycotoxin. Alternariol methyl ether altenuene, tenuazonic acid, and alternariol are the secondary metabolites of *Alternaria* sp., which are reported as mycotoxins and phytotoxins (Dall'Asta et al., 2014; Rajani et al., 2019). Alpha- and beta-glucosidase are the antifungal substances from endophytic bacterium *B. lentimorbus*, which exhibit a higher inhibitory effect on grey mould disease-causing *B. cinerea*. The application of *Bacillus cereus* CE3 formula enhances the shelf life of fruits by controlling *Endothia parasitica* (Murr) growth (Muthukumar et al., 2017). Fluorescent pseudomonads produce volatile/diffusible antifungal antibiotics and their root colonization was aggressive likewise Bacillus lipopeptides (LPs) have significant antagonistic activity and also stimulates defense mechanism in the host (Barquero, Terrón, Velázquez & González-Andrés, 2016). Most of the antimicrobial compounds studied are nonvolatile and their effect the pathogen's hyphae and cause cellular abnormality thereby interrupting growth (Ting, 2014).

Tenuazonic acid is the mycotoxin that binds to the ribosome to inhibit protein synthesis and this mycotoxin from endophytic fungus *Alternaria* sp. has capable to prohibit *S. rolfsii* growth. Extracellular metabolites secreted from bacterial antagonists suppress the plant pathogens and inhibit the pathogen's growth even at low concentrations (Rajani et al., 2019; Ulloa-Ogaz et al., 2015). Polyketide synthase (PKS) pathways are used to synthesis most of the secondary metabolites, particularly polyene, polyether, and macrolide compounds are synthesized by the catalytic action of type I PKSs (Liu, Dou, & Ma, 2016). Cytonic acids A, D, ecomycins B, C, fusicoccane diterpenes, munumbicins,*p*-amino acetophenonic acid, pyrrolidines A and B are the few active metabolites of endophytes reported to be active against pathogens (Indira Devi & Momota, 2015).

During the resting or stationary phase, microorganisms begin to produce secondary metabolites. Under stress conditions, the low molecular weight secondary metabolites from fungal sources help for the survival of the organism. This heterogeneous group of natural compounds is responsible for differentiation, competition, metal transport, and symbiosis. Methyl euginol [Benzene, 1,2- dimethoxy-4-(2-propenyl)] and other compounds with higher fatty acids and benzene group of *R. oryzae* contribute to its antibiotic activity against *R. solani*. (Inhibitory effect of 81.53%) (Peeran et al., 2018). In another study, *Phaeosphaeria nodorum* endophyte of *Prunus domestica* secretes inhibitory substances in growth medium to suppress the growth of *Colletotrichum gloeosporioides* and *Monilinia fructicola* which is the causative agent of anthracnose, blossom blight, brown rot, and twig blight on plum (Pimenta et al., 2012; Liarzi & Ezra, 2014). Munumbicins, munumbicin D is the active metabolite from *Streptomyces* NRRL 30562, (endohyte of *Kennedia nigriscans*) inhibits the development of *Bacillus anthracis*, *Mycobacterium tuberculosis*, *Plasmodium falciparum* (Indira Devi & Momota, 2015).

Volatile organic compounds are low water-soluble, and molecular weight compounds vaporize at normal atmospheric pressure and temperatures. About 300 VOCs identified from fungi and has a distinct odor. The chemical composition of VOC includes alcohols, aldehydes, simple hydrocarbons, phenols, ketones, thioalcohol, thioesters, heterocycles, and their derivatives (Bolívar-Anillo et al., 2020). In grass seeds, the clavicipitalean endophytes are transmitted vertically and systematically infect their hosts. The production of alkaloids (deterring metabolites) and toxic inside the host are linked with their protective actions (Witzell et al., 2014). The postharvest protection of food products, fruit, and vegetables by using VOCs and volatile emitting fungi is termed as "Mycofumigation". In *in vitro* studies, VOCs of *Muscodor albus* were found to be lethal for *B. cinerea*, *M. fructicola*, and *Penicillium expansum* (peach pathogens). *P. nodorum* is a fungal endophyte of *P. domestica* that secretes inhibitory VOCs against *M. fructicola* (Liarzi & Ezra, 2014). 3-methylbutan-1-ol, 3-methylbutyl acetate, and azulene derivatives are the two bioactive volatile metabolites of *Muscodor heveae* sp. nov. active against *Rigidoporus microporus* and *Phellinus noxius*, the pathogenic fungi that cause root disease in *H. brasiliensis* Müll.Arg (the rubber tree). In the Ascomycota phylum, the potent antimicrobial volatile metabolites are produced from fungi that belong to families like Hypocreaceae, Xylariaceae, and Diaporthaceae (Siri-udom, Suwannarach, & Lumyong, 2016).

3.7 Protection of a plant through the secretion of lytic enzymes

Endophytes use their enzymatic activities to enter and colonize the host plant, thereby the plants are protected or controlled by disease-causing phytopathogens. The cellulolytic enzymes discharged act on the cell wall material to dissolve pectin and cellulose so that the bacteria can easily invade inner plant tissue. Cell wall degrading enzymes are produced by endophytic bacteria. Cellulase and pectinases are the cellulolytic enzymes discharged and involved in the cellulolysis of a cell wall, so that the bacteria penetrate, localize, and disseminate in plant tissue (Dheeman et al., 2017). Lytic enzymes from BCA directly degrade cell wall material or disrupts the particular developmental stage of the pathogen (Ulloa-Ogaz et al., 2015). In the mechanism of Predation and Parasitism, few bacterial strains produce fungal cell wall lytic enzymes to lyse fungi and feed on dead materials. The endophytic bacteria isolated from potato roots produce high levels of hydrolytic enzymes like glucanase, chitinase, and cellulose. Chitinase is the lytic enzyme responsible for degrading fungal cell walls and it is well known that several chitinolytic endophytes are used for protection against plant diseases (Mercado-Blanco & Lugtenberg, 2014).

The elongation of the germ tube and spore germination of *B. cinerea* was inhibited through the chitinase enzyme produced by *Serratia plymuthica* (Indira Devi & Momota, 2015). Fungal strains harbored in various medicinal plants like *Catharanthus roseus*, *Alpinia calcarata*, *Calophyiium inophyllum*, and *Bixa orellana*, are involved in the production of enzymes like −1, 4- glucan, amylase, pectinase cellulose, lipase, and laccase. The polysaccharides present in the host plant are degraded by the endophytes enzymes (Sahoo et al., 2018). The effect of chitinase production of *Colletotrichum sublineolum* cell walls with its biocontrol potential by scanning electron microscopy was studied. This document confirmed the genetic correlation between the biocontrol effect of endophytic actinobacteria and chitinase production while studying with various phytopathogens. The actinobacteria have a wide diversity in their enzyme production and their use in agriculture and pharmaceutical industry is tremendous (Lacava & Azevedo, 2014).

3.8 Competition for niche and nutrition

Host plant serves as the shelter for endophytes in microbial competition and environmental stresses. As both plant pathogen and endophyte colonize similar niche, the latter can be the best in controlling pathogens either by direct antagonism or competitor. The development of *F. oxysporum* f. sp. Cubense was inhibited by 18.3% of bacterial strains isolated from surface-sterilized banana root tissues (Senthilkumar et al., 2009). The rhizosphere is the major route used by endophytes for plant colonization. Bacteria reach the rhizosphere by chemotaxis facing uproot exudate components and attaches. The exopolysaccharide lipopolysaccharide (LPS) and Type IV pili are the bacterial components involved in the attachment of endophytes toward plant tissue. The root hair zone and cell elongation zone are apical root zone with a thin-walled surface root layer as the preferential sites for attachment and entry to the plant. Likewise, intercellular spaces in the epidermis and differentiation zone of root regions are sites preferred for bacterial colonization. Once the endophytes exodermal crossed the barrier, it shifts and occupies the intercellular space of the cortex or stays at the entry site (Mercado-Blanco & Lugtenberg, 2014).

A specialized set of effectors are required for *P. indica* to execute host-dependent colonization strategies. Effectors are small molecules and proteins that increase microbial infections by repressing host defenses and changing host-cell structure and function (Card et al., 2016). The composition of the plant microbiome is determined by competitive elimination among microbes and the colonization of the host by pathogens was prohibited by the inhabiting endophytes via competition mechanism. The colonization of endophytic fungi occurs in various plant tissues systemically or locally and can be either in inter- or intracellular space. Due to accelerated colonization and scavenging of feasible nutrients, endophytes occupy the niche faster than the pathogenic organism (Latz et al., 2018). The high capability of bacterial endophytes to enter and colonize plants makes it the best protector of a host from phytopathogens among various beneficial microorganisms (Nigris et al., 2018). The mutualistic *Neotyphodium* endophytes provide competitive benefits on their hosts in both experimental and field studies (Bayman, 2007). The reduction of growth and survivorship of insects in endophyte-infected Grasses was reported in several studies. Endophytes have various effects on insect species that are closely related whereas the individual insect species also have a different mode of action toward different endophyte-infected grasses (Bayman, 2007).

In cucumber plants, Psl-induced disease was controlled by using antagonistic bacteria as BCA. The production of numerous bacteriocins and inhibitory substances and antagonistic activity of EB toward pathogen (Psl) colonization in the internal tissues are the major factors for disease suppression in cucumber (Akbaba & Ozaktan, 2018). Although the interaction between the host plant and endophytic microorganisms promotes the plant's immune response in the initial stage. The endophytes colonize the plant tissue by overcoming this immune response and act as natural vaccination or an immune stimulant (Nigris et al., 2018). When the organisms are involved in the competition for nutrients and niche for their survival. The growth and sporulation of the poor competitor are decreased during the competition. Slow-growing phytopathogens that are prevented from accessing particular host tissues required for their development are overgrown by many BCAs (Narayanasamy, 2013). *Mustela nivalis* var. neglecta 114 was the energetic competitor with *Ophiostoma novo-ulmi* for utilizing amino acids, alcohols, disaccharides, methyl-saccharides, phenolics and polysaccharides (Nigris et al., 2018). The colonization potential of strain toward inner plant tissue is very much essential to generate bacteria inocula for the agriculture field, before studying its biocontrol effect. The strain GL174 controls numerous plant pathogens by efficiently competing for iron nutrition with other microbes. The existence of genes that are related to the production and transport of siderophore may also influence the antagonistic effect of BCAs (Nigris et al., 2018).

Arthrobacter, *Cellulomonas*, *Microbacterium*, *Nesterenkonia*, *Streptomyces*, and *Propioni bacterium* are the various actinobacterial strains that are naturally capable to colonize tomato roots (Ting, 2014). *Clavibacter xyli* subsp. *cynodontis* is the xylem-inhabiting bacterial endophyte that can colonize many plant species.

These gram-positive bacteria received a gene that was derived from *Bacillus thuringiensis*. d-endotoxin produced by *Bacillus* was active against insects like Coleoptera and Lepidoptera and genetically modified bacterium that receives bacillus gene will be able to protect the plant against target insects attacks by toxin secretion inside the plant (Azevedo et al., 2000). Paratransgenesis is the genetic change of symbiotic microbes associated with insects. In symbiotic microbes, their genetic alterations are designed to improve their competitiveness within the insect vector at the loss of the pathogen (expense of the pathogen) (Lacava & Azevedo, 2014).

3.9 Induction of host resistance by endophytes

Systemic resistance is distinguished into two main types, which include "Induced Systemic Resistance (ISR)" and "Systemic Acquired Resistance (SAR)." Salicylic acid (SA) plays an outstanding role in the former SAR, whereas the latter requires ethylene and jasmonic acid. The cross-communication between these hormone signaling permits the plant to finely balance the defense response. The reinforcement of cell wall appositions and pathogen penetration is prevented by strengthening the plant's structural barriers (Latz et al., 2018). Inducing agents might be chemical compounds and living organisms. In the host, inducing agent triggers to forms translocatable signal and makes the host respond toward pathogen attack in a resistant manner. The signal formed will further trigger the differential expression of genes, specific metabolic changes, and synthesis of proteins. The plant metabolism changes make the plant unsuitable for the pathogens that are echoed in lowering the disease incidence. The term "Priming" refers to the protection that depends on activation of plant defense responses. Microbe-associated molecular patterns (MAMPs) or pathogen-associated molecular patterns (PAMPs) are compounds frequently derived from microbial origin. Plant receptors like-glucans and chitin recognize these compounds and induce PAMP/MAMP-triggered immunity (Latz et al., 2018). When compared with SAR, ISR varies in numerous biochemical and physiological phenotypes (Mercado-Blanco & Lugtenberg, 2014)

ISR of host plants can be indirectly influenced by biocontrol bacteria, and plant defense priming responses are activated by endophytes producing a plethora of metabolites. Numerous endophyte taxa are reported to produce LPs, an amphiphilic compound made of short cyclic oligopeptide connected with lipid tail. These molecules act as ISR elicitors in host plants and directly as antimicrobial compounds to control pathogens. Based on their chemical structure, LPs from fengycin, iturin, and surfactin families are the widely studied molecule (Nigris et al., 2018). In tobacco, LPSs from endophyte *Burkholderia cepaciae* act as an elicitor to induce plant defense responses (Alström & Van Vuurde, 2001). The primary infection of the pathogen activates the plant defense mechanisms, thereby develops SAR (Indira Devi & Momota, 2015).

Diverse signal molecules are used for the communication between BCAs and pathogenic microbes. During abiotic stress conditions, the production of abscisic acid, SA, and jasmonic acid is induced and these signal molecules are involved in defense against disease. The production of secondary metabolites and defense-related proteins are induced by jasmonic acid. SA plays a vital role in the biosynthesis of ethylene, growth and development, stromal behavior, and flowering. In the defense signaling process, seed dormancy was promoted by abscisic acid. Beneficial microbes stimulate ISR which is the plant immune response toward an infection. Protection can be enhanced when the plant produces an infection-induced immune response after the immunization process (Vijayabharathi, Sathya & Gopalakrishnan, 2016). Endophytes provoke the host defensive reactions to improve plant resistance. *Methylobacterium* sp. strain IMBG290, an endophytic bacteria isolated from potato shoot reported to induce resistance against the pathogen *Pectobacterium atrosepticum* in an inoculum density-dependent manner (Rabiey et al., 2019).

ISR is another mechanism involved in biocontrol. The induction of plant to produce resistant compounds like phenolics and the formation of lignin and glucan by endophytes increased the plant protection (Lacava & Azevedo, 2014). ISR is triggered by various bacterial traits such as the production of LPSs, siderophores, and flagellation (Indira Devi & Momota, 2015). LPs molecules, from diverse endophyte taxa, directly act as ISR elicitors and antimicrobial compound in host plants to biocontrol pathogenic infections. Short cyclic oligopeptides are associated with lipid tail to form these amphiphilic compounds (Blumenstein et al., 2015). Ashajyothi et al. (2020) confirmed extensive bacterial colonization in roots by examining fluorescence imaging of *P. putida* BP25 primed seedlings followed by qPCR and plate assay. The defense-related peroxidase (PO) and phenols activities are enhanced in primed plants. The expression of SAR related Os*PR1-1* (Pathogenesis related protein 1-1) was increased in the qPCR assay of rice defense and developmental genes and Os*PR3* was downregulated (Ashajyothi et al., 2020).

The endophytic *B. subtilis* (EPCO16, EPC5) along with rhizobacterium *P. fluorescens* applied to chilly plants and it effectively reduced the incidence of Fusarium wilt. Plants that are involved in the study treatment showed improved activities of phenolics, chitinase, β-1,3-glucanase, polyphenol oxidase (PPO) (PO), phenylalanine ammonia-lyase (PAL), and PO. This study is evidence of endophyte to suppress the disease through induced host resistance. The combination of endophyte-endophyte is a more effective multistrain approach rather than endophyte rhizobacteria mixtures (Ting, 2014). The treatments of plants with a mixture of endophyte inoculum combinations were reported to induce the production of defense-related enzymes such as PPO, PO, and PAL. The biosynthesis of phenolic compounds is up-regulated while supplementing the plants with a combination of strains, in which these phenolic compounds are considered to be a major part in promoting systemic resistance in opposition to Fusarium wilt of banana (Savani et al., 2020). *P. fluorescens* protects tomato against *F. oxysporum* f. sp. Radicislycopersici by inducing systemic resistance and in pea roots *F. oxysporum* f. sp. pisi is controlled by *B. pumilus* SE34 (Dheeman et al., 2017).

3.10 Indirect inhibition via siderophore production

Iron is an important element for the survival of microbes, and siderophores producing endophytes effectively utilize the bioavailable iron in the ferric ion. The disease-causing pathogen starves for iron which is essential for their survival (Latha, Karthikeyan, Rajeswari, 2019). Production of siderophore, an antibacterial compound, niche exclusion, ISR, and competition of nutrients are the several mechanisms demonstrated by endophytes during the biological control process (Akbaba & Ozaktan, 2018). The available iron was used up by the siderophores produced from bacteria and results in the inhibition of the phytopathogens growth. This action indirectly promotes the growth of the plant. The production of petrobactin type siderophores is recorded in different *Bacillus* species and pyoverdins in different *Pseudomonas* species. Both gram-negative (*Pseudomonas* sp.) and gram-positive (*Rhodococcus* sp. and *Bacillus* sp.) can produce siderophores. More than 50 structurally relevant siderophores are documented from structural studies and the production of pyoverdins was noted in different *Pseudomonas* species (Aswathy et al., 2013). Siderophore-mediated disease suppression has been demonstrated in many pathosystems that include Pythium damping-off of tomato, Fusarium wilt of radish (*Raphanus sativus* L.), and Fusarium wilt of carnation (*Dianthus caryophyllus* L.) (Mercado-Blanco & Lugtenberg, 2014).

3.11 Inhibition through phytohormone activity

The suppression of ethylene production in endogenous levels, solubilization of soil phosphorus and iron, nitrogen supply, and production of antibiotics and phytohormones like gibberellins, cytokinins, and auxins are involved in the enhancement of plant growth. The mechanisms expressed by endophytic bacteria to suppress plant pathogens *in vitro* may not be similar in natural conditions (Bacon & Hinton, 2006). Holobiont refers to plant and the microbiome associated with it. The plant genome and soil type influence the plant microbiome. Soil root interface was increased by biocontrol agents thereby it indirectly enhances the uptake of Fe and P nutrient in plant and improves plant growth. The plant ethylene levels were lowered by ACC deaminase to improve the stress tolerance ability of plants (Chhabra & Dowling, 2017). The stable microbial community is built by the interactions between a huge variety of microorganisms that occupy the plant interiors. The beneficial bacteria produce phytohormones to improve root formation to promote plant growth, this is an indirect mechanism for controlling pathogens attacks in hypocotyl and roots in the susceptible seedling stage (Alström & Van Vuurde, 2001). By producing ACC deaminase activity and siderophores, *Streptomyces virginiae* isolate Y30 and E36 are the two isolates that displayed favorable biocontrol activity against *Ralstonia solanacearum* (Ting, 2014). The uptake of phosphorus in plants is increased by phosphate solubilizing bacterial inoculants. Endophytic bacteria from the genus *Serratia*, *Bacillus*, *Enterobacter*, and *Pseudomonas* can solubilize the insoluble phosphate. *Achromobacter xyloxidans* and *Bacillus cereus* are the endophytes isolated from a potato plant root, while comparing with the control sample, the concentration of K, N and P, production of photosynthetic pigments, and plant vegetative growth are heightened when these strains are used as bio inoculants for potato tubers (Sandhya et al., 2017).

3.12 Conclusion and future perspective

Endophytes both directly and indirectly inhibit pathogens growth and reduce the disease incidence. Along with disease suppression, it also promotes plant growth by producing phytohormones. These directly inhibit pathogens development by producing antibiotic compounds, lytic enzymes, and mycoparasitic activity. Systemic resistance was triggered by endophytic microbe to enhance the plant tolerance toward pathogenic attack. Endophytes compete with pathogens for nutrition, niche, and other essential elements for survival. It is proven that control of plant diseases using endophyte will be the best alternative to future agriculture in the situation where excessive agrochemical use results in environmental pollution and destruction. Despite the challenges associated with the biocontrol of plant diseases, research shows that endophyte treatments can be successfully implemented as biocontrol agent in the future. Further, it is important to determine the exact mechanism of endophytes involved in preventing and controlling plant disease-causing pathogens. The efficacy of the biocontrol method can be enhanced by a combination of microbes BCA with proper agriculture practices. In this chapter, we discussed the current state of knowledge regarding endophytic microbes as BCAs for plants. This relatively unexplored field is, therefore, seen as an interesting source of BCA particularly in disease suppression and improving crop yield.

Acknowledgments

The authors thank the management of Vellore Institute of Technology, Vellore, for their kind support and encouragement.

References

Akbaba, M., & Ozaktan, H. (2018). Biocontrol of angular leaf spot disease and colonization of cucumber (*Cucumis sativus* L.) by endophytic bacteria. *Egyptian Journal of Biological Pest Control*, *28*, 1–10. Available from https://doi.org/10.1186/s41938-017-0020-1.

Alström, S., & Van Vuurde, J. W. L. (2001). Endophytic bacteria and biocontrol of plant diseases. In S. H. De Boer (Ed.), *Plant Pathogenic Bacteria* (pp. 60–67). Dordrecht: Springer. Available from https://doi.org/10.1007/978-94-010-0003-1_11.

Ashajyothi, M., Kumar, A., Sheoran, N., Ganesan, P., Gogoi, R., Subbaiyan, G. K., & Bhattacharya, R. (2020). Black pepper (*Piper nigrum* L.) associated endophytic *Pseudomonas putida* BP25 alters root phenotype and induces defense in rice (*Oryza sativa* L.) against blast disease incited by *Magnaporthe oryzae*. *Biological Control*, *143*, 104181. Available from https://doi.org/10.1016/j.biocontrol.2019.104181.

Azevedo, J. L., Maccheroni, W., Jr., Pereira, J. O., & de Araújo, W. L. (2000). Endophytic microorganisms: a review on insect control and recent advances on tropical plants. *Electronic Journal of Biotechnology*, *3*, 15–16.

Bacon, C. W., & Hinton, D. M. (2006). Bacterial endophytes: The endophytic niche, its occupants, and its utility. In S. S. Gnanamanickam (Ed.), *Plant-Associated Bacteria* (pp. 155–194). Dordrecht: Springer. Available from https://doi.org/10.1007/978-1-4020-4538-7_5.

Barquero, M., Terrón, A., Velázquez, E., & González-Andrés, F. (2016). Biocontrol of *Fusarium oxysporum* f.sp. *phaseoli* and *Phytophthora capsici* with autochthonous endophytes in common bean and pepper in Castilla y León (Spain). In F. González-Andrés, & E. James (Eds.), *Biological Nitrogen Fixation and Beneficial Plant-Microbe Interaction* (pp. 221–235). Cham: Springer. Available from https://doi.org/10.1007/978-3-319-32528-6_19.

Bayman, P. (2007). Fungal endophytes. In C. Kubicek, & I. Druzhinina (Eds.), *Environmental and Microbial Relationships. The Mycota* (vol. 4, pp. 213–227). Berlin, Heidelberg: Springer. Available from https://doi.org/10.1007/978-3-540-71840-6_13.

Blumenstein, K., Albrectsen, B. R., Martín, J. A., Hultberg, M., Sieber, T. N., Helander, M., & Witzell, J. (2015). Nutritional niche overlap potentiates the use of endophytes in biocontrol of a tree disease. *BioControl*, *60*, 655–667. Available from https://doi.org/10.1007/s10526-015-9668-1.

Bolívar-Anillo, H. J., Garrido, C., & Collado, I. G. (2020). Endophytic microorganisms for biocontrol of the phytopathogenic fungus *Botrytis cinerea*. *Phytochemistry Reviews*, *19*, 721–740. Available from https://doi.org/10.1007/s11101-019-09603-5.

Card, S., Johnson, L., Teasdale, S., & Caradus, J. (2016). Deciphering endophyte behaviour: The link between endophyte biology and efficacious biological control agents. *FEMS Microbiology Ecology*, *92*(8), fiw114. Available from https://doi.org/10.1093/femsec/fiw114.

Chhabra, S., & Dowling, D. N. (2017). Endophyte-promoted nutrient acquisition: Phosphorus and iron. In S. Doty (Ed.), Functional Importance of the Plant Microbiome (pp. 21–42). Cham: Springer. Available from https://doi.org/10.1007/978-3-319-65897-1_3.

Chowdhary, K., & Sharma, S. (2020). Plant growth promotion and biocontrol potential of fungal endophytes in the inflorescence of *Aloe vera* L. *Proceedings of the National Academy of Sciences, India Section B: Biological Sciences*, *90*, 1045–1055. Available from https://doi.org/10.1007/s40011-020-01173-3.

Compant, S., Saikkonen, K., Mitter, B., Campisano, A., & Mercado-Blanco, J. (2016). Editorial special issue: Soil, plants and endophytes. *Plant and Soil*, *405*, 1–11. Available from https://doi.org/10.1007/s11104-016-2927-9.

Dall'Asta, C., Cirlini, M., & Falavigna, C. (2014). Mycotoxins from Alternaria: toxicological implications. *Advances in Molecular Toxicology*, *8*, 107–121. Available from https://doi.org/10.1016/B978-0-444-63406-1.00003-9.

de Carvalho, C. R., Ferreira-D'Silva, A., Wedge, D. E., Cantrell, C. L., & Rosa, L. H. (2018). Antifungal activities of cytochalasins produced by *Diaporthe miriciae*, an endophytic fungus associated with tropical medicinal plants. *Canadian Journal of Microbiology*, *64*, 1–25.

Dheeman, S., Maheshwari, D. K., & Baliyan, N. (2017). Bacterial endophytes for ecological intensification of agriculture. In D. Maheshwari (Ed.), *Endophytes: Biology and Biotechnology, Sustainable Development and Biodiversity* (vol. 15, pp. 193–231). Cham: Springer. Available from https://doi.org/10.1007/978-3-319-66541-2_9.

El-Katatny, M. H., Somitsch, W., Robra, K. H., El-Katatny, M. S., & Gübitz, G. M. (2000). Production of chitinase and β-1,3-glucanase by *Trichoderma harzianum* for control of the phytopathogenic fungus *Sclerotium rolfsii*. *Food Technology and Biotechnology.*, *38*, 173–180.

Indira Devi, S., & Momota, P. (2015). Plant-endophyte interaction and its unrelenting contribution towards plant health. In N. Arora (Ed.), *Plant Microbes Symbiosis: Applied Facets* (pp. 147–162). New Delhi: Springer. Available from https://doi.org/10.1007/978-81-322-2068-8_7.

Jasim, B., Joseph, A. A., John, C. J., Mathew, J., & Radhakrishnan, E. K. (2014). Isolation and characterization of plant growth promoting endophytic bacteria from the rhizome of *Zingiber officinale*. *3 Biotech*, *4*, 197–204. Available from https://doi.org/10.1007/s13205-013-0143-3.

Lacava, P. T., & Azevedo, J. L. (2014). Biological Control of Insect-Pest and Diseases by Endophytes. In V. Verma, & A. Gange (Eds.), *Advances in Endophytic Research* (pp. 231–256)). New Delhi: Springer. Available from https://doi.org/10.1007/978-81-322 1575-2_13.

Latha, P., Karthikeyan, M., & Rajeswari, E. (2019). Endophytic bacteria: Prospects and applications for the plant disease management. In R. Ansari, & I. Mahmood (Eds.), *Plant Health Under Biotic Stress* (pp. 1–50). Singapore: Springer. Available from https://doi.org/10.1007/978-981-13-6040-4_1.

Latz, M. A. C., Jensen, B., Collinge, D. B., & Jørgensen, H. J. L. (2018). Endophytic fungi as biocontrol agents: Elucidating mechanisms in disease suppression. *Plant Ecology Diversity*, *11*(5-6), 555–567. Available from https://doi.org/10.1080/17550874.2018.1534146.

Liarzi, O., & Ezra, D. (2014). Endophyte-mediated biocontrol of herbaceous and non-herbaceous plants. In V. Verma, & A. Gange (Eds.), *Advances in Endophytic Research* (pp. 335–369). New Delhi: Springer. Available from https://doi.org/10.1007/978-81-322-1575-2_18.

Liu, X., Dou, G., & Ma, Y. (2016). Potential of endophytes from medicinal plants for biocontrol and plant growth promotion. *Journal of General Plant Pathology*, *82*, 165–173. Available from https://doi.org/10.1007/s10327-016-0648-9.

McKinnon, A. C., Saari, S., Moran-Diez, M. E., Meyling, N. V., Raad, M., & Glare, T. R. (2017). Beauveria bassiana as an endophyte: A critical review on associated methodology and biocontrol potential. *BioControl*, *62*, 1–17. Available from https://doi.org/10.1007/s10526-016-9769-5.

Mercado-Blanco, J., & Lugtenberg, B. J. J. (2014). Biotechnological applications of bacterial endophytes. *Current Biotechnology*, *3*, 60–75. Available from https://doi.org/10.2174/22115501113026660038.

Muthukumar, A., Udhayakumar, R., & Naveenkumar, R. (2017). Role of bacterial endophytes in plant disease control. In D. Maheshwari, & K. Annapurna (Eds.), *Endophytes: Crop Productivity and Protection, Sustainable Development and Biodiversity* (vol. 16, pp. 133–161). Cham: Springer. Available from https://doi.org/10.1007/978-3-319-66544-3_7.

Narayanasamy, P. (2013). *Detection and identification of fungal biological control agents. Biological Management of Diseases of Crops. Progress in Biological Control* (vol. 15, pp. 9–98). Dordrecht: Springer. Available from https://doi.org/10.1007/978-94-007-6380-7_2.

Nigris, S., Baldan, E., Tondello, A., Zanella, F., Vitulo, N., Favaro, G., ... Baldan, B. (2018). Biocontrol traits of *Bacillus licheniformis* GL174, a culturable endophyte of *Vitis vinifera* cv. Glera. *BMC Microbiology*, *18*, 133. Available from https://doi.org/10.1186/s12866-018-1306-5.

Pang, F., Wang, T., Zhao, C., Tao, A., Yu, Z., Huang, S., & Yu, G. (2016). Novel bacterial endophytes isolated from winter wheat plants as biocontrol agent against stripe rust of wheat. *BioControl*, *61*, 207–219. Available from https://doi.org/10.1007/s10526-015-9708-x.

Paul, N. C., Deng, J. X., Sang, H. K., Choi, Y. P., & Yu, S. H. (2012). Distribution and antifungal activity of endophytic fungi in different growth stages of chili pepper (*Capsicum annuum* L.) in Korea. *The Plant Pathology Journal*, *28*, 10–19. Available from https://doi.org/10.5423/PPJ.OA.07.2011.0126.

Peeran, M. F., Prasad, L., & Kamil, D. (2018). Characterization of secondary metabolites from *Rhizopus oryzae* and its effect on plant pathogens. *International Journal of Current Microbiology and Applied Sciences*, *7*(3), 705–710. Available from https://doi.org/10.20546/ijcmas.2018.703.082.

Pimenta, S. R., Moreira, J. F., Silva, D. A., Buyer, J. S., & Janisiewicz, W. J. (2012). Endophytic fungi from plums (*Prunus domestica*) and their antifungal activity against *Monilinia fructicola. Journal of Food Protection*, *75*, 1883–1889. doi:10.4315/0362-028X.JFP-12-156.

Pinto, C., Sousa, S., Froufe, H., Egas, C., Clément, C., Fontaine, F., & Gomes, A. C. (2018). Draft genome sequence of *Bacillus amyloliquefaciens* subsp. *plantarum* strain Fito-F321, an endophyte microorganism from *Vitis vinifera* with biocontrol potential. *Standards in Genomic Sciences*, *13*, 30. Available from https://doi.org/10.1186/s40793-018-0327-x.

Rabiey, M., Hailey, L. E., Roy, S. R., Grenz, K., Al-Zadjali, M. A. S., Barrett, G. A., & Jackson, R. W. (2019). Endophytes vs tree pathogens and pests: Can they be used as biological control agents to improve tree health? *European Journal of Plant Pathology*, *155*, 711–729. Available from https://doi.org/10.1007/s10658-019-01814-y.

Rajani, P., Aiswarya, H., Vasanthakumari, M. M., Jain, S. K., Bharate, S. B., Rajasekaran, C., ... Uma Shaanker, R. (2019). Inhibition of the collar rot fungus, *Sclerotium rolfsii* Sacc. by an endophytic fungus *Alternaria* sp.: Implications for biocontrol. *Plant Physiology Reports*, *24*, 521–532. Available from https://doi.org/10.1007/s40502-019-00484-6.

Sahoo, S., Sarangi, S., & Kerry, R. G. (2018). Bioprospecting of endophytes for agricultural and environmental sustainability. In J. Patra, C. Vishnuprasad, & G. Das (Eds.), Microbial Biotechnology (pp. 429–458). Singapore: Springer. Available from https://doi.org/10.1007/978-981-10-6847-8_19.

Sandhya, V., Shrivastava, M., Ali, S. Z., & Sai Shiva Krishna Prasad, V. (2017). Endophytes from maize with plant growth promotion and biocontrol activity under drought stress. *Russian Agricultural Sciences*, *43*, 22–34. Available from https://doi.org/10.3103/s1068367417010165.

Savani, A. K., Bhattacharyya, A., & Baruah, A. (2020). Endophyte mediated activation of defense enzymes in banana plants pre-immunized with covert endophytes. *Indian Phytopathology*, *73*, 433–441. Available from https://doi.org/10.1007/s42360-020-00245-8.

Senthilkumar, M., Swarnalakshmi, K., Govindasamy, V., Lee, Y. K., & Annapurna, K. (2009). Biocontrol potential of soybean bacterial endophytes against charcoal rot fungus, *Rhizoctonia bataticola*. *Current Microbiology*, *58*, 288. Available from https://doi.org/10.1007/s00284-008-9329-z.

Siri-udom, S., Suwannarach, N., & Lumyong, S. (2016). Existence of *Muscodor vitigenus*, *M. equiseti* and *M. heveae* sp. nov. in leaves of the rubber tree (*Hevea brasiliensis* Müll.Arg.), and their biocontrol potential. *Annals of Microbiology*, *66*, 437–448. Available from https://doi.org/10.1007/s13213-015-1126-x.

Terhone, E., Kovalchuk, A., Zarsav, A., & Asiegbu, F. O. (2018). Biocontrol Potential of Forest Tree Endophytes. In A. Pirttilä, & A. Frank (Eds.), *Endophytes of Forest Trees, Forestry Sciences* (vol. 86, pp. 283–318). Cham: Springer International Publishing. Available from https://doi.org/10.1007/978-3-319-89833-9_13.

Ting, A. S. Y. (2014). Biosourcing endophytes as biocontrol agents of wilt diseases. In V. Verma, & A. Gange (Eds.), *Advances in Endophytic Research* (pp. 283–300). New Delhi: Springer. Available from https://doi.org/10.1007/978-81-322-1575-2_15.

Troskie, A. M., de Beer, A., Vosloo, J. A., Jacobs, K., & Rautenbach, M. (2014). Inhibition of agronomically relevant fungal phytopathogens by tyrocidines, cyclic antimicrobial peptides isolated from Bacillus aneurinolyticus. *Microbiology (Reading)*, *160* (Pt 9), 2089–2101. Available from https://doi.org/10.1099/mic.0.078840-0.

Ulloa-Ogaz, A. L., Muñoz-Castellanos, L., & Nevárez-Moorillón, G. (2015). Biocontrol of phytopathogens: Antibiotic production as mechanism of control. In A. Méndez-Vilas (Ed.), The Battle Against Microbial Pathogens: Basic Science, Technological Advances and Educational Programs (1st ed., pp. 305–309). Formatex.

Vijayabharathi, R., Sathya, A., & Gopalakrishnan, S. (2016). A renaissance in plant growth- promoting and biocontrol agents by endophytes. In D. Singh, H. Singh, & R. Prabha (Eds.), *Microbial Inoculants in Sustainable Agricultural Productivity* (pp. 37–60). New Delhi: Springer. Available from https://doi.org/10.1007/978-81-322-2647-5_3.

Witzell, J., Martín, J. A., & Blumenstein, K. (2014). Ecological aspects of endophyte-based biocontrol of forest diseases. In V. Verma, & A. Gange (Eds.), *Advances in Endophytic Research* (pp. 1–454). New Delhi: Springer. Available from https://doi.org/10.1007/978-81-322-1575-2_17.

CHAPTER

Antimicrobial metabolites from endophytic microorganisms and its mode of action

4

Saranya Shankar, Gayathri Segaran and Mythili Sathiavelu
School of Bioscience and Technology, Vellore Institute of Technology, Vellore, India

4.1 Introduction

Biological control becomes an environment-friendly option for reducing plant diseases that are caused by plant pathogens. Biocontrol against plant diseases is not only used as a substitute for chemically synthesized pesticides but also supports the plant by controlling the diseases that cannot govern by some other approaches (Iqrar, Shinwari, El-sayed, & Ali, 2019). In developing countries, to feed the growing society the production of crop improvement is very necessary. So, they rely on the usage of chemical fertilizers, which leads to adverse conditions on the habitat. A rise in groundwater and soil pollution, and leaching of nitrogen and phosphorous into groundwater are some of the harmful effects. To reduce the need for chemical fertilizers, we should use active, nutrient-providing microorganisms that will enhance the sustainability of agricultural practices (ALKahtani et al., 2020). Endophytes are defined as microorganisms that live inside the tissues of the host plant for at least one period of their life cycle without causing any noticeable symptoms to their host. Various works reveal that endophytes serve as a reservoir of natural products for a variety of distinct classes of secondary metabolites in the field of medicine and agriculture (Conti et al., 2012). They are universal and have been isolated from almost all parts of the plants. Their association may be facultative or obligate and causes no negative effects on the plants. They reveal complex interactions that include antagonism and mutualism with their hosts. The growth of the endophytes is limited strictly by the plant, and these endophytes gently adapt to their environments by using several mechanisms. Endophytes produce various potential bioactive compounds that help to grow the plants and serve them to adapt better to the surroundings to sustain strong symbiosis (Nair & Padmavathy, 2014). The sources of endophytes may be leaf, seed, root, stem, fruit, and tuber, both in the intercellular space or inside the cells along with conducting vessels. The plant tissue is penetrated by microorganisms through natural openings such as stomata, hydathodes, or through wounds that are caused

Biocontrol Mechanisms of Endophytic Microorganisms. DOI: https://doi.org/10.1016/B978-0-323-88478-5.00001-8

by the friction of growing roots or the appearance of secondary roots in the soil (Souza et al., 2016). Better understanding about endophytes, their roles, and significance may help the researcher for the development of endophyte resources that bring active and novel biologically active metabolites that cannot be synthesized by chemical reactions (Siddiqui & Shaukat, 2003).

Endophytic microorganisms research has stable academic interests in addition to economic conditions, regarding the discovery of unique microbial species during tropical hosts (Azevedo, 2000). The United States of America's Food and Drug Administration has certified that 47% of drugs were developed from medicinal plants. From that 3% of drugs are antimicrobials. The phytoconstituents such as flavonoids, steroids, and saponins produced are reported to be responsible for the pharmaceutical activities of medicinal plants that lead to the development of antimicrobial drugs (Photolo, Mavumengwana, Sitole, & Tlou, 2020). Pathogens that affect plant tissue, insect invasion, and herbivores are the causes of host damage. Endophytes with parasitic, mutualistic, and communalistic relationships safeguard the plants from these damages. They contain resistance mechanisms by producing antimicrobial metabolites to protect their host plant from pathogenic attack (Suresha & Jayashankar, 2019). Secondary metabolites are not only used for the growth of an organism but also in ecological interactions and environmental stresses. It also plays a flexible role in working as the signaling and defense compound. Endophytes are the microorganisms that produce various secondary metabolites such as phytohormones, antimicrobial compounds, and vitamins like B12 and B1 with low molecular weight (Singh, Kumar, Singh, & Deo, 2017). To control the pathogenic invasion, the endophyte develops a resistance mechanism by producing secondary metabolites like phenols, flavonoids, alkaloids, peptides etc. (Singh, Kumar, et al., 2017). All over the world, food loss is caused by two major factors through pests and plant diseases. Plant infections caused by pathogens are about 20%–30% loss of $40 billion worldwide annually (Bolívar-Anillo, Hernando Jose Garrido, & Collado, 2019).

4.2 Importance of endophytic microorganisms as biocontrol agents

Endophytic microorganisms can act as biocontrol agents because they colonize environmental niches as same as that of few phytopathogens. Their efficacy is based on several factors such as patterns of the colonization, growth phase, host specificity, capacity to activate systemic resistance, and physiological state of the plant. To control plant diseases, microorganisms and their compounds obtain huge importance when they do not exhibit harmful effects on human and animal health (Bolívar-Anillo et al., 2019). Biocontrol mechanisms include hyperparasitism, systemic resistance in plants, competition for a substrate, predation, and allelochemical production like lytic enzymes, siderophores, and antibiotics. When compared to antibiosis and induced systemic resistance (ISR) mechanisms for substrates in endophytes, the competition and parasitism mechanisms are less powerful. Minerals solubilization, phytohormones production, and biological nitrogen fixation are the several mechanisms that endophytic microorganisms carried out to promote the growth of the host plant (Bolívar-Anillo et al., 2019). Various microorganisms have been considered desirable biocontrol agents like *Pseudomonas fluorescens* and a few *Streptomyces* species. Several fungal and bacterial infections are controlled by *Ulocladium atrum* and *Trichoderma*, as they exhibit 30% and 50% of control efficacy. Due to their strong, resistant endospores and antibiotics production, the *Bacillus* sp. turn into interesting biocontrol agents (Shafi, Tian, & Ji, 2017). Commonly used antimicrobial compounds become resistant to many microbes present in agriculture. To overcome this problem nowadays researchers are searching for new ecofriendly agents by a natural way to control the pathogen. For example, endophytes produced metabolites like amides and amines that are toxic to insects but not for mammals (Singh, Kumar, et al., 2017). Some of the antimicrobial metabolites from endophytic microorganisms and their mode of action are represented in Table 4.1.

4.3 Endophytic bacteria

Many studies reported that microorganisms become resistant to antibiotics by molecular mechanisms like drug modification and the prevention of drug targets approach. This is the major reason to develop new antibiotics from endophytic bacteria, especially from medicinal plants (Suresha & Jayashankar, 2019). Endophytic bacteria from the plant have feasible applications in the pharmaceutical, agriculture, and food industry because of their novel biomolecule production, and also they can inhibit disease development in plants (Suresha & Jayashankar, 2019). As an indirect stimulation to improve the plant's development and growth, the bacterial endophytes possess acceleration of digestion, phytohormones protection, and biological nitrogen fixation, etc., that promote resistance to biotic factors. Endophytes act indirectly on the development of absorption of water, nutrients, and minerals. Abiotic factors cause various stress such as risk to xenobiotics, heavy metals, and osmotic stress to the host. They also act in the biocontrol of phytopathogens by eliminating the harmful microbes (Langner et al., 2018). Bacterial endophytes are located in different environments like temperate, aquatic, coastal forests, and geothermal soils (Singh, Kumar, et al., 2017). Plant growth–promoting rhizobacteria and biocontrol agents such as *Azotobacter*, *Clostridium*, *Bacillus*, *Enterobacter*, *Pseudomonas*, *Serratia*, and *Azoarcus* were reported in earlier studies (Waheda et al., 2018).

4.4 Endophytic fungi

The endophytic fungi not only improve plant growth by producing hormones that resist abiotic stress but also produce bioactive compounds related to their host

Table 4.1 Antimicrobial metabolites from endophytic microorganisms from plants and their mode of its action.

Host plant	Endophytic microorganisms	Phytopathogens inhibited	Antimicrobial metabolites	Mode of action	Reference
Lycopersicon esculentum	*Bacillus halotolerans, Bacillus amyloliquefaciens*	*Botrytis cinerea*	Fengycin	Antibiosis	Bolívar-Anillo et al. (2019)
Persea indica	*Hypoxylon* sp.	*B. cinerea*	1,8-Cineole, 1-Methyl-1,4-cyclohexadiene, Alpha-methylene-alpha-fenchocamphorone	Antibiosis and antibiotics suppressing pathogens	Tomsheck et al. (2016)
Cinnamomum zeylanicum	*Muscodor albus*	Aspergillus sp., Colletotrichum sp., Geotrichum sp.	*N*-methyl-*N*-nitrosoisobutyramide	Complete inhibition of mycelium growth by antibiosis	Kaddes, Fauconnier, Sassi, and Nasraoui (2019)
Glycyrrhiza uralensis	*Bacillus atrophaeus*	*Verticillium dahliae*	1,2-Benzenedicarboxylic acid, Bis(2-methylpropyl) ester, 9,12-Octadecadienoic acid (Z,Z)-methyl ester, 9-Octadecenoic acid, methyl ester, Decanedioic acid, Bis(2-ethylhexyl) ester	Antibiosis	Mohamad et al. (2018)
Dendrobium nobile	*Trichoderma longibrachiatum*	*Bacillus subtilis*, *Bacillus mycoides*, *Staphylococcus* sp.	Dendrobine	Antibiosis and antibiotics suppressing pathogens	Sarsaiya et al. (2020)
Leptospermum scoparium	*Pseudomonas* sp.	*Pseudomonas syringae pv. actinidiae*	Hydrogen cyanide	Competition	Wicaksono et al. (2018)
C. zeylanicum	*M. albus*	*Ustilago hordei*	1-Butanol, 3-methyl-, acetate	Antibiosis	Strobel, Dirkse, Sears, and Markworth (2001)

L. esculentum	*B. halotolerans*, *B. subtilis*	*B. cinerea*	Surfactin	Antibiosis and antibiotics suppressing pathogens	Bolívar-Anillo et al. (2019)
Aloe dhufarensis Lavranos	*Sarocladium kiliense*	*Fusarium sp.*, *Cladosporium* sp.	2,3-Butanediol	Antibiosis	Khuseib et al. (2020)
L. esculentum	*B. amyloliquefaciens*	*B. cinerea*	Bacillomycin D	Antibiosis	Bolívar-Anillo et al. (2019)
Paris polyphylla var. yunnanensis Hand.-Mazz	*Fusarium* sp.	*B. subtilis*, *Staphylococcus haemolyticus*, *Agrobacterium tumefaciens*, *Pseudomonas lachrymans*, *Xanthomonas vesicatoria*, *Magnaporthe oryzae*	5α, 8α-Epidioxyergosta-6, 22-dien-3β-ol, Butanedioic acid	Antibiosis	Huang et al. (2009)
L. esculentum	*B. subtilis*	*B. cinerea*	Iturin	Antibiosis and antibiotics suppressing pathogens	Bolívar-Anillo et al. (2019)
Aloe dhufarensis Lavranos	*Penicillium oxalicum*	*Fusarium* sp., *Cladosporium* sp.	Tetradecanoic acid, Dodecanoic acid, 2-Furanmethanol	Competition and antibiosis	Khuseib et al. (2020)
Cucumis sativus	*Trichoderma harzianum*	*Fusarium oxysporum*	Diterpene	Inhibition of mycelium growth	Kaddes et al. (2019)
Hedera helix	*B. amyloliquefaciens*	*B. cinerea*	Iturin, Fengycin, Surfactin, Bacillomycin D	Antibiosis and antibiotics suppressing pathogens	Bolívar-Anillo et al. (2019)

(Continued)

Table 4.1 Antimicrobial metabolites from endophytic microorganisms from plants and their mode of its action. *Continued*

Host plant	Endophytic microorganisms	Phytopathogens inhibited	Antimicrobial metabolites	Mode of action	Reference
Urospermum picroides	*Ampelomyces* sp.	*Staphylococcus aureus*, *Staphylococcus epidermidis*, *Enterococcus faecalis*	3-*O*-Methylalaternin, Altersolanol	Antibiosis	Pavithra, Bindal, Rana, and Srivastava (2020)
Maize	*Acremonium zeae*	*Aspergillus flavus*, *Fusarium verticillioides*	Pyrrocidines	Antibiosis	Pavithra et al. (2020)
Olea europaea L.	*Daldinia cf. concentric*	*B. cinerea*	Transoct-2-enal	Antibiosis	Liarzi, Bar, Lewinsohn, and Ezra (2016)
Solanum lycopersicum	*T. harzianum*	*B. cinerea*	Diterpene	Induced systemic resistance and antibiosis	Kaddes et al. (2019)
Theobroma cacao	*Pseudomonas aeruginosa*, *Chryseobacterium proteolyticum*	*Phytophthora palmivora*	Eicosane, 1-phenanthrenecarboxylic acid, Hexatriacontane, Tetratetracontane, Heneicosane, Phenol, 2,4-bis(1,1-dimethylethyl)	Induced systemic resistance and antibiosis	Alsultan, Vadamalai, Khairulmazmi, and Saud (2019)
Panax ginseng	*Phoma terrestris*	*B. cinerea*	*N*-Amino-3-hydroxy-6-methoxyphthalim-ide, 3-Methylthiobenzothiophene, 5-Hydroxy-dodecanoic acid lactone, 5-(Methoxycar-bonyloxy) pent-3-yn-2-ol, 2-Phenylindole	Antibiosis and antibiotics suppressing pathogens	Bolívar-Anillo et al. (2019)

Cassia spectabilis	*Phomopsis cassiae*	*Cladosporium sphaerospermum*, *Cladosporium cladosporioides*	3,11,12-Trihydroxy-cadalene	Antibiosis	Gao, Dai, and Liu (2010)
Catalpa ovata	*Bacillus velezensis*	*B. cinerea*	2,5-Dimethylpyrazine, 4-Chloro-3-methylphenol, Benzothiazole 2,4-Bis (1,1-dimethylethyl) phenol	Antibiosis and antibiotics suppressing pathogens	Gao, Zhang, Liu, Han, and Zhang (2016)
Taxus cuspidata	*Periconia* sp.	*B. subtilis*, *Klebsiella pneumonia*, *S. aureus*, *Salmonella typhimurium*	Fusicoccane diterpenes	Antibiosis	Pavithra et al. (2020)
Taxus mairei	*Aspergillus clavatonanicus*	*B. cinerea*	Clavatol, Patulin	Antibiosis and antibiotics suppressing pathogens	Zhang et al. (2008)
Black pepper	*Pseudomonas putida*	*Rhizoctonia solani*, *Colletotrichum gloeosporioides*, *Athelia rolfsii*, *Gibberella moniliformis*, *Magnaporthe oryzae*, *Ralstonia pseudosolanacearum*	2,5-Dimethyl pyrazine, 2-Methyl pyrazine, Dimethyl trisulfide, 2-Ethyl 5-methyl pyrazine, 2-Ethyl 3,6-dimethyl pyrazine	Competition	Agisha, Kumar, Eapen, and Suseelabhai (2019)
Sorghum	*Streptomyces* sp.	*Rhizoctonia solani*	2-(Chloromethyl)-2-cyclopropyloxirane, 2,4-Ditert-butylphenol, 1-Ethylthio-3-methyl-1,3-butadiene	Direct inhibition of phytopathogens by antibiosis	Patel, Madaan, and Archana (2018)
Melia azedarach	*Aspergillus fumigatus*	*B. cinerea*	Fumitremorgin B, Helvolic acid, 12β-hydroxy-13α-methoxyverruculogen	Antibiosis	Bolívar-Anillo et al. (2019)

(*Continued*)

Table 4.1 Antimicrobial metabolites from endophytic microorganisms from plants and their mode of its action. *Continued*

Host plant	Endophytic microorganisms	Phytopathogens inhibited	Antimicrobial metabolites	Mode of action	Reference
Allium fistulosum	*Streptomyces* sp.	*Alternaria brassicicola*	Fistupyrone	Antibiosis	Ayswaria, Vasu, and Krishna (2020)
Zingiber officinale	*Nocardiopsis* sp.	*Pythium myriotylum*	Phenol, 2,4-bis (1,1-dimethylethyl), Trans cinnamic acid	Antibiosis and antibiotics suppressing pathogens	Sabu and Radhakrishnan (2017)
Physalis ixocarpa	*Pseudomonas stutzeriStenotrophomonas maltophilia*	*B. cinerea*	Dimethyl disulfide	Direct contact with the phytopathogen	Rojas-solís et al. (2018)
Pterocarpus santalinus	*Trichoderma* sp.	*Sclerotinia sclerotiorum Sclerotium rolfsii F. oxysporum*	Hydrocarbons, alcohols, ketones, aldehydes, esters, acids, ethers, and terpenes	Antibiosis	Rajani et al. (2020)
Ginkgo biloba	*Chaetomium globosum*	*Fusarium graminearum*	1,2-Benzenedicarboxaldehyde-3,4,5-trihydroxy-6-methyl (flavipin)	Antibiosis	Xiao et al. (2013)
Z. officinale	*Paraconiothyrium* sp.	*P. myriotylum*	Danthron	Antibiosis	Radhakrishnan (2018)
Capsicum frutescens	*Trichoderma viride*	*B. cinerea F. oxysporum*	6-Pentyl-2H-Pyran-2-one	Inhibition of mycelium growth	Kaddes et al. (2019)
Gossypium hirsutum	*Phomopsis* sp.	*B. cinerea*	*Cytochalasin N, Cytochalasin H, Epoxycytochalasin H*	Antibiosis	Fu, Zhou, Li, Ye, and Guo (2011)

plant (An et al., 2020). It sustains its antagonistic environmental conditions to protect the host plant from herbivory and phytopathogens (Obare, Indieka, & Matasyoh, 2020). By producing novel secondary metabolites, the endophytic fungi are confirmed as a potential source. Because of this reason, they gain an advantage to find novel structures and functions obtaining secondary metabolites (Singh, Katoch, et al., 2017). The investigation of each plant species confirms the presence of endophytic fungi. Based on species, host plant chemistry, physiological state, environmental stress, etc., differ the connection between host and their endophytes from mutualistic to parasitic. In every terrestrial ecosystem, these fungi play a crucial role in the community structure, evolution, and biogeography of the plant due to the strength and survival of the plants (Macías-rubalcava, Hernández-bautista, Oropeza, & Anaya, 2010). On the other hand, in the forest community, endophytic fungi reside in the various plant species and also provide diversity to the natural ecosystems. Some endophytic fungi studies reported that they have promising applications in resistance and biocontrol by producing several bioactive chemicals. In natural ecosystems, they play a significant role in recycling nutrients (Gashgari, Gherbawy, Ameen, & Alsharari, 2016).

4.5 Endophytic actinomycetes

Actinomycetes are considered the best creative microbes when compared to other microorganisms producing bioactive compounds. Medicinal plants serve as a potent resource for isolating novel actinobacteria that produce bioactive compounds (Ningthoujam, 2017). These biologically active compounds are used as therapeutic agents and for crop protection. More than 140 genera of actinomycetes are reported to produce important antibiotics till date. Various types of secondary metabolites from actinomycetes exhibited several bioactivities to develop therapeutic agents (Singh, Kumar, et al., 2017). Both medicinal and crop plants are responsible for rising endophytic actinobacterial isolates. Various mechanisms were followed by endophytic actinobacteria to overcome disease symptoms caused by plant pathogens and develop the plant growth, secondary metabolites production initiates systemic acquired in plants, and changes in host physiology (Ningthoujam, 2017).

4.6 Endophytic microorganisms from the plant as a resource of secondary metabolites

In recent years, natural products have played a powerful act in the invention and improvement of the drug. Endophytic microorganisms from medicinal plants lead to the discovery of potential natural products. By enhancing separation, isolation techniques, and screening programs resulted in a finding of more than 1 million natural compounds. From these, 5% are obtained from the microbial origin and 50%–60% from plant origin. Bioactivities were demonstrated for 20%–25% of the reported natural compounds. Actinomycetes, bacteria, and fungi are the potential resources for producing 45% of bioactive compounds still now. Secondary metabolites from the plant have a vast range of medicinal properties and they are explored for various applications, including industrial, pharmacological, and agricultural. Several metabolites were determined as lead compounds for drug discovery (Tikole, Tarate, & Shelar, 2018).

4.7 Effects of phytopathogens on plant community

Plant pathogens not only disturb the growth of the plant and seedling survival but also decrease the quality and production of a crop that leads to huge economic loss in agricultural ecosystems. By modifying fitness differences, pathogens may affect the composition and dynamics of the community. Ecosystem destabilization occurs when the pathogen removes species in communities rather than developing species coexistence. The diversity of species influenced by pathogens results in the pathogen intensity change. In theoretical aspects, plant diversity damage can either increase or decrease the disease's existence (Chen & Nan, 2015). Environmental conditions, fertilization, and plant community diversity were examined through long-term manipulative analysis, which alters the plant quality and pathogen or herbivore disturbance in usual prairie legume *Lespedeza capitata* (Jeger, Salama, Shaw, & van Den, 2014).

4.8 Conclusion

Endophytes are a poorly examined group of microorganisms that provide a rich source of bioactive novel compounds. They are used in a wide range of pharmaceutical and agricultural fields, etc. Endophytes may have the capacity to produce the same or similar bioactive compounds as host plants because of their symbiotic relationships and intergeneric exchange of genetic information with host plants. Interactions of plant and endophyte may result in the promotion of plant health and plays an important role in low input sustainable agriculture applications for both food and nonfood crops. In this chapter, we have tabulated the endophytic microorganisms producing antimicrobial metabolites and their mode of action for protecting the host against phytopathogens. Endophytic microorganisms are significant sources of secondary metabolites and valuable for further analysis. Broad studies are needed for understanding and characterizing the mode of action of these biocontrol agents.

Acknowledgments

The authors thank the management of Vellore Institute of Technology, Vellore for their kind support and encouragement.

References

Agisha, V. N., Kumar, A., Eapen, S. J., & Suseelabhai, R. (2019). Broad-spectrum antimicrobial activity of volatile organic compounds from endophytic *Pseudomonas putida* BP25 against diverse plant pathogens. *Biocontrol Science and Technology*, *29*(11), 1–21. Available from https://doi.org/10.1080/09583157.2019.1657067.

ALKahtani, M. D. F., Fouda, A., Attia, K. A., Al-Otaibi, F., Eid, A. M., Ewais, E. E.-D., & Abdelaal, K. A. A. (2020). Isolation and characterization of plant growth promoting endophytic bacteria from desert plants and their application as bioinoculants for sustainable agriculture. *Agronomy*, *10*(1325), 1–18.

Alsultan, W., Vadamalai, G., Khairulmazmi, A., & Saud, H. M. (2019). Isolation, identification and characterization of endophytic bacteria antagonistic to *Phytophthora palmivora* causing black pod of cocoa in Malaysia. *European Journal of Plant Pathology*, *155*, 1077–1091.

An, C., Ma, S., Shi, X., Xue, W., Liu, C., & Ding, H. (2020). Diversity and antimicrobial activity of endophytic fungi isolated from *Chloranthus japonicus* Sieb in Qinling Mountains, China. *International Journal of Molecular Sciences*, *21*, 1–15.

Ayswaria, R., Vasu, V., & Krishna, R. (2020). Critical reviews in microbiology diverse endophytic *Streptomyces* species with dynamic metabolites and their meritorious applications: A critical review. *Critical Reviews in Microbiology*, *46*(6), 750–758. Available from https://doi.org/10.1080/1040841X.2020.1828816.

Azevedo, J. L. (2000). Endophytic microorganisms: A review on insect control and recent advances on tropical plants. *EJB Electronic Journal of Biotechnology*, *3*(1), 41–65.

Bolívar-Anillo., Hernando Jose Garrido, C., & Collado, I. G. (2019). Endophytic microorganisms for biocontrol of the phytopathogenic fungus *Botrytis cinerea*. *Phytochemistry Reviews*, *19*(3). Available from https://doi.org/10.1007/s11101-019-09603-5.

Chen, T., & Nan, Z. (2015). Acta ecologica sinica effects of phytopathogens on plant community dynamics: A review. *Acta Ecologica Sinica*, *35*(6), 177–183. Available from https://doi.org/10.1016/j.chnaes.2015.09.003.

Conti, R., Cunha, I. G. B., Siqueira, V. M., Souza-motta, C. M., Amorim, E. L. C., & Janete, M. (2012). Endophytic microorganisms from leaves of *Spermacoce verticillata* (L.): Diversity and antimicrobial activity. *Journal of Applied Pharmaceutical Science*, *2*(12), 17–22. Available from https://doi.org/10.7324/JAPS.2012.21204.

Fu, J., Zhou, Y., Li, H., Ye, Y., & Guo, J. (2011). Endophytic fungus in *Gossypium hirsutum*. *African Journal of Microbiology Research*, *5*(10), 1231–1236. Available from https://doi.org/10.5897/AJMR11.272.

Gao, F., Dai, C., & Liu, X. (2010). Mechanisms of fungal endophytes in plant protection against pathogens. *African Journal of Microbiology Research*, *4*(13), 1346–1351.

Gao, Z., Zhang, B., Liu, H., Han, J., & Zhang, Y. (2016). Identification of endophytic *Bacillus velezensis* ZSY-1 strain and antifungal activity of its volatile compounds against *Alternaria solani* and *Botrytis cinerea. Biological Control*, *105*, 27–39. Available from https://doi.org/10.1016/j.biocontrol.2016.11.007.

Gashgari, R., Gherbawy, Y., Ameen, F., & Alsharari, S. (2016). Molecular characterization and analysis of antimicrobial activity of endophytic—From medicinal plants in Saudi Arabia. *Jundishapur Journal of Microbiology*, *9*(1), 1–8. Available from https://doi.org/10.5812/jjm.26157.

Huang, Y., Zhao, J., Zhou, L., Wang, M., Wang, J., Li, X., & Chen, Q. (2009). Antimicrobial compounds from the endophytic fungus *Fusarium* sp. Ppf4 isolated from the medicinal plant *Paris polyphylla var. yunnanensis. Natural Product Communications*, *4*, 1455–1458.

Iqrar, I., Shinwari, Z.K., El-sayed, A., & Ali, G.S. (2019). *Bioactivity-driven high throughput screening of microbiomes of medicinal plants for discovering new biological control agents*, bioRxiv, 1–26.

Jeger, M. J., Salama, N. K. G., Shaw, M. W., & van Den, B. E. (2014). Effects of plant pathogens on population dynamics and community composition in grassland ecosystems: Two case studies. *European Journal of Plant Pathology*, *138*, 513–527. Available from https://doi.org/10.1007/s10658-013-0325-1.

Kaddes, A., Fauconnier, M., Sassi, K., & Nasraoui, B. (2019). Endophytic fungal volatile compounds as solution for sustainable agriculture. *Molecules (Basel, Switzerland)*, *24*, 1–16. Available from https://doi.org/10.3390/molecules24061065.

Khuseib, F., Al-rashdi, H., Al-sadi, A. M., Al-riyamy, B. Z., Maharachchikumbura, S. S. N., Al-sabahi, J. N., & Velazhahan, R. (2020). Endophytic fungi from the medicinal plant *Aloe dhufarensis* Lavranos exhibit antagonistic potential against phytopathogenic fungi. *South African Journal of Botany*, 1–8. Available from https://doi.org/10.1016/j.sajb.2020.05.022, https://www.sciencedirect.com/science/article/abs/pii/S0254629920309340.

Langner, M., Berlitz, D. L., Leticia, S., Wiest, F., Schünemann, R., Knaak, N., & Fiuza, L. M. (2018). Benefits associated with the interaction of endophytic bacteria and plants. *Brazilian Archives of Biology and Technology*, *61*, 1–11.

Liarzi, O., Bar, E., Lewinsohn, E., & Ezra, D. (2016). Use of the endophytic fungus *Daldinia cf. concentrica* and Its volatiles as bio-control agents. *PLoS One*, *11*(12), 1–18. Available from https://doi.org/10.1371/journal.pone.0168242.

Macías-rubalcava, M. L., Hernández-bautista, B. E., Oropeza, F., & Anaya, A. L. (2010). Allelochemical effects of volatile compounds and organic extracts from muscodor yucatanensis, a tropical endophytic fungus from *Bursera simaruba. Journal of Chemical Ecology*, *36*, 1122–1131. Available from https://doi.org/10.1007/s10886-010-9848-5.

Mohamad, O. A. A., Li, L., Ma, J.-B., Hatab, S., Xu, L., Jian-We, G., ... Li, W. (2018). Evaluation of the antimicrobial activity of endophytic bacterial populations from Chinese traditional medicinal plant licorice and characterization of the bioactive secondary metabolites produced by *Bacillus atrophaeus* against *Verticillium dahliae. Frontiers in Microbiology*, *9*(924), 1–14. Available from https://doi.org/10.3389/fmicb.2018.00924.

Nair, D. N., & Padmavathy, S. (2014). Impact of endophytic microorganisms on plants, environment and humans. *The Scientific World Journal*, *2014*.

Ningthoujam, D. S. (2017). Biocontrol and PGP potential of endophytic actinobacteria from selected ethnomedicinal plants in. *Journal of Bacteriology & Mycology*, *4*(6), 1–8. Available from https://doi.org/10.15406/jbmoa.2017.04.00112.

Obare, R. M., Indieka, S. A., & Matasyoh, J. (2020). Antibacterial activity of endophytic fungi isolated from leaves of medicinal Plant *Leucas martinicensis* L. growing in a Kenyan tropical forest. *African Journal of Biochemistry Research*, *14*(3), 81–91. Available from https://doi.org/10.5897/AJBR2020.1055.

Patel, J. K., Madaan, S., & Archana, G. (2018). Antibiotic producing endophytic *Streptomyces* spp. colonize above-ground plant parts and promote shoot growth in multiple healthy and pathogen-challenged cereal crops. *Microbiological Research*, *215*, 36–45. Available from https://doi.org/10.1016/j.micres.2018.06.003.

Pavithra, G., Bindal, S., Rana, M., & Srivastava, S. (2020). Asian journal of plant sciences review article role of endophytic microbes against plant pathogens: A review. *Asian Journal of Plant Sciences*, *19*, 54–62. Available from https://doi.org/10.3923/ajps.2020.54.62.

Photolo, M. M., Mavumengwana, V., Sitole, L., & Tlou, M. G. (2020). Antimicrobial and antioxidant properties of a bacterial endophyte, methylobacterium radiotolerans MAMP 4754, isolated from *Combretum erythrophyllum* seeds. *International Journal of Microbiology*, *2020*, 1–11.

Radhakrishnan, C. A. P. S. E. K. (2018). Endophytic *Paraconiothyrium* sp. from *Zingiber officinale* Rosc. displays broad-spectrum antimicrobial activity by production of Danthron. *Current Microbiology*, *75*(3), 343–352. Available from https://doi.org/10.1007/s00284-017-1387-7.

Rajani, P., Rajasekaran, C., Vasanthakumari, M. M., Olsson, S. B., Ravikanth, G., & Shaanker, R. U. (2020). Inhibition of plant pathogenic fungi by endophytic *Trichoderma* spp. through mycoparasitism and volatile organic compounds. *Microbiological Research*, *242*, 126595. Available from https://doi.org/10.1016/j.micres.2020.126595.

Rojas-solís, D., Zetter-salmón, E., Contreras-pérez, M., Rocha-granados, C., Macías-rodríguez, L., & Santoyo, G. (2018). Biocatalysis and agricultural biotechnology *Pseudomonas stutzeri* E25 and *Stenotrophomonas maltophilia* CR71 endophytes produce antifungal volatile organic compounds and exhibit additive plant growth-promoting effects. *Biocatalysis and Agricultural Biotechnology*, *13*, 46–52. Available from https://doi.org/10.1016/j.bcab.2017.11.007.

Sabu, R., & Radhakrishnan, K. R. S. E. K. (2017). Endophytic *Nocardiopsis* sp. from *Zingiber officinale* with both antiphytopathogenic mechanisms and antibiofilm activity against clinical isolates. *3 Biotech*, *7*(115), 1–13. Available from https://doi.org/10.1007/s13205-017-0735-4.

Sarsaiya, S., Jain, A., Fan, X., Jia, Q., Xu, Q., Shu, F., & Chen, J. (2020). New insights into detection of a dendrobine compound from a novel endophytic *Trichoderma longibrachiatum* strain and its toxicity against phytopathogenic bacteria. *Frontiers in Microbiology*, *11*(337), 1–12. Available from https://doi.org/10.3389/fmicb.2020.00337.

Shafi, J., Tian, H., & Ji, M. (2017). *Bacillus* species as versatile weapons for plant pathogens: A review. *Biotechnology & Biotechnological Equipment*, *31*(3), 446–459. Available from https://doi.org/10.1080/13102818.2017.1286950.

Siddiqui, I. A., & Shaukat, S. S. (2003). Endophytic bacteria: Prospects and opportunities for the biological control of plant-parasitic nematodes. *Nematologia Mediterranea*, *31*, 111–120.

Singh, G., Katoch, A., Razak, M., Kitchlu, S., Goswami, A., & Katoch, M. (2017). Bioactive and biocontrol potential of endophytic fungi associated with *Brugmansia aurea* Lagerh. *FEMS Microbiology Letters*, *364*, 1–10. Available from https://doi.org/10.1093/femsle/fnx194.

Singh, M., Kumar, A., Singh, R., & Deo, K. (2017). Endophytic bacteria: A new source of bioactive compounds. *3 Biotech*, *2*. Available from https://doi.org/10.1007/s13205-017-0942-z.

Souza, I. F. A. C., Napoleão, T. H., De Sena, K. X. R. F., Paiva, P. M. G., De Araújo, J. M., & Coelho, L. C. B. B. (2016). Endophytic microorganisms in leaves of *Moringa oleifera* collected in three localities at pernambuco state, Northeastern Brazil. *British Microbiology Research Journal*, *13*(5), 1–7. Available from https://doi.org/10.9734/BMRJ/2016/24722.

Strobel, G. A., Dirkse, E., Sears, J., & Markworth, C. (2001). Volatile antimicrobials from *Muscodor albus*, a novel endophytic fungus. *Microbiology (Reading, England)*, *147*, 2943–2950.

Suresha, S, & Jayashankar, M (2019). Antimicrobial activity of endophytic bacteria isolated from few plants of muthathi wildlife sanctuary mandya, Karnataka. *International Journal of Pharmaceutical Sciences and Research*, *10*(5), 2523–2527. Available from https://doi.org/10.13040/IJPSR.0975-8232.10(5).2523-27.

Tikole, S. S., Tarate, B., & Shelar, P. (2018). Endophytic micro-organisms as emerging trend of secondary metabolite. *International Journal of Pharmaceutical Sciences and Research*, *9*(12), 211–214.

Tomsheck, A. R., Strobel, G. A., Booth, E., Geary, B., Spakowicz, D., Knighton, B., & Ezra, D. (2016). *Hypoxylon* sp., an endophyte of *Persea indica*, producing 1, 8-cineole and other bioactive volatiles with fuel potential. *Microbial Ecology*, *60*(4), 903–914. Available from https://doi.org/10.1007/S00248-0.

Waheda, M., Ansary, R., Rezwan, F., Prince, K., Haque, E., Sultana, F., & Akanda, A. M. (2018). Endophytic *Bacillus* spp. from medicinal plants inhibit mycelial growth of *Sclerotinia sclerotiorum* and promote plant growth. *Zeitschrift für Naturforschung C*, *73*.

Wicaksono, W. A., Jones, E. E., Casonato, S., Monk, J., Hayley, J., & Monk, J. (2018). Biological control of *Pseudomonas syringae pv. actinidiae* (Psa), the causal agent of bacterial canker of kiwifruit, using endophytic bacteria recovered from a medicinal plant. *Biological Control*, *116*, 103–112. Available from https://doi.org/10.1016/j.biocontrol.2017.03.003.

Xiao, Y., Li, H., Li, C., Wang, J., Li, J., Wang, M., & Ye, Y. (2013). Antifungal screening of endophytic fungi from *Ginkgo biloba* for discovery of potent anti-phytopathogenic fungicides. *FEMS Microbiology Letters*, *339*, 130–136. Available from https://doi.org/10.1111/1574-6968.12065.

Zhang, C., Zheng, B., Lao, J., Mao, L., Chen, S., Kubicek, C. P., & Lin, F. (2008). Clavatol and patulin formation as the antagonistic principle of *Aspergillus clavatonanicus*, an endophytic fungus of *Taxus mairei*. *Applied Microbiology and Biotechnology*, *78*, 833–840. Available from https://doi.org/10.1007/s00253-008-1371-z.

CHAPTER

Induction of plant defense response by endophytic microorganisms

5

Aswani R[1], Roshmi Thomas[2] and Radhakrishnan E.K.[1]
[1]*School of Biosciences, Mahatma Gandhi University, Kottayam, India*
[2]*Sanatana Dharma College, Alappuzha, India*

5.1 Introduction

Agrochemicals are globally used to enhance the agricultural productivity and yield. This is due to the fact that these chemicals can effectively act against pathogens and protect plants from both stress conditions. But due to the severe negative effects of these agrochemicals, it has been advised to develop sustainable agricultural practices and products by Food and Agriculture Organization and European Union (FAO, 2013; Westman, Kloth, & Hanson, 2019). Here comes the relevance of exploration of agriculturally important microorganisms especially those which are plant associated like the endophytes. The term endophyte was introduced by De Bary in 1866 and these are microorganisms present within the plant tissues which execute multifaceted interactions with the host plant. In addition, they play a dynamic role in the growth and development of plants. Hence these microorganisms are generally referred to as the second genome of plants as they promote plant growth, provide disease protection, and elicit defense response against biotic and abiotic stress factors. Endophytic microorganisms are distributed abundantly in plants and are widely reported to provide protection to plants from pathogens multimechanistically. The endophytic microorganisms themselves produce bioactive metabolites such as phenolic acids, alkaloids, quinones, steroids, saponins, tannins, and terpenoids, which make them more resistant to stress conditions. The colonization ability of endophytes with in the plants marks them as a promising biocontrol agent over other microorganisms. These features indicate the need to explore endophytic microbes as an environment-friendly tool in the development of bio-inoculant for sustainable agricultural practices. A detailed understanding on the molecular mechanisms and adaptations of endophytic microorganisms is highly essential to utilize its application for sustainable agricultural practices. Therefore, the applications of endophytic microorganisms for plant growth and disease management are discussed well in this chapter. Mechanisms involved in plant innate immunity and the role of endophytes to activate plant

Biocontrol Mechanisms of Endophytic Microorganisms. DOI: https://doi.org/10.1016/B978-0-323-88478-5.00002-X

defense has also been described. Hence this chapter provides insight into importance of plant–microbe interactions which will enable to design or develop suitable bioinoculants for field applications.

5.2 Endophytic microorganisms

Endophytic microorganisms are the microbes that inhabit within the internal parts of a plant without harming the host plant (Yadav, Kumar, & Dhaliwal, 2018). The interaction of endophytes with its host plants support the plant growth and defense against pathogens both directly and indirectly. Endophytes directly improves the plant growth through the secretion of phytohormones, nutrient mobilization like phosphate solubilization, 1-aminocyclopropane-1-carboxylate (ACC) deaminase production and indirectly by the protection of plants from phytopathogens (Jasim, Joseph, & John, 2014; Jasim, Mathew, & Radhakrishnan, 2016; Jayakumar, Krishna, & Mohan, 2019; Sabu, Aswani, & Jishma, 2019). Many studies have already been demonstrated the inoculation of endophytic microorganisms to result in improved plant growth and enhanced resistance to plant diseases (Aswathy, Jasim, Jyothis, & Radhakrishnan, 2013; Jasim, Geethu, Mathew, & Radhakrishnan, 2015; Jasim, Mathew et al., 2016; Jasim, Sreelakshmi, Mathew, & Radhakrishnan, 2016; Jimtha, Smitha, & Anisha, 2014; Rohini, Aswani, & Kannan, 2018). Hence, the endophytic microorganisms have emerging agricultural, biotechnological, and industrial applications. They have immense promises as biofertilizer, and biocontrol agents, antioxidants, and antibiotics because of their ability to produce various secondary metabolites (Yadav et al., 2018). Wide range applications of endophytic microorganisms ranging from medicine to agriculture has already been reported (Christina, Christapher, & Bhore, 2013; Nair & Padmavathy, 2014; Singh, Gaba, & Yadav, 2016).

The word endophyte itself indicates the meaning "within the plant" and these organisms are established generally within the plant through the invasion from phyllospheric and rhizospheric regions (Verma, Yadav, & Kumar, 2017). At the same time, there are increasing number of reports showing the distribution of endophytes within seeds. Hence endophytes can be considered to get transmitted both horizontally and vertically. Generally, microbes associated with plants can be endophytes, epiphytes, or pathogenic microorganisms (Brader, Corretto, & Sessitsch, 2017). The endophytic microorganisms generally include bacteria, fungi, actinomycetes, and archea, which inhabit within plant tissues with beneficial and mutualistic relationship with their host plant. There are various reports on the isolation and identification of endophytic microorganisms from diverse plant sources such as ginger, rice, turmeric, wheat, tomato, cowpea, maize, strawberry, chickpea, mustard, chili, citrus, soybean, cotton, brahmi, etc. (Jasim et al., 2014, 2015; Jayakumar et al., 2019; Jimtha et al., 2014; Rohini et al., 2018; Sabu et al., 2019; Verma et al., 2017). The distribution of endophytes within different species of plants or plant parts hence can be a universal phenomenon.

5.3 Colonization of endophytic microorganisms

The colonization and existence of endophytic microorganisms within the plants are considered to be favored by different factors. Once the endophytes invade in to the host tissue, it is recognized by the plant and initiate cross talk signals (Khare, Mishra, & Arora, 2018). For example, the root exudates produced from the host plants are rich source of biomolecules that attracts microbes through chemotactic response (Compant, Clément, & Sessitsch, 2010; Rosenblueth & Martínez-Romero, 2006). Flavonoids are one such metabolite characterized as a chemoattractant produced by many plants that play a vital role in the plant–endophytic interactions. In a previous study, strigolactone secreted from *Arabidopsis thaliana* has been described to have significant role as a signal molecule in its interactions with endophytic *Mucor* sp. (Rozpadek, Domka, & Nosek, 2018). The treatment with strigolactone was also found to provoke the signaling pathway in the plants (López-Ráez, Shirasu, & Foo, 2017). Likewise, the root exudates, including sugars, amino acids, organic acids, and other metabolites, produced from the plants are known to influence the mutualistic interaction between plants and microorganisms (Badri & Vivanco, 2009; Weert, Vermeiren, & Mulders, 2002). All these interactions help endophytic microbes to enter into the plant tissue.

5.4 Association of endophytes with plants

The association of endophytic microorganisms with plants is generally mutualistic relationship and can happens on a short- or the long-term basis. Based on taxonomy, endophytic lifestyle, mode of transmission, and functional diversity, endophytic microorganisms can be generally classified as (1) systemic or true endophytes and (2) nonsystemic or transient endophytes (Mostert, Crous, & Petrini, 2000; Wani & Ashraf, 2015). Systemic endophytes are proposed to be evolved with the host plant and hence did not cause any infections to the future generations. As these endophytes coevolved with the plant itself, they exhibit vertical gene transfer mechanisms and form the core endobiome (Mostert et al., 2000; Wani & Ashraf, 2015). Instead, nonsystemic endophytes are facultative and short-term associated microorganisms. Furthermore, the diversity and the number of microbial communities are dependent on various abiotic and biotic factors such as host plant, environmental conditions, and the pathogenic stress. In addition to the vertically transmitted endophytes, other possible ways of endophytic microbial transmission in plants occur through the penetration and colonization of microbes from the rhizosphere and phyllosphere and eventually they form either systemic (long-term) or nonsystemic (short-term) association.

5.5 Identification of endophytic microbial diversity

The identification of complete microbial communities in the host plant is mainly carried out through the culture- and culture-independent approaches. Studies have also reported the combined use of these methods to have the promises to provide better characterization of the whole microbiome composition rather than the individual approach (Anguita-Maeso, Olivares-García, & Haro, 2020; Berg, Grube, Schloter, & Smalla, 2014; Turner, James, & Poole, 2013).

5.5.1 Culture-dependent methods

The conventional method for the identification of endophytic microorganisms includes the culture-based approaches (Bell, Dickie, Harvey, & Chan, 1995; Stoltzfus, So, & Malarvithi, 1997). As per this method, endophytic microorganisms can be isolated from the plant tissue by culturing under the laboratory conditions. Here, the isolation of cultivable endophytic microorganisms involves a series of procedures like surface sterilization, maceration, serial dilution, and plating or imprinting the surface-sterilized materials on to the growth medium (Jasim et al., 2014; Jasim, Mathew et al., 2016; Jayakumar et al., 2019; Nxumalo, Ngidi, Shandu, & Maliehe, 2020; Sabu et al., 2019). Surface sterilization is usually carried out using the sterilizing agents like sodium hypochlorite and can remove the epiphytic and other surface attached contaminants and thus improve the efficiency of sterilization (Lodewyckx, Vangronsveld, & Porteous, 2002; Romero, Carrión, & Rico-Gray, 2001; Schulz, Wanke, Draeger, & Aust, 1993). The residual chemicals can then be removed by subsequent wash with sterile distilled water for several times. Here, the total number of endophytic microorganisms which can be isolated through this method is mainly determined by the selection of growth media, growth conditions, and the type and concentrations of the surface sterilizing agent used. The cultivable bacteria obtained could be further purified by means of their differences in morphological, physiological, biochemical characteristics. The distinct bacterial colonies are basically identified by 16S rRNA-based molecular method and the fungal endophytes with internal transcribed spacer-based sequencing (Johnson, Spakowicz, & Hong, 2019; Kaul, Sharma, & Dhar, 2016; Srinivasan, Karaoz, & Volegova, 2015). The molecular techniques are considered to be more accurate and it helps to identify the microorganisms with their phylogenetic relationship (Srinivasan et al., 2015). As the culture-dependent methods identify only the cultivable bacteria, the obtained isolates could represent only 0.001%–10% of the actual endophytic microbial communities. However, identification of whole endophytic microbial communities using culture-independent methods are highly significant to get an over all idea about types of endophytes distributed in various plants (Alain & Querellou, 2009; Torsvik & Øvreås, 2002).

5.5.2 Culture-independent methods

The diversity of endophytic microbial populations in a plant can be studied using culture-independent methods that are mainly based on the extraction of the meta DNA from host plants. For determining the endophytic microbial communities, the source material is primarily surface sterilized, homogenized and used for the meta DNA extraction (Sessitsch, Reiter, Pfeifer, & Wilhelm, 2002). The obtained DNA can further be analyzed by using different molecular techniques. Among them, the 16S rRNA (rDNA) gene is considered the widely used phylogenetic marker for the characterization of bacterial diversity (Dunbar, Takala, & Barns, 1999; Felske, Wolterink, Van Lis, & Akkermans, 1998). Also, several 16S rDNA-based methods have previously been carried out to identify the total microbiome. Various molecular techniques used to study bacterial diversity include denaturing gradient gel electrophoresis, temperature gradient gel electrophoresis (Muyzer, 1999; Sabu, Aswani, & Prabhakaran, 2018), restriction fragment length polymorphism (RFLP) (Laguerre, Allard, Revoy, & Amarger, 1994), terminal RFLPs (Dunbar, Ticknor, & Kuske, 2000; Su, Lei, & Duan, 2012), single-strand conformation polymorphism (Lee, Zo, & Kim, 1996), and 16S rDNA sequencing, etc. (Dunbar et al., 1999; Felske, Wolterink, & van Lis, 1999). In addition to these, several other methods have also been reported to identify the microorganisms present in the certain ecosystem that include fluorescence in situ hybridization (Bottari, Ercolini, & Gatti, 2006) and microarray (Bodrossy & Sessitsch, 2004).

Recent advances in high-throughput next-generation sequencing (NGS) approaches help to sequence the complete microbiome or metagenome and thus provide in-depth insight into their taxonomical, phylogenetical, and evolutionary classification (Chaudhry, Sharma, Bansal, & Patil, 2017; Kaul et al., 2016). NGS-based platforms are used for the identification whole microbial communities which includes whole genome sequencing, metagenomics (Xu, 2006), metatranscriptomics, etc. (Poretsky, Gifford, & Rinta-Kanto, 2009; Wilmes & Bond, 2004). From sequence analysis of endophytic microorganisms, it is possible to predict the genetic features that directly or indirectly involved in the colonization, mobility, bioactivity and defense response. This method aids in the identification of particular genes involved in plant growth promotion, secondary metabolites production, secretory systems, transport systems, and other metabolic processes (Kaul et al., 2016). In addition, the genes encoding for endophytic lifestyle can be identified by genomic analysis (Hardoim, van Overbeek, & Berg, 2015). In the basic metagenomic study, it involves the DNA extraction from the complete bacterial population followed by the sequence analysis of PCR amplicon of 16S rDNA (V3–V4 regions) by NGS (Akinsanya, Goh, Lim, & Ting, 2015; Beckers, Op De Beeck, & Thijs, 2016; Giangacomo, Mohseni, Kovar, & Wallace, 2020). The metagenomic study thus provides information about the microbial diversity and its functional and metabolic features. But the major limitation of this method is the inability to differentiate the expressed and the nonexpressed genes. Whereas the metatranscriptomic approaches can identify the RNA-based regulation and predict the expressed biological signatures in the host (Su et al., 2012), the transcriptome analysis hence can detect more comprehensive picture of the plant–microbe interaction which can further led to the discovery of key plant and microbial genes that characterize the plant–microbe interaction and disease resistance (Nobori, Velásquez, & Wu, 2018; Schenk, Carvalhais, & Kazan, 2012).

5.6 Mechanisms of endophytic microorganisms in plant disease management

Endophytic microorganisms exhibit several direct and indirect mechanisms to boost plant resistance against various biotic and abiotic stress factors. Here we discuss detailed mechanistic aspects of various endophytic fungal and bacterial mechanisms which support plant growth under stressful conditions.

5.6.1 Direct mechanisms of plant disease protection by endophytes

Endophytic microorganisms have the potential to exert direct activity against pathogenic microorganisms and thereby protecting the plant from the pathogen attack. The activity of endophytic microorganisms against diverse phytopathogens has further revealed the phytovaccinating potential of endophytes which could be explored for the management of plant diseases. The simple and the most common method used directly to analyze the antagonistic activity is the in vitro screening by dual culture and the in vivo study by comparing the infected and survived plants (Rohini et al., 2018; Sabu et al., 2019). Different mechanisms expressed by endophytes during the pathogen interaction include (1) antibiotic production, (2) lytic enzymes production, (3) siderophore production, and (4) competition with pathogens.

5.6.1.1 Production of antibiotics

Most of the endophytes are reported to have the ability to produce diverse bioactive metabolites that helps in the inhibition of plant pathogenic microorganisms (Gunatilaka, 2006). Several bioactive metabolites have already been reported from the endophytic microorganisms such as lipopeptide compounds, phenazine 1 carboxylic acid, piperine, danthrone, gliotoxin, and camptothecin (Anisha & Radhakrishnan, 2015; Anisha, Sachidanandan, & Radhakrishnan, 2018; Aswani, Jasim, & Arun Vishnu, 2020; Jasim, Sreelakshmi et al., 2016, Chithra et al., 2014). The production of secondary metabolites to manage plant diseases can either occurs in the endophytes or it can induce the production of bioactive metabolites from the host plant (Kusari, Hertweck, & Spiteller, 2012). Interestingly, endophytes are also known to have the ability to synthesize host-specific metabolites due to its mutualistic association (Aswani et al., 2020; Chithra, Jasim, & Anisha, 2014). Liu, Zou, Lu, and Tan (2001) have previously demonstrated the antifungal activity of endophytic fungi against the most threatening plant pathogens *Gaeumannomyces graminis* var. *tritici*, *Rhizoctonia cerealis*, *Helminthosporium sativum*, *Fusarium graminearum*, *Gerlachia nivalis*, and *Phytophthora capsici* through the production of various biochemicals. Likewise, various volatile organic compounds produced by endophytes have also been reported to have promising antagonistic activity against diverse plant pathogens

(Sánchez-Fernández, Diaz, & Duarte, 2016; Wonglom, Ito, & Sunpapao, 2020; Xie, Liu, & Gu, 2020).

5.6.1.2 Secretion of lytic enzymes

The endophytic microbes are found to produce diverse lytic enzymes for their successful colonization (Gao, Dai, & Liu, 2010). In addition, these hydrolytic enzymes can also be active against the fungal pathogen indirectly through the cell wall degradation. There are various reports on the secretion of lytic enzymes like lipase, proteinase, chitinases, cellulases, and 1,3-glucanases from endophytes (Desire, Bernard, & Forsah, 2014; Fadiji & Babalola, 2020). The lytic enzymes reported from the *Streptomyces* were found to have activity against cacao witches' broom disease which indicate the role of lytic enzymes in the plant disease management (Macagnan, Romeiro, Pomella, & deSouza, 2008). Gao et al. (2010) have previously described the mutagenesis to 1,3-glucanase gene of *Lysobacter enzymogenes* to result in reduced antifungal activity against *Pythium* infection. Although the lytic enzyme production cannot be used merely as a strong antifungal agent, the antifungal activity can be increased when it synergistically acts with other mechanisms.

5.6.1.3 Siderophore production

Siderophores are iron-chelating compounds that make iron available for the plant (Yadav, 2018). As well, the siderophore production helps in the nitrogen fixation by diazotrophic microorganisms since it requires Fe^{2+} and Mo factors for the synthesis and functioning of nitrogenase (Kraepiel, Bellenger, Wichard, & Morel, 2009). Interestingly, some siderophores like hydroxymate, phenolate, and catecholate are reported to confer antimicrobial activities also (Rajkumar, Ae, Prasad, & Freitas, 2010). The mechanisms involved in iron chelation include the oxidation of Fe^{2+} to form Fe^{3+}siderophore complex in the bacterial membrane, which is later introduced into the plant cell by endophytes through gating mechanism (Gao et al., 2010). Sharma et al. (2003) have also described the siderophore producing endophytic *Pseudomonas* strain GRP3 to have the ability to reduce chlorotic symptoms on *Vigna radiata*. Similarly, many endophytic microorganisms such as *Bacillus* sp., *Pseudomonas* sp., *Stenotrophomonas* sp., *Streptomyces* sp., and *Nocardia* sp. are also known to produce siderophores (Jasim et al., 2014; Singh & Dubey, 2018).

5.6.1.4 Competition with pathogens

Competition by endophytes is considered the strong mechanism in plant protection by preventing the colonization of pathogens into the host (Martinuz, Schouten, & Sikora, 2012). Instead of using independent biocontrol activity, the competition by most endophytes generally takes place in combination with other mechanisms. The protection in the leaves of cacao tree from *Phytophthora* sp. infection through competition suggested the significance of this mechanism as one of the important methods in plant disease suppression. However, in the presence of high pathogenic stress, the competition-mediated biocontrol activity may be inactive and hence the method encountered a major drawback in its application (Lahlali, McGregor, & Song, 2014). Also, the hyperparasitism and predation are the characteristic features of endophytes against pathogenic fungi (Fadiji & Babalola, 2020). Hyperparasitism is another mechanism used by endophytic microorganisms for protecting the plant from pathogens. Here, the endophytes target pathogens or their propagules (Tripathi et al., 2008). In a previous study, Grosch, Scherwinski, Lottmann, and Berg (2006) have demonstrated the endophytic *Trichoderma* sp. to have the ability to capture and penetrate the hyphae of *Rhizoctonia solani* and thereby exhibit biocontrol activity through cell wall disruption. Similarly, the microbial predation is another mechanism of plant protection from pathogens. Most of the endophytes exhibit predatory characteristics under nutrient-deficient conditions.

5.6.2 Indirect mechanisms of plant disease protection by endophytes

Plants are known to employ several mechanisms to withstand biotic and abiotic stress conditions. The general response developed during disease resistance includes hypersensitive response, cellular necrosis, and phytoalexin production. These general responses may act against broad range of pathogens and protect plants from infection. Endophytes increase the plant defense mechanisms through the modulation of secondary metabolite production and/or through the induction of plant disease resistance.

5.6.2.1 Modulation of plant secondary metabolites

Secondary metabolites produced from plants are of great significance in the adaptation of plants to different environmental conditions (Bourgaud, Gravot, Milesi, & Gontier, 2001). The different secondary metabolitesinclude terpenoids, flavonoids, and many others (Gao et al., 2010). The production of metabolites of phytoalexin group was first demonstrated by a French botanist Noel Bernard in the *Orchis morio* and *Loroglossum hircinum* in response to a fungal infection. The phytoalexin production can also be modulated through the application of elicitors. The modulation in production of triterpene and dipertene compounds in the cell suspensions of *Euphorbia pekinensis* by *Fusarium* E5 elicitors further confirms the effect of elicitors in the metabolite enhancement. Similarly, in the *Taxus cuspidata*, the inoculation of endophytes was found to result in the enhanced production of paclitaxel (Li & Tao, 2009). Studies have also described the co-culturing of endophytes with various elicitors to result in increased antifungal activity and secondary metabolite production (Aswani et al., 2020; Ravi, Rajan, & Khalid, 2021).

5.6.2.2 Induction of plant resistance

Plants generally respond to pathogens and other biotic stress through induced systemic resistance (ISR) and systemic acquired resistance (SAR). The ISR responses are generally induced by nonpathogenic rhizobacteria and it is moderated by ethylene (ET) or jasmonic acid (JA) and finally lead to the enhancement of defense genes, while the SAR response is associated with infections from pathogens and it is mediated by salicylic acid (SA) and finally lead to the building up of pathogenesis-related (PR) proteins (Tripathi, Kamal, & Sheramati, 2008). These PR proteins help in the lysis of invading cells through the secretion of lytic enzymes such as 1,3-glucanases and chitinases (Gao et al., 2010). In a previous study, endophytic *Fusarium solani* has been reported to have the potential to increase ISR response against *Septoria lycopersici* and protect the plant from infection through the activation of PR genes (Kavroulakis, Ntougias, & Zervakis, 2007). Redman, Freeman, and Clifton (1999) have previously confirmed the inoculation of *Colletotrichum magna* on *Cucumis sativus*, and *Citrullus lanatus* to result in enhancement of peroxidase , lignin deposition, and phenylalanine ammonia lyase and thereby protecting the plant from disease caused by *Fusarium oxysporum* and *Colletotrichum orbiculare*. The enhanced activities of PO and superoxide dismutase are also reported to cause the protection of *Neotyphodium lolii* from various pathogens (Tian, Nan, Li, & Spangenberg, 2008). In a recent study, accumulation of defense-related enzymes like peroxidase, phenylalanine ammonia lyase, and polyphenol oxidase have been reported to get increased in the susceptible cv. Rasthali planlets when treated with endophytic microbial consortia containing *Trichoderma reesei*, *Polyporus vinctus*, and *Sphingobacterium tabacisoli* (Savani, Bhattacharyya, & Baruah, 2020).

5.6.2.3 Plant growth promotion

Plant growth-promoting properties such as phytohormone production, phosphate solubilization, and nitrogen fixation of endophytic microorganisms are already known to increase the growth and development of plant. As the growth of plant increases, it also develops vigor and improved resistance to pathogens (Kuldau & Bacon, 2008). The inoculation of endophytic microorganisms with indole acetic acid (IAA) production, ACC deaminase production, nitrogen fixation, and phosphate solubilization has also been reported to increase the plant growth in various experimental models (Jasim, Mathew et al., 2016; Jayakumar et al., 2019; Sabu et al., 2019). Under the growth promotion provided by the beneficial microorganisms, plant may able to execute its defense mechanisms in a better way.

5.7 Modulation of plant immune system by endophytic and pathogenic microorganisms

The plant immune responses toward different types of microorganisms have already been described (Azmina, Malik, & Kumar, 2020).

The innate immune mechanisms of plants toward infection involve two branches of defense responses, the first line and second line of defense. The first line of defense response represents the preformed barriers at the cell surface which prevent the pathogen invasion. The plants also recognize the highly conserved molecular patterns called microbe or pathogen associated molecular patterns (MAMPs or PAMPs) derived from endophytic and other microorganisms as part of second line defense. These include flagellin, elongation factor TU, peptidoglycan, lipopolysaccharides, bacterial proteins, β-glycans, β-glucans, chitin, etc. (Chinchilla, Bauer, & Regenass, 2006; Gust, Biswas, & Lenz, 2007; Newman, Sundelin, Nielsen, & Erbs, 2013). In addition to this, plants can also recognize endogenous signals released by themselves under the biotic or abiotic stress conditions and these stress-related patterns are called damage-associated molecular patterns (Boller & Felix, 2009; Choi & Klessig, 2016; Klarzynski & Fritig, 2001). All these conserved patterns are recognized by pattern recognition receptors (PRRs) present at the cell surface. PRRs are generally plasma membrane-localized receptor-like kinases or receptor-like proteins with modular functional domains (De Lorenzo, Ferrari, Cervone, & Okun, 2018; Newman et al., 2013). Also, the PRR-mediated defense responses are commonly referred to as MAMP-triggered immunity (MTI) or PAMP-triggered immunity. Therefore endophytes with the potential to cross the first-line defense can only enter and colonizes into the plant tissue. For example, in the case of fungal endophytes, the cell wall component chitin is recognized by chitin-specific receptors present in the plants and activates the defense mechanisms toward these endophytes. Here the proposed mechanisms of fungal endophytes to protect themselves from plant defense system involve the synthesis of chitin deacetylases that further modify the chitin or consequently form oligomers and thereby hide this chitin from the plant immune system (Cord-Landwehr, Melcher, Kolkenbrock, & Moerschbacher, 2016; Sánchez-Vallet, Mesters, & Thomma, 2015). In the case of endophytic bacterial interaction with plant receptors, it is different from those toward pathogenic interactions (Vandenkoornhuyse, Quaiser, & Duhamel, 2015). This is because the perception of plants toward MAMPs from the endophytes is different in pathogenic microbial interaction. Trdá, Fernandez et al. (2014), Trdá, Boutrot et al. (2015) have also reported the differences in perception of pathogenic and beneficial bacteria by plants and their influence on host immunity. The study proposed the flagellin from the endophytic *Burkholderia phytofirmans* to be different from the flagellin of pathogens like *Pseudomonas aeruginosa* or *Xanthomonas campestris*. Furthermore, endophytic microorganisms were described to have the ability to protect themselves from the released reactive oxygen species as part of plant defense mechanism through the production of various defense enzymes which scavenge the released reactive oxygen species and in turn avoid the cellular damage (Khare et al., 2018). Additionally, the type III secretion system (T3SS) present in the pathogenic bacteria can also trigger the plant immune system. These secretory systems are essential to deliver the effector proteins from the pathogenic microorganisms into the host cell. Interestingly, in the case of endophytic or mutualistic microorganisms, these secretion systems are absent or exist in low quantity (Wang, Yang, & Zhang, 2020). Bernal, Llamas, and Filloux (2018) have reported that the Type VI secretion system (T6SS) is notably present in both pathogenic and plant-associated bacteria and is associated with important functions like fitness and colonization advantages in the plants rather than the virulence properties.

The interaction of molecular patterns with plant surface initiates the signal transmission and induces a cascade of reactions, including the activation of defense genes (Azmina et al., 2020; Nadarajah & Kassim, 2009). These effector molecules finally result in effector-triggered immunity (ETI) in the host plant (Jones & Dangl, 2006).

In the case of ETI, the effector molecules act as a major indicator of pathogens and further activates the ETI-mediated defense response to protect plants from pathogens. Here, the recognition of effector molecules take place directly or indirectly with the help of plant resistance (R) genes. When the effector molecules come in contact, the R proteins present in the plants immediately induce its immune response through the activation of ETI. The ETI response is hence considered an accelerated response in plants to deliver disease resistance (Newman et al., 2013). The mechanisms involved in the plant disease resistance upon the perception of elicitor molecules include the increased oxidative burst, hormone accumulation, mitogen-activated protein kinases activation, bioactive metabolite production, and the expression of PR proteins (Sanabria, Huang, & Dubery, 2010; Stael, Kmiecik, & Willems, 2015; Villena, Kitazawa, & Van Wees, 2018). The activation of ETI finally result in the cell death in the host infected area and also prevents infection spread to other parts through hypersensitive response. The survival of plants from the infection at one site further develops enhanced resistance to get protection from the subsequent pathogen attack. The transmission of resistance from one parts to the other parts of the plant resulting in the enhanced resistance throughout the plant is called SAR (Azmina et al., 2020; Nadarajah, Kasim, & Fui, 2009).

There are several other signaling molecules also involved in the plant–pathogen interactions which include SA, methyl salicylate, JA, methyl jasmonate, and ET. These plant hormones are well-recognized signaling molecules and play a major role in the activation of defense response against the biotrophic pathogens. In addition to its role in plant defense mechanisms, they also play an important role in the modulation of secondary metabolite production in the plants (Yamamoto, Iwanaga, Al-Busaidi, & Yamanaka, 2020; Yamamoto, Ma et al., 2020).

The salicylic acid-mediated response mainly targets the activation of hypersensitive (HR) response toward the biotrophic pathogens while the jasmonic acid -mediated response mainly targets the herbivorous insects and necrotrophic pathogens. Smith, De Moraes, and Mescher (2009) have also described the effect of increased SA concentration in tomato plant and showed HR-like responses near to the *Cuscuta pentagona* infected area which further revealed the SA-mediated signaling response against the pathogens. The biosynthesis of SA occurs through the phenylpropanoid or isochorismate pathway (Sendon, Seo, & Song, 2011). Interestingly, both the endogenous and exogenous occurrences of SA have been reported to elicit PR proteins and enzymes against a wide range of pathogens in diverse plant species (Clarke, Volko, & Ledford, 2000; Enyedi, Yalpani, Silverman, & Raskin, 1992; Yalpani, Silverman, & Wilson, 1991). In addition, several studies have reported the exogenous applications of SA to result in plant growth, biocontrol and reduced pesticide toxic effects in several model plant systems (Fatma, Kamal, & Srivastava, 2018; Panichikkal, Prathap, Nair, & Krishnankutty, 2021). The hormone-mediated regulation of disease resistance primarily achieved through the effects on gene transcription. Here, the expression or repression of genes could be accomplished through the *trans*-acting proteins and *cis*-acting DNA elements. Even though the transcription factors and coregulators can be controlled themselves at the transcriptional level, they are also subjected to posttranslational modification through various processes like oxidation, reduction, sequestration, phosphorylation, degradation, or its interaction with other transcription or cofactors (Moore, Loake, & Spoel, 2011). In the SA-mediated defense signaling, the transcriptional and post-translational regulatory mechanisms are important. In SA signaling pathway, the NPR1 was identified as a major transcriptional coregulator that regulates the SA-dependent genes and triggers transcriptional reprogramming and resistance toward a broad range of pathogens. SA facilitates the translocation of NPR1 through cellular redox reactions instead of its direct binding. Generally, in the absence of pathogen interaction, the NPR1 is retained as an oligomer through redox-sensitive intermolecular disulfide bonds in the cytoplasm. During the pathogen challenge, SA induces changes in the cellular redox state that can be detected by NPR1, which further mediates the reduction of intermolecular disulfide bonds. This result in the conversion of NPR1 oligomer into monomers with the help of thioredoxins TRX-h5 and TRX-h3 and finally get released into the nucleus. In the nucleus, the NPR1 function as a cofactor for transcription factor-like TGAs and induce defense-related genes (Fu & Dong, 2013). Fu, Yan, and Saleh (2012) have described the regulation of NPR1 to be mainly depend on the concentration of SA and SA-dependent binding of NPR1 homologs NPR3 and NPR4 with NPR1. Because, NPR3 and NPR4 can act as CUL3 ligase adapter proteins in proteasome-mediated degradation of NPR1 based on the available SA concentrations. When the SA levels are low, NPR4 interacts with NPR1 mediates its degradation leading to the transcriptional inactivation. Similarly, high SA levels facilitate the binding of NPR3 with NPR1 and leading to the degradation of NPR1. However, when the SA levels are intermediate, it prevents the binding of NPR3 with NPR1 and therefore the free NPR1 activates SA-dependent defense response to act against the pathogen (Fu et al., 2012). Many endophytic microorganisms are likely to have influence on these signaling pathways and many plant associated bacteria have also been described to have the ability to produce salicylic acid.

JA is known to play an important role in the plant processes such as photosynthesis, flower, and fruit development, senescence, root growth and in defense response (Turner, Ellis, & Devoto, 2002). JA-associated defense response protect plant from insects and necrotrophic pathogens (Farmer, Alméras, &

Krishnamurthy, 2003; Halim, Vess, Scheel, & Rosahl, 2006; Heil & Bostock, 2002; Walling, 2000). The synthesis of JA and its derivatives occur through the octadecanoid pathway (Balbi & Devoto, 2008; Wasternack, 2007). During the wound formation by the insect, the host plant release linolenic acid from the membrane lipids and it enters into the octadecanoid pathway. The linolenic acid is enzymatically converted into an active jasmonate that triggers the expression of defense genes and pathways (Smith et al., 2009). JA plays an important role in the induced long-term systemic defense response called ISR. Under stress conditions the JA usually cause expression of physiological and molecular responses. The physiological responses include the activation of antioxidant systems like superoxide and PO (Karpets, Kolupaev, Lugovaya, & Oboznyi, 2014), accumulation of amino acids and soluble sugars, and the regulation of stomatal opening and closure (Acharya & Assmann, 2009; Wasternack, 2014). The molecular responses include the expression of JA-associated genes like JAZ, AOS1, AOC, LOX2, and COI1 (Hu, Jiang, & Han, 2017; Robson, Okamoto, & Patrick, 2010), interactions with other plant hormones (abscisic acid, ET, SA, and indole-3-acetic acid) (Yang, Duan, & Li, 2019) and interactions with transcription factors such as MaMYC2 and osbHLH148 (Seo, Joo, & Kim, 2011; Zhao, Wang, & Shan, 2013). Here, the F-box protein COI1 and the JAZ repressor proteins are known as the major regulators of JA signaling pathway. In the presence of JA and its bioactive forms, COI1 initiates the binding with JA and targets the degradation of JAZ repressor proteins through proteasome-assisted degradation. These interactions also facilitate the release of transcriptional activators that further lead to the activation of JA-responsive genes. Plant microbiome can also expect to have impact on JA-mediated signaling. However, the detailed understanding on the same is limited.

5.8 Priming methods and applications of endophytes in agriculture

Use of appropriate method for the application of endophytic microorganisms is an important criterion to improve biocontrol efficiency. Different methods such as seed priming, soil drenching, stem injection, and foliar spraying are tested in different crops using various microbial cultures (Fahey, Dimock, & Tomasino, 1991). Several chemical-based priming methods have already been developed to increase seed quality. However, the negative environmental and health issues have limited its use and this demand the need for biological alternatives. Here, the seed priming with endophytic microorganisms with inherent plant growth promoting properties such as phytohormone production, induction of resistance, and tolerance to abiotic and biotic stresses has been described to have application as environment-friendly approach to mitigate the plant productivity and yield (Kumar, Droby, & White, 2020). The commonly reported endophytic microorganisms for seed biopriming belong to the genus *Pseudomonas*, *Bacillus*, *Burkholderia*, *Stenotrophomonas*, *Micrococcus*, *Rhizobium*, *Pantoea*, and *Microbacterium* (Jayakumar et al., 2019; Romero, Marina, & Pieckenstain, 2014; Rosenblueth & Martínez-Romero, 2006; Sabu et al., 2019; Seghers, Wittebolle, & Top, 2004). Musson, McInroy, and Kloepper (1995) have previously evaluated the efficiency of different methods of application for the successful delivering of microbial endophytes into plant. These include stem inoculation, seed coating, seed soaking, leaf inoculation, furrow application of granules comprised of microbial consortia, vacuum infiltration, and pruned-root dip application. Among the seven methods, the stem inoculation was found to be as the best method even though it is labor intensive and forms wound in the plants. To enhance the shelf life and compatibility, studies also recommend the application of differentiated pressure for drying the seeds treated with endophytic microbial suspension (Latha, Karthikeyan, & Rajeswari, 2019). Also, the addition of certain nutrients or elicitors has also been reported to enhance the survival of the bacteria in the formulations.

Filippi, da Silva, and Silva-Lobo (2011) have reported the soil drenching with the antagonistic bacteria Rizo-55 to result in *Magnaporthe oryzae* inhibition and thereby reduce the leaf blight in rice. Similarly, combined soil application of *Bacillus* and *Pseudomonas* has reported to result in significant reduction of fusarium wilt in cv. Grand Naine Banana (Raman & Muthukathan, 2015). The antagonistic bacteria can also be applied as a talc-based formulation in which the microbial suspension is mixed with talc powder (Basheer, Ravi, Mathew, & Krishnankutty, 2019). In a study, soil drenching method was reported as a more efficient method in the plant disease management than the talc-based formulation. This is demonstrated with the inhibition of *R. solani* damping-off in cotton where the application of the bacterial strains as a soil drench was found to be more efficient in the disease suppression than the talc-based bioformulation (Selim, Gomaa, & Essa, 2017). Talc-based formulations of *Pseudomonas fluorescens* and *Bacillus subtilis* was also found to provide enhanced disease protection from fusarium wilt in chile pepper (Sundaramoorthy, Raguchander, Ragupathi, & Samiyappan, 2012). This in turn suggest the efficiency of application method is mainly dependent on the type of microbial culture.

Knudsen and Spurr (1987) have previously demonstrated about the spray application of lyophilized bacteria in dust formulations or suspensions on the fruits and flowers. The foliar application of *Bacillus amyloliquefaciens* was also reported to protect strawberry from Colletotrichum gloeosporioides anthracnose (Yamamoto, Shiraishi, & Suzuki, 2015). Among the different methods for plant protection, the seed priming method is considered as the economically dependable and efficient method (Latha, Karthikeyan, & Rajeswari, 2019). The seed inoculation by endophytic *Achromobacter xylosoxidans* has also shown to result in enhanced yield and disease resistance in rice further confirms the seed priming method as an effective method in the sustainable agricultural practices (Joe, Islam, & Karthikeyan, 2012). Additionally, the combined application of seed treatment, soil drenching, and foliar application was also reported to enhance the colonization ability of endophytic microorganisms and thereby their beneficial effects.

5.9 Conclusion

The diverse mechanisms used by endophytes for the protection of plants from various pathogens makes them highly sought after microbial communities for the sustainable agriculture. Endophytic microorganisms improve the plant growth and aid plant protection through several direct and indirect mechanisms. The use of endophytic microorganism for the agriculture can significantly reduce the agrochemical input and thereby limit its toxic side effects. Endophytic microorganisms protect plants from pathogens either through the production of bioactive metabolites or through the induction of plant immunity. However, for the successful suppression of plant diseases, the selection of suitable endophytic microorganism is important. A detailed insight into the mechanisms of plant disease management by endophytic microorganisms as described in the chapter further helps in the better understanding of antimicrobial mechanisms in the plant–endophyte, plant–pathogen, and endophyte–pathogen interactions. The detailed knowledge can helps in the designing and development of suitable microbial agents and method of application for the improved agricultural practices.

Acknowledgment

The authors are thankful to Kerala State Council for Science, Technology and Environment—Kerala Biotechnology Commission—Young Investigator Programme in Biotechnology (673/2017/KSCSTE dated October 13, 2017) and JAIVAM project, Mahatma Gandhi University, Kottayam (3972/AD A7/2019 dated August 17, 2019). The authors also acknowledge the Kerala state plan fund project and DST-PURSE P II Programme.

References

Acharya, B. R., & Assmann, S. M. (2009). Hormone interactions in stomatal function. *Plant Molecular Biology*, *69*, 451–462. Available from https://doi.org/10.1007/s11103-008-9427-0.

Akinsanya, M. A., Goh, J. K., Lim, S. P., & Ting, A. S. Y. (2015). Metagenomics study of endophytic bacteria in *Aloe vera* using next-generation technology. *Genomics Data*, *6*, 159–163. Available from https://doi.org/10.1016/j.gdata.2015.09.004.

Alain, K., & Querellou, J. (2009). Cultivating the uncultured: Limits, advances and future challenges. *Extremophiles: Life Under Extreme Conditions*, *13*, 583–594. Available from https://doi.org/10.1007/s00792-009-0261-3.

Anguita-Maeso, M., Olivares-García, C., Haro, C., et al. (2020). Culture-dependent and culture-independent characterization of the olive xylem microbiota: Effect of sap extraction methods. *Frontiers in Plant Science*, *10*, 1708. Available from https://doi.org/10.3389/fpls.2019.01708.

Anisha, C., & Radhakrishnan, E. K. (2015). Gliotoxin-producing endophytic Acremonium sp. from *Zingiber officinale* found antagonistic to soft rot pathogen *Pythium myriotylum*. *Applied Biochemistry and Biotechnology*, *175*, 3458–3467. Available from https://doi.org/10.1007/s12010-015-1517-2.

Anisha, C., Sachidanandan, P., & Radhakrishnan, E. K. (2018). Endophytic paraconiothyrium sp. from *Zingiber officinale* Rosc. displays broad-spectrum antimicrobial activity by production of danthron. *Current Microbiology*, *75*, 343–352. Available from https://doi.org/10.1007/s00284-017-1387-7.

Aswani, R., Jasim, B., Arun Vishnu, R., et al. (2020). Nanoelicitor based enhancement of camptothecin production in fungi isolated from Ophiorrhiza mungos. *Biotechnology Progress*, *36*, e3039. Available from https://doi.org/10.1002/btpr.3039.

Aswathy, A. J., Jasim, B., Jyothis, M., & Radhakrishnan, E. K. (2013). Identification of two strains of Paenibacillus sp. as indole 3 acetic acid-producing rhizome-associated endophytic bacteria from Curcuma longa. *3 Biotech*, *3*, 219–224. Available from https://doi.org/10.1007/s13205-012-0086-0.

Azmina N., Malik A., & Kumar I.S. (2020) *Elicitor and receptor molecules: Orchestrators of plant defense and immunity*.

Badri, D. V., & Vivanco, J. M. (2009). Regulation and function of root exudates. *Plant, Cell & Environment*, *32*, 666–681. Available from https://doi.org/10.1111/j.1365-3040.2008.01926.x.

Balbi, V., & Devoto, A. (2008). Jasmonate signalling network in *Arabidopsis thaliana*: Crucial regulatory nodes and new physiological scenarios. *The New Phytologist*, *177*, 301–318. Available from https://doi.org/10.1111/j.1469-8137.2007.02292.x.

Basheer, J., Ravi, A., Mathew, J., & Krishnankutty, R. E. (2019). Assessment of plant-probiotic performance of novel endophytic Bacillus sp. in talc-based formulation. *Probiotics Antimicrob Proteins*, *11*, 256–263. Available from https://doi.org/10.1007/s12602-018-9386-y.

Beckers, B., Op De Beeck, M., Thijs, S., et al. (2016). Performance of 16s rDNA primer pairs in the study of rhizosphere and endosphere bacterial microbiomes in metabarcoding studies. *Frontiers in Microbiology*, *7*, 650. Available from https://doi.org/10.3389/fmicb.2016.00650.

Bell, C. R., Dickie, G. A., Harvey, W. L. G., & Chan, J. W. Y. F. (1995). Endophytic bacteria in grapevine. *Canadian Journal of Microbiology*, *41*, 46–53. Available from https://doi.org/10.1139/m95-006.

Berg, G., Grube, M., Schloter, M., & Smalla, K. (2014). Unraveling the plant microbiome: Looking back and future perspectives. *Frontiers in Microbiology*, *5*, 148. Available from https://doi.org/10.3389/fmicb.2014.00148.

Bernal, P., Llamas, M. A., & Filloux, A. (2018). Type VI secretion systems in plant-associated bacteria. *Environmental Microbiology*, *20*, 1–15. Available from https://doi.org/10.1111/1462-2920.13956.

Bodrossy, L., & Sessitsch, A. (2004). Oligonucleotide microarrays in microbial diagnostics. *Current Opinion in Microbiology*, *7*, 245–254. Available from https://doi.org/10.1016/j.mib.2004.04.005.

Boller, T., & Felix, G. (2009). A renaissance of elicitors: Perception of microbe-associated molecular patterns and danger signals by pattern-recognition receptors. *Annual Review of Plant Biology*, *60*, 379–406. Available from https://doi.org/10.1146/annurev.arplant.57.032905.105346.

Bottari, B., Ercolini, D., & Gatti, M. (2006). Application of FISH technology for microbiological analysis: Current state and prospects. *Applied Microbiology and Biotechnology*, *73*, 485–494. Available from https://doi.org/10.1007/s00253-006-0615-z.

Bourgaud, F., Gravot, A., Milesi, S., & Gontier, E. (2001). Production of plant secondary metabolites: A historical perspective. *Plant Science (Shannon, Ireland)*, *161*, 839–851. Available from https://doi.org/10.1016/S0168-9452(01)00490-3.

Brader, G., Corretto, E., & Sessitsch, A. (2017). Metagenomics of plant microbiomes. *Functional metagenomics: Tools and applications*, 179–200. Available from https://doi.org/10.1007/978-3-319-61510-3.

Chaudhry, V., Sharma, S., Bansal, K., & Patil, P. B. (2017). Glimpse into the genomes of rice endophytic bacteria: Diversity and distribution of firmicutes. *Frontiers in Microbiology*, *7*, 2115. Available from https://doi.org/10.3389/fmicb.2016.02115.

Chinchilla, D., Bauer, Z., Regenass, M., et al. (2006). The arabidopsis receptor kinase FLS2 binds flg22 and determines the specificity of flagellin perception. *The Plant Cell*, *18*, 465–476. Available from https://doi.org/10.1105/tpc.105.036574.

Chithra, S., Jasim, B., Anisha, C., et al. (2014). LC–MS/MS based identification of piperine production by endophytic Mycosphaerella sp. PF13 from *Piper nigrum*. *Applied Biochemistry and Biotechnology*, *173*, 30–35. Available from https://doi.org/10.1007/s12010-014-0832-3.

Choi, H. W., & Klessig, D. F. (2016). DAMPs, MAMPs, and NAMPs in plant innate immunity. *BMC Plant Biology*, *16*, 1–10. Available from https://doi.org/10.1186/s12870-016-0921-2.

Christina, A., Christapher, V., & Bhore, S. J. (2013). Endophytic bacteria as a source of novel antibiotics: An overview. *Pharmacognosy Reviews*, *7*, 11–16. Available from https://doi.org/10.4103/0973-7847.112833.

Clarke, J. D., Volko, S. M., Ledford, H., et al. (2000). Roles of salicylic acid, jasmonic acid, and ethylene in cpr-induced resistance in arabidopsis. *The Plant Cell*, *12*, 2175–2190. Available from https://doi.org/10.1105/tpc.12.11.2175.

Compant, S., Clément, C., & Sessitsch, A. (2010). Plant growth-promoting bacteria in the rhizo- and endosphere of plants: Their role, colonization, mechanisms involved and prospects for utilization. *Soil Biology & Biochemistry*, *42*, 669–678. Available from https://doi.org/10.1016/j.soilbio.2009.11.024.

Cord-Landwehr, S., Melcher, R., Kolkenbrock, S., & Moerschbacher, B. (2016). A chitin deacetylase from the endophytic fungus *Pestalotiopsis* sp. efficiently inactivates the elicitor activity of chitin oligomers in rice cells. *Scientific Reports*, *6*, 38018. Available from https://doi.org/10.1038/srep38018.

De Lorenzo, G., Ferrari, S., Cervone, F., & Okun, E. (2018). Extracellular DAMPs in plants and mammals: Immunity, tissue damage and repair. *Trends in Immunology*, *39*, 937–950. Available from https://doi.org/10.1016/j.it.2018.09.006.

Desire, M. H., Bernard, F., Forsah, M. R., et al. (2014). Enzymes and qualitative phytochemical screening of endophytic fungi isolated from *Lantana camara* Linn. *Leaves*. Available from https://doi.org/10.7324/JABB.2014.2601.

Dunbar, J., Takala, S., Barns, S. M., et al. (1999). Levels of bacterial community diversity in four arid soils compared by cultivation and 16S rRNA gene cloning. *Applied and*

Environmental Microbiology, *65*, 1662–1669. Available from https://doi.org/10.1128/AEM.65.4.1662-1669.1999.

Dunbar, J., Ticknor, L. O., & Kuske, C. R. (2000). Assessment of microbial diversity in four Southwestern United States soils by 16S rRNA gene terminal restriction fragment analysis. *Applied and Environmental Microbiology*, *66*, 2943–2950. Available from https://doi.org/10.1128/AEM.66.7.2943-2950.2000.

Enyedi, A. J., Yalpani, N., Silverman, P., & Raskin, I. (1992). Localization, conjugation, and function of salicylic acid in tobacco during the hypersensitive reaction to tobacco mosaic virus. *Proceedings of the National Academy of Sciences of the United States of America*, *89*, 2480–2484.

Fadiji, A. E., & Babalola, O. O. (2020). Elucidating mechanisms of endophytes used in plant protection and other bioactivities with multifunctional prospects. *Front Bioeng Biotechnol*, *8*, 467. Available from https://doi.org/10.3389/fbioe.2020.00467.

Fahey, J. W., Dimock, M. B., Tomasino, S. F., et al. (1991). Genetically engineered endophytes as biocontrol agents: A case study from industry. In J. H. Andrews, & S. S. Hirano (Eds.), *Microbial ecology of leaves Brock/Springer series in contemporary bioscience* (pp. 401–411). New York: Springer. Available from https://doi.org/10.1007/978-1-4612-3168-4_20.

FAO (2013) *International code of conduct on the distribution and use of pesticides annotated list of technical guidelines for the implementation of the international code of conduct on the distribution and use of pesticides.*

Farmer, E. E., Alméras, E., & Krishnamurthy, V. (2003). Jasmonates and related oxylipins in plant responses to pathogenesis and herbivory. *Current Opinion in Plant Biology*, *6*, 372–378. Available from https://doi.org/10.1016/s1369-5266(03)00045-1.

Fatma, F., Kamal, A., & Srivastava, A. (2018). Exogenous application of salicylic acid mitigates the toxic effect of pesticides in *Vigna radiata* (L.) Wilczek. *Journal of Plant Growth Regulation*, *37*, 1185–1194. Available from https://doi.org/10.1007/s00344-018-9819-6.

Felske, A., Wolterink, A., Van Lis, R., & Akkermans, A. D. L. (1998). Phylogeny of the main bacterial 16S rRNA sequences in drentse a grassland soils (The Netherlands). *Applied and Environmental Microbiology*, *64*, 871–879. Available from https://doi.org/10.1128/AEM.64.3.871-879.1998.

Felske, A., Wolterink, A., van Lis, R., et al. (1999). Searching for predominant soil bacteria: 16S rDNA cloning vs strain cultivation. *FEMS Microbiology Ecology*, *30*, 137–145. Available from https://doi.org/10.1111/j.1574-6941.1999.tb00642.x.

Filippi, M. C. C., da Silva, G. B., Silva-Lobo, V. L., et al. (2011). Leaf blast (*Magnaporthe oryzae*) suppression and growth promotion by rhizobacteria on aerobic rice in Brazil. *Biological Control*, *58*, 160–166. Available from https://doi.org/10.1016/j.biocontrol.2011.04.016.

Fu, Z. Q., & Dong, X. (2013). Systemic acquired resistance: Turning local infection into global defense. *Annual Review of Plant Biology*, *64*, 839–863. Available from https://doi.org/10.1146/annurev-arplant-042811-105606.

Fu, Z. Q., Yan, S., Saleh, A., et al. (2012). NPR3 and NPR4 are receptors for the immune signal salicylic acid in plants. *Nature*, *486*, 228–232. Available from https://doi.org/10.1038/nature11162.

Gao, F., Dai, C., & Liu, X. (2010). Mechanisms of fungal endophytes in plant protection against pathogens. *African Journal of Microbiology Research*, *4*, 1346–1351.

Giangacomo, C., Mohseni, M., Kovar, L., & Wallace, J. G. (2020). Comparing DNA extraction and 16s amplification methods for plant-associated bacterial communities. *bioRxiv*. Available from https://doi.org/10.1101/2020.07.23.217901.

Grosch, R., Scherwinski, K., Lottmann, J., & Berg, G. (2006). Fungal antagonists of the plant pathogen Rhizoctonia solani: Selection, control efficacy and influence on the indigenous microbial community. *Mycological Research*, *110*, 1464–1474. Available from https://doi.org/10.1016/j.mycres.2006.09.014.

Gunatilaka, A. A. L. (2006). Natural products from plant-associated microorganisms: Distribution, structural diversity, bioactivity, and implications of their occurrence. *Journal of Natural Products*, *69*, 509–526. Available from https://doi.org/10.1021/np058128n.

Gust, A. A., Biswas, R., Lenz, H. D., et al. (2007). Bacteria-derived peptidoglycans constitute pathogen-associated molecular patterns triggering innate immunity in arabidopsis. *The Journal of Biological Chemistry*, *282*, 32338–32348. Available from https://doi.org/10.1074/jbc.M704886200.

Halim, V. A., Vess, A., Scheel, D., & Rosahl, S. (2006). The role of salicylic acid and jasmonic acid in pathogen defence. *Plant Biology (Stuttgart, Germany)*, *8*, 307–313. Available from https://doi.org/10.1055/s-2006-924025.

Hardoim, P. R., van Overbeek, L. S., Berg, G., et al. (2015). The hidden world within plants: Ecological and evolutionary considerations for defining functioning of microbial endophytes. *Microbiology and Molecular Biology Reviews: MMBR*, *79*, 293–320. Available from https://doi.org/10.1128/MMBR.00050-14.

Heil, M., & Bostock, R. M. (2002). Induced systemic resistance (ISR) against pathogens in the context of induced plant defences. *Annals of Botany*, *89*, 503–512. Available from https://doi.org/10.1093/aob/mcf076.

Hu, Y., Jiang, Y., Han, X., et al. (2017). Jasmonate regulates leaf senescence and tolerance to cold stress: Crosstalk with other phytohormones. *Journal of Experimental Botany*, *68*, 1361–1369. Available from https://doi.org/10.1093/jxb/erx004.

Jasim, B., Geethu, P. R., Mathew, J., & Radhakrishnan, E. K. (2015). Effect of endophytic Bacillus sp. from selected medicinal plants on growth promotion and diosgenin production in Trigonella foenum-graecum. *Plant Cell, Tissue and Organ Culture*, *122*, 565–572. Available from https://doi.org/10.1007/s11240-015-0788-1.

Jasim, B., Joseph, A. A., John, C. J., et al. (2014). Isolation and characterization of plant growth promoting endophytic bacteria from the rhizome of *Zingiber officinale*. *3 Biotech*, *4*, 197–204. Available from https://doi.org/10.1007/s13205-013-0143-3.

Jasim, B., Mathew, J., & Radhakrishnan, E. K. (2016). Identification of a novel endophytic Bacillus sp. from *Capsicum annuum* with highly efficient and broad spectrum plant probiotic effect. *Journal of Applied Microbiology*, *121*, 1079–1094. Available from https://doi.org/10.1111/jam.13214.

Jasim, B., Sreelakshmi, K. S., Mathew, J., & Radhakrishnan, E. K. (2016). Surfactin, Iturin, and Fengycin Biosynthesis by endophytic Bacillus sp. from *Bacopa monnieri*. *Microbial Ecology*, *72*, 106–119. Available from https://doi.org/10.1007/s00248-016-0753-5.

Jayakumar, A., Krishna, A., Mohan, M., et al. (2019). Plant growth enhancement, disease resistance, and elemental modulatory effects of plant probiotic endophytic Bacillus sp. Fcl1. *Probiotics Antimicrob Proteins*, *11*, 526–534. Available from https://doi.org/10.1007/s12602-018-9417-8.

Jimtha, J. C., Smitha, P. V., Anisha, C., et al. (2014). Isolation of endophytic bacteria from embryogenic suspension culture of banana and assessment of their plant growth promoting properties. *Plant Cell, Tissue and Organ Culture*, *118*, 57–66. Available from https://doi.org/10.1007/s11240-014-0461-0.

Joe, M. M., Islam, M. R., Karthikeyan, B., et al. (2012). Resistance responses of rice to rice blast fungus after seed treatment with the endophytic *Achromobacter xylosoxidans* AUM54 strains. *Crop Protection (Guildford, Surrey)*, *42*, 141–148. Available from https://doi.org/10.1016/j.cropro.2012.07.006.

Johnson, J. S., Spakowicz, D. J., Hong, B. Y., et al. (2019). Evaluation of 16S rRNA gene sequencing for species and strain-level microbiome analysis. *Nature Communications*, *10*, 1–11. Available from https://doi.org/10.1038/s41467-019-13036-1.

Jones, J. D. G., & Dangl, J. L. (2006). The plant immune system. *Nature*, *444*, 323–329.

Karpets, Y. V., Kolupaev, Y. E., Lugovaya, A. A., & Oboznyi, A. I. (2014). Effect of jasmonic acid on the pro-/antioxidant system of wheat coleoptiles as related to hyperthermia tolerance. *Russian Journal of Plant Physiology*, *61*, 339–346. Available from https://doi.org/10.1134/S102144371402006X.

Kaul, S., Sharma, T., & Dhar, M. K. (2016). "Omics" tools for better understanding the plant-endophyte interactions. *Frontiers in Plant Science*, *7*, 955. Available from https://doi.org/10.3389/fpls.2016.00955.

Kavroulakis, N., Ntougias, S., Zervakis, G. I., et al. (2007). Role of ethylene in the protection of tomato plants against soil-borne fungal pathogens conferred by an endophytic *Fusarium solani* strain. *Journal of Experimental Botany*, *58*, 3853–3864. Available from https://doi.org/10.1093/jxb/erm230.

Khare, E., Mishra, J., & Arora, N. K. (2018). Multifaceted interactions between endophytes and plant: Developments and prospects. *Frontiers in Microbiology*, *9*, 1–12. Available from https://doi.org/10.3389/fmicb.2018.02732.

Klarzynski, O., & Fritig, B. (2001). Stimulation des défenses naturelles des plantes. *Comptes Rendus de l'Académie des Sciences - Series III—Sciences de la Vie*, *324*, 953–963. Available from https://doi.org/10.1016/S0764-4469(01)01371-3.

Knudsen, G. R., & Spurr, H. W. J. (1987). Field persistence and efficacy of five bacterial preparations for control of peanut leaf spot. *Plant Disease*, *71*, 442–445. Available from https://doi.org/10.1094/pd-71-0442.

Kraepiel, A. M. L., Bellenger, J. P., Wichard, T., & Morel, F. M. M. (2009). Multiple roles of siderophores in free-living nitrogen-fixing bacteria. *Biometals: An International Journal on the Role of Metal Ions in Biology, Biochemistry, and Medicine*, *22*, 573. Available from https://doi.org/10.1007/s10534-009-9222-7.

Kuldau, G., & Bacon, C. (2008). Clavicipitaceous endophytes: Their ability to enhance resistance of grasses to multiple stresses. *Biological Control*, *46*, 57–71. Available from https://doi.org/10.1016/j.biocontrol.2008.01.023.

Kumar, A., Droby, S., White, J. F., et al. (2020). Endophytes and seed priming: Agricultural applications and future prospects. *Microbial Endophytes*, 107–124. Available from https://doi.org/10.1016/b978-0-12-819654-0.00005-3.

Kusari, S., Hertweck, C., & Spiteller, M. (2012). Chemical ecology of endophytic fungi: Origins of secondary metabolites. *Chemistry & Biology*, *19*, 792–798. Available from https://doi.org/10.1016/j.chembiol.2012.06.004.

Laguerre, G., Allard, M.-R., Revoy, F., & Amarger, N. (1994). Rapid identification of Rhizobia by restriction fragment length polymorphism analysis of PCR-amplified 16S rRNA genes. *Applied and Environmental Microbiology*, *60*, 56–63.

Lahlali, R., McGregor, L., Song, T., et al. (2014). Heteroconium chaetospira induces resistance to clubroot via upregulation of host genes involved in jasmonic acid, ethylene, and auxin biosynthesis. *PLoS One*, *9*, e94144.

Latha, P., Karthikeyan, M., & Rajeswari, E. (2019). Endophytic bacteria: Prospects and applications for the plant disease management. In R. Ansari, & I. Mahmood (Eds.), *Plant health under biotic stress*. Singapore: Springer.

Lee, D. H., Zo, Y. G., & Kim, S. J. (1996). Nonradioactive method to study genetic profiles of natural bacterial communities by PCR-single-strand-conformation polymorphism. *Applied and Environmental Microbiology*, *62*, 3112–3120.

Li, Y.-C., & Tao, W.-Y. (2009). Paclitaxel-producing fungal endophyte stimulates the accumulation of taxoids in suspension cultures of Taxus cuspidate. *Scientia Horticulturae*, *121*, 97–102. Available from https://doi.org/10.1016/j.scienta.2009.01.016.

Liu, C. H., Zou, W. X., Lu, H., & Tan, R. X. (2001). Antifungal activity of Artemisia annua endophyte cultures against phytopathogenic fungi. *Journal of Biotechnology*, *88*, 277–282. Available from https://doi.org/10.1016/S0168-1656(01)00285-1.

Lodewyckx, C., Vangronsveld, J., Porteous, F., et al. (2002). Endophytic bacteria and their potential applications. *Critical Reviews in Plant Sciences*, *21*, 583–606. Available from https://doi.org/10.1080/0735-260291044377.

López-Ráez, J. A., Shirasu, K., & Foo, E. (2017). Strigolactones in plant interactions with beneficial and detrimental organisms: The yin and yang. *Trends in Plant Science*, *22*, 527–537. Available from https://doi.org/10.1016/j.tplants.2017.03.011.

Macagnan, D., Romeiro, Rd. S., Pomella, A. W. V., & deSouza, J. T. (2008). Production of lytic enzymes and siderophores, and inhibition of germination of basidiospores of Moniliophthora (ex Crinipellis) perniciosa by phylloplane actinomycetes. *Biological Control*, *47*, 309–314. Available from https://doi.org/10.1016/j.biocontrol.2008.08.016.

Martinuz, A., Schouten, A., & Sikora, R. A. (2012). Systemically induced resistance and microbial competitive exclusion: Implications on biological control. *Phytopathology*, *102*, 260–266. Available from https://doi.org/10.1094/PHYTO-04-11-0120.

Moore, J. W., Loake, G. J., & Spoel, S. H. (2011). Transcription dynamics in plant immunity. *The Plant Cell*, *23*, 2809–2820. Available from https://doi.org/10.1105/tpc.111.087346.

Mostert L., Crous P. W., & Petrini O. (2000). *Endophytic fungi associated with shoots and leaves of Vitis vinifera, with specific reference to the Phomopsis viticola complex.*

Musson, G., McInroy, J. A., & Kloepper, J. W. (1995). Development of delivery systems for introducing endophytic bacteria into cotton. *Biocontrol Science and Technology*, *5*, 407–416. Available from https://doi.org/10.1080/09583159550039602.

Muyzer, G. (1999). DGGE/TGGE a method for identifying genes from natural ecosystems. *Current Opinion in Microbiology*, *2*, 317–322. Available from https://doi.org/10.1016/S1369-5274(99)80055-1.

Nadarajah, K., Kasim, N. M., & Fui, V. V. (2009). The modulation of abiotic stresses by mitogen-activated protein kinase in rice. *The Journal of Biological Sciences*, *9*, 402–412.

Nadarajah, K., & Kassim, N. M. (2009). Effect of signal molecules and hormones on the expression of protein kinase gene OrMKK1 in rice. *Journal of Plant Sciences*, *4*, 32–42.

Nair, D. N., & Padmavathy, S. (2014). Impact of endophytic microorganisms on plants, environment and humans. *Scientific World Journal*, *2014*, 250693. Available from https://doi.org/10.1155/2014/250693.

Newman, M.-A., Sundelin, T., Nielsen, J., & Erbs, G. (2013). MAMP (microbe-associated molecular pattern) triggered immunity in plants. *Frontiers in Plant Science*, *4*, 139. Available from https://doi.org/10.3389/fpls.2013.00139.

Nobori, T., Velásquez, A. C., Wu, J., et al. (2018). Transcriptome landscape of a bacterial pathogen under plant immunity. *Proceedings of the National Academy of Sciences of the United States of America*, *115*, E3055–E3064. Available from https://doi.org/10.1073/pnas.1800529115.

Nxumalo, C. I., Ngidi, L. S., Shandu, J. S. E., & Maliehe, T. S. (2020). Isolation of endophytic bacteria from the leaves of *Anredera cordifolia* CIX1 for metabolites and their biological activities. *BMC Complementary Medicine and Therapies*, *20*, 1–11. Available from https://doi.org/10.1186/s12906-020-03095-z.

Panichikkal, J., Prathap, G., Nair, R. A., & Krishnankutty, R. E. (2021). Evaluation of plant probiotic performance of *Pseudomonas* sp. encapsulated in alginate supplemented with salicylic acid and zinc oxide nanoparticles. *International Journal of Biological Macromolecules*, *166*, 138–143. Available from https://doi.org/10.1016/j.ijbiomac.2020.10.110.

Poretsky, R. S., Gifford, S., Rinta-Kanto, J., et al. (2009). Analyzing gene expression from marine microbial communities using environmental transcriptomics. *Journal of Visualized Experiments*, *18*, e1086. Available from https://doi.org/10.3791/1086.

Rajkumar, M., Ae, N., Prasad, M. N. V., & Freitas, H. (2010). Potential of siderophore-producing bacteria for improving heavy metal phytoextraction. *Trends in Biotechnology*, *28*, 142–149. Available from https://doi.org/10.1016/j.tibtech.2009.12.002.

Raman, T., & Muthukathan, G. (2015). Field suppression of Fusarium wilt disease in banana by the combined application of native endophytic and rhizospheric bacterial isolates possessing multiple functions. *Phytopathologia Mediterranea*, *54*, 241–252. Available from https://doi.org/10.14601/Phytopathol.

Ravi, A., Rajan, S., Khalid, N. K., et al. (2021). Impact of supplements on enhanced activity of Bacillus amyloliquefaciens BmB1 against pythium aphanidermatum through lipopeptide modulation. *Probiotics Antimicrob Proteins*, *13*, 367–374. Available from https://doi.org/10.1007/s12602-020-09707-x.

Redman, R. S., Freeman, S., Clifton, D. R., et al. (1999). Biochemical analysis of plant protection afforded by a nonpathogenic endophytic mutant of *Colletotrichum magna*. *Plant Physiology*, *119*, 795–804. Available from https://doi.org/10.1104/pp.119.2.795.

Robson, F., Okamoto, H., Patrick, E., et al. (2010). Jasmonate and phytochrome a signaling in arabidopsis wound and shade responses are integrated through JAZ1 stability. *The Plant Cell*, *22*, 1143–1160. Available from https://doi.org/10.1105/tpc.109.067728.

Rohini, S., Aswani, R., Kannan, M., et al. (2018). Culturable endophytic bacteria of ginger rhizome and their remarkable multi-trait plant growth-promoting features. *Current Microbiology*, *75*, 505–511. Available from https://doi.org/10.1007/s00284-017-1410-z.

Romero, A., Carrión, G., & Rico-Gray, V. (2001). Fungal latent pathogens and endophytes from leaves of *Parthenium hysterophorus* (Asteraceae). *Fungal Divers*, *7*, 81–87.

Romero, F. M., Marina, M., & Pieckenstain, F. L. (2014). The communities of tomato (*Solanum lycopersicum* L.) leaf endophytic bacteria, analyzed by 16S-ribosomal RNA gene pyrosequencing. *FEMS Microbiology Letters*, *351*, 187–194. Available from https://doi.org/10.1111/1574-6968.12377.

Rosenblueth, M., & Martínez-Romero, E. (2006). Bacterial endophytes and their interactions with hosts. *Molecular Plant-Microbe Interactions*, *19*, 827–837. Available from https://doi.org/10.1094/MPMI-19-0827.

Rozpadek, P., Domka, A. M., Nosek, M., et al. (2018). The role of strigolactone in the cross-talk between *Arabidopsis thaliana* and the endophytic fungus *Mucor* sp. *Frontiers in Microbiology*, *9*, 1–14. Available from https://doi.org/10.3389/fmicb.2018.00441.

Sabu, R., Aswani, R., Jishma, P., et al. (2019). Plant growth promoting endophytic Serratia sp. ZoB14 protecting ginger from fungal pathogens. *Proceedings of the National Academy of Sciences, India Section B: Biological Sciences*, *89*, 213–220. Available from https://doi.org/10.1007/s40011-017-0936-y.

Sabu, R., Aswani, R., Prabhakaran, P., et al. (2018). Differential modulation of endophytic microbiome of ginger in the presence of beneficial organisms, pathogens and both as identified by DGGE analysis. *Current Microbiology*, *75*, 1033–1037. Available from https://doi.org/10.1007/s00284-018-1485-1.

Sanabria, N. M., Huang, J.-C., & Dubery, I. A. (2010). Self/nonself perception in plants in innate immunity and defense. *Self Nonself*, *1*, 40–54. Available from https://doi.org/10.4161/self.1.1.10442.

Sánchez-Fernández, R. E., Diaz, D., Duarte, G., et al. (2016). Antifungal volatile organic compounds from the endophyte Nodulisporium sp. strain GS4d2II1a: A qualitative change in the intraspecific and interspecific interactions with *Pythium aphanidermatum*. *Microbial Ecology*, *71*, 347–364. Available from https://doi.org/10.1007/s00248-015-0679-3.

Sánchez-Vallet, A., Mesters, J. R., & Thomma, B. P. H. J. (2015). The battle for chitin recognition in plant-microbe interactions. *FEMS Microbiology Reviews*, *39*, 171–183. Available from https://doi.org/10.1093/femsre/fuu003.

Savani, A. K., Bhattacharyya, A., & Baruah, A. (2020). Endophyte mediated activation of defense enzymes in banana plants pre-immunized with covert endophytes. *Indian Phytopathology*, *73*, 433–441. Available from https://doi.org/10.1007/s42360-020-00245-8.

Schenk, P. M., Carvalhais, L. C., & Kazan, K. (2012). Unravelling plant-microbe interactions: Can multi-species transcriptomics help. *Trends in Biotechnology*, *30*.

Schulz, B., Wanke, U., Draeger, S., & Aust, H.-J. (1993). Endophytes from herbaceous plants and shrubs: Effectiveness of surface sterilization methods. *Mycological Research*, *97*, 1447–1450. Available from https://doi.org/10.1016/S0953-7562(09)80215-3.

Seghers, D., Wittebolle, L., Top, E. M., et al. (2004). Impact of agricultural practices on the Zea mays L. endophytic community. *Applied and Environmental Microbiology*, *70*, 1475–1482. Available from https://doi.org/10.1128/AEM.70.3.1475-1482.2004.

Selim, H. M. M., Gomaa, N. M., & Essa, A. M. M. (2017). Application of endophytic bacteria for the biocontrol of *Rhizoctonia solani* (Cantharellales: ceratobasidiaceae) damping-off disease in cotton seedlings. *Biocontrol Science and Technology*, *27*, 81–95. Available from https://doi.org/10.1080/09583157.2016.1258452.

Sendon, P. M., Seo, H. S., & Song, J. T. (2011). Salicylic acid signaling: Biosynthesis, metabolism, and crosstalk with jasmonic acid. *Journal of the Korean Society for*

Applied Biological Chemistry, *54*, 501–506. Available from https://doi.org/10.3839/jksabc.2011.077.

Seo, J.-S., Joo, J., Kim, M.-J., et al. (2011). OsbHLH148, a basic helix-loop-helix protein, interacts with OsJAZ proteins in a jasmonate signaling pathway leading to drought tolerance in rice. *The Plant Journal: For Cell and Molecular Biology*, *65*, 907–921. Available from https://doi.org/10.1111/j.1365-313X.2010.04477.x.

Sessitsch, A., Reiter, B., Pfeifer, U., & Wilhelm, E. (2002). Cultivation-independent population analysis of bacterial endophytes in three potato varieties based on eubacterial and actinomycetes-specific PCR of 16S rRNA genes. *FEMS Microbiology Ecology*, *39*, 23–32. Available from https://doi.org/10.1111/j.1574-6941.2002.tb00903.x.

Sharma, A., & Johri, B. N. (2003). Growth promoting influence of siderophore-producing Pseudomonas strains GRP3A and PRS9 in maize (Zea mays L.) under iron limiting. *conditions.Microbiological*, *research,158(3)*, 243–248.

Singh, R., & Dubey, A. K. (2018). Diversity and applications of endophytic actinobacteria of plants in special and other ecological niches. *Frontiers in Microbiology*, *9*, 1767. Available from https://doi.org/10.3389/fmicb.2018.01767.

Singh, R. N., Gaba, S., Yadav, A. N., et al. (2016). First high quality draft genome sequence of a plant growth promoting and cold active enzyme producing psychrotrophic *Arthrobacter agilis* strain L77. *Standards in Genomic Sciences*, *11*, 54. Available from https://doi.org/10.1186/s40793-016-0176-4.

Smith, J. L., De Moraes, C. M., & Mescher, M. C. (2009). Jasmonate- and salicylate-mediated plant defense responses to insect herbivores, pathogens and parasitic plants. *Pest Management Science*, *65*, 497–503. Available from https://doi.org/10.1002/ps.1714.

Srinivasan, R., Karaoz, U., Volegova, M., et al. (2015). Use of 16S rRNA gene for identification of a broad range of clinically relevant bacterial pathogens. *PLoS One*, *10*, e0117617. Available from https://doi.org/10.1371/journal.pone.0117617.

Stael, S., Kmiecik, P., Willems, P., et al. (2015). Plant innate immunity – sunny side up? *Trends in Plant Science*, *20*, 3–11. Available from https://doi.org/10.1016/j.tplants.2014.10.002.

Stoltzfus, J. R., So, R., Malarvithi, P. P., et al. (1997). Isolation of endophytic bacteria from rice and assessment of their potential for supplying rice with biologically fixed nitrogen. *Plant and Soil*, *194*, 25–36. Available from https://doi.org/10.1007/978-94-011-5744-5_4.

Su, C., Lei, L., Duan, Y., et al. (2012). Culture-independent methods for studying environmental microorganisms: Methods, application, and perspective. *Applied Microbiology and Biotechnology*, *93*, 993–1003. Available from https://doi.org/10.1007/s00253-011-3800-7.

Sundaramoorthy, S., Raguchander, T., Ragupathi, N., & Samiyappan, R. (2012). Combinatorial effect of endophytic and plant growth promoting rhizobacteria against wilt disease of Capsicum annum L. caused by *Fusarium solani*. *Biological Control*, *60*, 59–67. Available from https://doi.org/10.1016/j.biocontrol.2011.10.002.

Tian, P., Nan, Z., Li, C., & Spangenberg, G. (2008). Effect of the endophyte Neotyphodium lolii on susceptibility and host physiological response of perennial ryegrass to fungal pathogens. *European Journal of Plant Pathology*, *122*, 593–602. Available from https://doi.org/10.1007/s10658-008-9329-7.

Torsvik, V., & Øvreås, L. (2002). Microbial diversity and function in soil: From genes to ecosystems. *Current Opinion in Microbiology*, *5*, 240–245. Available from https://doi.org/10.1016/s1369-5274(02)00324-7.

Trdá, L., Boutrot, F., Claverie, J., et al. (2015). Perception of pathogenic or beneficial bacteria and their evasion of host immunity: Pattern recognition receptors in the frontline. *Frontiers in Plant Science*, *6*, 219. Available from https://doi.org/10.3389/fpls.2015.00219.

Trdá, L., Fernandez, O., Boutrot, F., et al. (2014). The grapevine flagellin receptor VvFLS2 differentially recognizes flagellin-derived epitopes from the endophytic growth-promoting bacterium *Burkholderia phytofirmans* and plant pathogenic bacteria. *The New Phytologist*, *201*, 1371–1384. Available from https://doi.org/10.1111/nph.12592.

Tripathi, S., Kamal, S., Sheramati, I., et al. (2008). Mycorrhizal fungi and other root endophytes as biocontrol agents against root pathogens. In A. Varma (Ed.), *Mycorrhiza* (pp. 281–306). Berlin, Heidelberg: Springer. Available from https://doi.org/10.1007/978-3-540-78826-3_14.

Turner, J. G., Ellis, C., & Devoto, A. (2002). The Jasmonate signal pathway. *The Plant Cell*, *14*, S153–S164. Available from https://doi.org/10.1105/tpc.000679.

Turner T. R., James E. K., & Poole P.S. (2013) *The plant microbiome*. 1–10.

Vandenkoornhuyse, P., Quaiser, A., Duhamel, M., et al. (2015). The importance of the microbiome of the plant holobiont. *The New Phytologist*, *206*, 1196–1206. Available from https://doi.org/10.1111/nph.13312.

Verma, P., Yadav, A. N., Kumar, V., et al. (2017). Beneficial plant–microbes interactions: Biodiversity of microbes from diverse extreme environments and its impact for crop improvement. In D. P. Singh, H. B. Singh, & R. Prabha (Eds.), *Plant–microbe interactions in agro-ecological perspectives: Volume 2: Microbial interactions and agro-ecological impacts* (pp. 543–580). Singapore: Springer Singapore.

Villena, J., Kitazawa, H., Van Wees, S. C. M., et al. (2018). Receptors and signaling pathways for recognition of bacteria in livestock and crops: Prospects for beneficial microbes in healthy growth strategies. *Frontiers in Immunology*, *9*, 2223. Available from https://doi.org/10.3389/fimmu.2018.02223.

Walling, L. L. (2000). The myriad plant responses to herbivores. *Journal of Plant Growth Regulation*, *19*, 195–216. Available from https://doi.org/10.1007/s003440000026.

Wang, W., Yang, J., Zhang, J., et al. (2020). An arabidopsis secondary metabolite directly targets expression of the bacterial type III secretion system to inhibit bacterial virulence. *Cell Host & Microbe*, *27*, 601–613. Available from https://doi.org/10.1016/j.chom.2020.03.004, e7.

Wani Z. A., & Ashraf N. (2015) *Plant-endophyte symbiosis, an ecological perspective*. 30:177–184. Available from https://doi.org/10.1007/s00253-015-6487-3.

Wasternack, C. (2007). Jasmonates: An update on biosynthesis, signal transduction and action in plant stress response, growth and development. *Annals of Botany*, *100*, 681–697. Available from https://doi.org/10.1093/aob/mcm079.

Wasternack, C. (2014). Action of jasmonates in plant stress responses and development—applied aspects. *Biotechnology Advances*, *32*, 31–39. Available from https://doi.org/10.1016/j.biotechadv.2013.09.009.

Wasternack, C., & Hause, B. (2013). Jasmonates: Biosynthesis, perception, signal transduction and action in plant stress response, growth and development. An update to the

2007 review in Annals of Botany. *Annals of Botany*, *111*, 1021–1058. Available from https://doi.org/10.1093/aob/mct067.

Weert, S. D., Vermeiren, H., Mulders, I. H. M., et al. (2002). Flagella-driven chemotaxis towards exudate components is an important trait for tomato root colonization by *Pseudomonas fluorescens*. *Molecular Plant-Microbe Interactions: MPMI*, *15*, 1173–1180.

Westman, S. M., Kloth, K. J., Hanson, J., et al. (2019). Defence priming in Arabidopsis – a *Meta*-Analysis. *Scientific Reports*, *9*, 1–13. Available from https://doi.org/10.1038/s41598-019-49811-9.

Wilmes, P., & Bond, P. L. (2004). The application of two-dimensional polyacrylamide gel electrophoresis and downstream analyses to a mixed community of prokaryotic microorganisms. *Environmental Microbiology*, *6*, 911–920. Available from https://doi.org/10.1111/j.1462-2920.2004.00687.x.

Wonglom, P., Ito, S., & Sunpapao, A. (2020). Volatile organic compounds emitted from endophytic fungus Trichoderma asperellum T1 mediate antifungal activity, defense response and promote plant growth in lettuce (Lactuca sativa). *Fungal Ecology*, *43*, 100867. Available from https://doi.org/10.1016/j.funeco.2019.100867.

Xie, S., Liu, J., Gu, S., et al. (2020). Antifungal activity of volatile compounds produced by endophytic *Bacillus subtilis* DZSY21 against *Curvularia lunata*. *Annals of Microbiology*, *70*, 2. Available from https://doi.org/10.1186/s13213-020-01553-0.

Xu, J. (2006). Microbial ecology in the age of genomics and metagenomics: Concepts, tools, and recent advances. *Molecular Ecology*, *15*, 1713–1731. Available from https://doi.org/10.1111/j.1365-294X.2006.02882.x.

Yadav, A. N. (2018). Biodiversity and biotechnological applications of host-specific endophytic fungi for sustainable agriculture and allied sectors. *Acta Scientific Microbiology*, *1*, 44.

Yadav, A. N., Kumar, V., Dhaliwal, H. S., et al. (2018). Chapter 15—Microbiome in crops: Diversity, distribution, and potential role in crop improvement. In R. Prasad, S. S. Gill, & N. Tuteja (Eds.), *Crop improvement through microbial biotechnology* (pp. 305–332). Elsevier.

Yalpani, N., Silverman, P., Wilson, T. M., et al. (1991). Salicylic acid is a systemic signal and an inducer of pathogenesis-related proteins in virus-infected tobacco. *The Plant Cell*, *3*, 809–818. Available from https://doi.org/10.1105/tpc.3.8.809.

Yamamoto, F., Iwanaga, F., Al-Busaidi, A., & Yamanaka, N. (2020). Roles of ethylene, jasmonic acid, and salicylic acid and their interactions in frankincense resin production in *Boswellia sacra* Flueck. trees. *Scientific Reports*, *10*, 16760. Available from https://doi.org/10.1038/s41598-020-73993-2.

Yamamoto, R., Ma, G., Zhang, L., et al. (2020). Effects of salicylic acid and methyl jasmonate treatments on flavonoid and carotenoid accumulation in the juice sacs of satsuma mandarin in vitro. *Applied Sciences*, *10*, 1–13. Available from https://doi.org/10.3390/app10248916.

Yamamoto, S., Shiraishi, S., & Suzuki, S. (2015). Are cyclic lipopeptides produced by *Bacillus amyloliquefaciens* S13–3 responsible for the plant defence response in strawberry against *Colletotrichum gloeosporioides*? *Letters in Applied Microbiology*, *60*, 379–386. Available from https://doi.org/10.1111/lam.12382.

Yang, J., Duan, G., Li, C., et al. (2019). The crosstalks between jasmonic acid and other plant hormone signaling highlight the involvement of jasmonic acid as a core component in plant response to biotic and abiotic stresses. *Frontiers in Plant Science*, *10*, 1349. Available from https://doi.org/10.3389/fpls.2019.01349.

Zhao, M.-L., Wang, J.-N., Shan, W., et al. (2013). Induction of jasmonate signalling regulators MaMYC2s and their physical interactions with MaICE1 in methyl jasmonate-induced chilling tolerance in banana fruit. *Plant, Cell & Environment*, *36*, 30–51. Available from https://doi.org/10.1111/j.1365-3040.2012.02551.x.

Further reading

Chithra, S., Jasim, B., Mathew, J., & Radhakrishnan, E. K. (2017). Endophytic Phomopsis sp. colonization in Oryza sativa was found to result in plant growth promotion and piperine production. *Physiologia Plantarum*, *160*, 437–446. Available from https://doi.org/10.1111/ppl.12556.

CHAPTER

Plant disease management through microbiome modulation

6

Aswani R and Radhakrishnan E.K.
School of Biosciences, Mahatma Gandhi University, Kottayam, India

6.1 Introduction

Plant-associated microorganisms play an important role in the plant growth and development. The interaction between the host and the associated microorganisms forms an assemblage of species known as holobiont. The term "holobiont" was introduced in 1991 by Lynn Margulis (Margulis & Fester, 1991). Generally, the plant holobiont comprises all types of microorganisms, including bacteria, fungi, and actinomycetes, termed plant microbiota and the microbial genomes (microbiome) which are found in the phyllosphere, rhizosphere, and endosphere regions of the plant. At the same time, the plant diseases caused by pathogenic microorganisms cause a major challenge to the agricultural productivity. Here comes the importance of plant-associated microorganisms with multiple plant beneficial features for the enhancement of plant growth, health, and stress tolerance. Due to the vital role of plant-associated microorganisms in the growth, health, and productivity of plants they are also considered as the plant's second genome (Berendsen, Pieterse, & Bakker, 2012). The beneficial features of plant microbiota, include production of phytohormones, mobilization of phosphorus and minerals, nitrogen fixation, protection from biotic and abiotic factors through the production of bioactive metabolites and these advantages offer promises for their agricultural applications. The environment friendly nature of these microbial agents, in turn makes them as effective alternative to the currently used agrochemicals (Mendes, Garbeva, & Raaijmakers, 2016; Mitter, Pfaffenbichler, & Sessitsch, 2016). Therefore, a deeper understanding on the functionality of plant–microbe interactions and the factors involved is essential for the selection of suitable candidate organisms for the sustainable agricultural application (Hardoim et al., 2015). Due to the immense promises of plant associated bacteria and fungi, the present chapter focuses on these components of plant microbiota with their plant growth promoting and biocontrol promises.

Plant microbiota consists of both below-ground and above-ground microbiota. For the below-ground microbiota, plants actively recruit microorganisms from the surrounding soil reservoir to make use of their plant growth promoting mechanisms.

Biocontrol Mechanisms of Endophytic Microorganisms. DOI: https://doi.org/10.1016/B978-0-323-88478-5.00011-0

The rhizosphere represents one of the most complex ecosystems of the soil environment (Igiehon & Babalola, 2017) and is considered the hot spot of microbial activity (Berendsen et al., 2012). Here the microorganisms are horizontally transferred from the soil environment to the plant roots and are generally dominated by the phylum acidobacteria, bacteroidetes, proteobacteria, and planctomycetes. In addition to this, the production and composition of root exudates composed of organic acids, amino acids, fatty acids, phenolics, plant growth regulators, sugars, sterols, and vitamins are known to drive the microbial composition of the rhizosphere through rhizosphere effect. The rhizospheric microbes improve the plant growth, development, and nutrient absorption through mineralization of organic substances, N_2 fixation, potassium and phosphate solubilization, indole-3-acetic acid (IAA) production, and 1-aminocyclopropane-1-carboxylate (ACC) deaminase activity (Meena et al., 2017; Zhang et al., 2017). These microbes are also found to exhibit disease-protective effect against broad range of phytopathogens through the production of antimicrobial metabolites and also by eliciting the plant defense responses.

The above-ground microbiota includes endophytic and phyllospheric microbial communities. The endophytic microorganisms generally inhabit within the plant tissues and also support plant growth, health, and disease protection without causing any harm to the host (Farrar, Bryant, & Cope-Selby, 2014; Kumar, Droby, Singh, Singh, & White, 2020; Truyens, Weyens, Cuypers, & Vangronsveld, 2015). Most endophytes can spread systemically via the xylem to distinct compartments of the plants (Tian et al., 2017) and its entry into the plant tissues can also be through the aerial parts such as flowers and fruits (Yeoh et al., 2017). Additionally, the seeds also form an important source of endophytic microorganisms where the bacteria could be transmitted vertically. A diverse range of endophytic microorganisms (root endosphere) also colonizes plant roots internally. The entry of endophytes into the root tissues often occurs through the passive processes such as root cracks, and also through the colonization via lateral roots (Shi et al., 2018). However, the colonization and transmission of endophytes within plants depend on many factors such as the allocation of plant resources and their ability for colonization. The endophytic microorganisms which reside within plants are also able to produce IAA, ACC deaminase, siderophores and solubilize phosphate, zinc, and potassium, which eventually promote the plant growth and development (Kumar et al., 2016; Zhao et al., 2018).

6.2 Priority effects in plant microbiome assembly

The plant microbiome assembly is highly influenced by the differences in the order of species arrival and its early colonization processes. This, in turn, could lead to a larger differences in the community structure of the plant termed priority effects (Braun-Kiewnick, Jacobsen, & Sands, 2000). Based on the priority effects, the plant microbiome management can be executed through the (1) recruitment of functional species, (2) assembly of those with the potential to act against pathogen/pest, and (3) core reinforcement. The recruitment of functional species preferentially involves the recruitment of beneficial microbial populations into the plant system with desirable physiological functions (Freilich et al., 2011). These microorganisms can be found in the seeds or seedlings as pioneer species or it can be from the native pools of potential microbial symbionts. The assembly of potential microbial communities in plants also provides protection from various pathogens and pests . Here, the priority effect enables to prevent the entry of antagonistic late colonizers using the resident or early colonizing microorganisms (Arnold et al., 2003; Wei et al., 2015). The identification of core microbiome that form strong facilitative and mutualistic interactions with plants further strengthen the recruitment and infection control functioning of beneficial microorganisms.

6.3 Core microbiome of plants

Microorganisms that are tightly associated with plants and independent of soil and environmental conditions with multiple plant beneficial features are defined as the core plant microbiome (Toju et al., 2018). The core plant microbiome, thus, comprises keystone microbial taxa with essential functional genes important for the fitness of plant holobiont. This may be established through the evolutionary mechanisms of selection and enrichment (Jones, Garcia, Furches, Tuskan, & Jacobson, 2019). In a previous study, Pfeiffer et al. (2017) have identified *Bradyrhizobium*, *Sphingobium*, and *Microvirga* as the core microbiome of *Solanum tuberosum*. Similarly, Edwards et al. (2015) have identified bacteria belonging to the genus *Deltaproteobacteria*, *Alphaproteobacteria*, and *Actinobacteria* as the core microbiome in rice. The analysis of core microbiome of 21 *Salvia miltiorrhiza* seeds using 16S rDNA and internal transcribed spacer (ITS) amplicon sequencing has reported the identification of Gammaproteobacteria, Alphaproteobacteria, Betaproteobacteria, Sphingobacteria, and Actinobacteria as the core microbiome (Chen et al., 2018).

6.4 Beneficial features of plant microbiome

The plant microbiome consists of beneficial, neutral, or pathogenic microorganisms in the rhizosphere, phyllosphere, and endosphere region. Remarkably, the plant growth–promoting bacteria (PGPBs) form a large proportion in the plant microbiome and help to promote plant growth and protection. The various mechanisms of PGPB for plant growth promotion involve nitrogen fixation, phosphate solubilization, phytohormone production, ACC deaminase activity, and biocontrol properties (Fig. 6.1). Nitrogen is one of the major nutrients essential for the plant

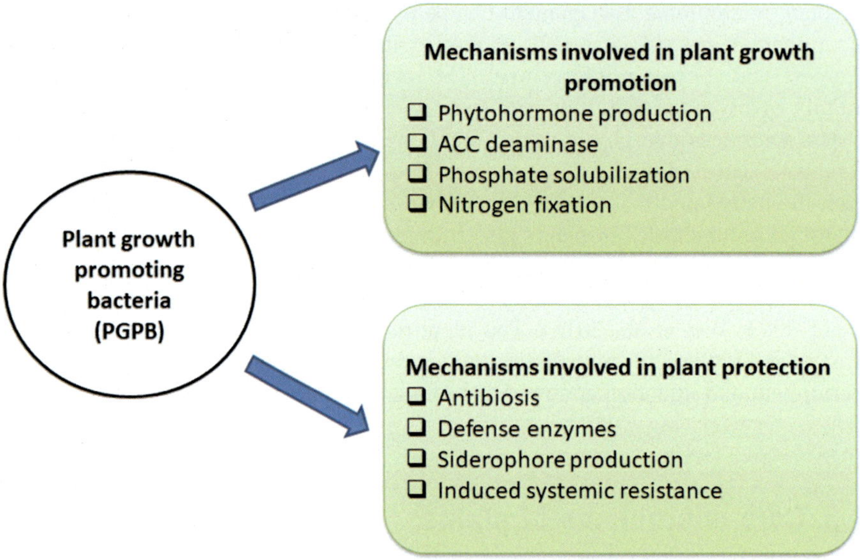

FIGURE 6.1

Beneficial features of plant-associated microorganisms.

growth. Due to the increased nitrogen demand in agriculture, synthetic fertilizers have been extensively used to meet the required nitrogen balance. However, this has led to the challenges like nitrate pollution that suggest the need for suitable alternative methods for sustainable agriculture productivity and to maintain the healthy ecosystem functioning. Here, biological nitrogen fixation (BNF), the microbial conversion of atmospheric nitrogen to a plant-usable form, is highly significant as it is a natural method of nitrogen fixation. BNF by leguminous and nonleguminous (associative, endosymbiotic, and endophytic) systems have been reported to have beneficial impact by increasing the nutrient content and thereby reducing the need for synthetic nitrogen fertilizers in the field (Mahmud, Makaju, Ibrahim, & Missaoui, 2020). Different bacterial species are reported to have the ability to fix nitrogen. In the case of *B. japonicum*, it has been described to fix atmospheric N_2 in to the plant's available form and thereby promote the growth and productivity in soybean (Ronner et al., 2016).

Phosphorus (P) is one of the essential elements in plants responsible for the photosynthesis, maturation, and stress tolerance (Gyaneshwar, Kumar, Parekh, & Poole, 2002; Vance, Uhde-Stone, & Allan, 2003). But P always forms complexe with other compounds in the soil (Theodorou & Plaxton, 1993). These insoluble complexes further make the P to be inaccessible for the plants and hence result in severe P deficiency in both acidic and alkaline soils (Shen et al., 2011). To overcome the P deficiency in agriculture, various phosphatic fertilizers have been used which pose a serious threat to the environment. Therefore alternate strategies are being required for the enhancement of agricultural productivity. Many studies have reported the use of microorganisms to solubilize the insoluble phosphates to plant available form and thus provide an effective replacement to phosphatic fertilizers (Abdelrahman et al., 2016; Sudisha, Mostafa, Phan, & Shin-Ichi, 2018). Phosphate solubilizing bacteria (PSB) secrete organic acids that dissolve unavailable P (PO_4^{3-}) to plant's available forms such as HPO_4^{2-} and $H_2PO_4^-$ (Richardson & Simpson, 2011). Among the organic acids, gluconic acid is reported as the most frequent organic acid produced in the bacteria by an enzyme glucose dehydrogenase mediated by a cofactor pyrroloquinoline quinone (PQQ) (Choi et al., 2008; Deppenmeier, Hoffmeister, & Prust, 2002). The cloning and expression of genes coding for PQQ biosynthesis reported in the previous study further demonstrated the importance of gluconic and 2-ketogluconic acid production in phosphate solubilization (Han et al., 2008). The application of PSB as biofertilizers have also shown to have promising enhancement in grain yield of cereals and other crops (Egamberdiyeva & Höflich, 2004; Saharan & Vibha, 2011). Similarly, *Pseudomonas* and *Bacillus* strains have also been reported to increase yield of wheat through the P solubilization and organic acid production (Egamberdiyeva & Höflich, 2004).

Some PGPB produce phytohormones like auxin, cytokinin, and gibberellin which can have a major impact on plant metabolism and growth by modulating the endogenous hormone levels. The auxin is an important phytohormone that supports the growth and developmental processes such as cell division, elongation, and differentiation (Asgher, Khan, Anjum, & Khan, 2015). Many studies have also identified the indole 3 acetic acid (IAA) producing microorganisms from plants with suggested role in plant growth promotion (Jasim, Jimtha John, Shimil, Jyothis, & Radhakrishnan, 2014). The exogenous application of IAA in combination with beneficial microorganisms has also reported to have an impact on enhanced growth in various model plant systems which further described the role of microbially produced IAA in plant growth and development (Jimtha et al., 2014; Rohini, Aswani, Kannan, Sylas, & Radhakrishnan, 2018; Sabu, Aswani, Jishma, et al., 2019). Cytokinins (CK) are important plant hormones involved in the cellular proliferation and differentiation. It can also help to prevent senescence and, thereby inhibit premature leaf senescence (Schmülling, 2002). The balance between the auxin and cytokinin has also been reported to determine the events in plant physiology such as meristem functioning, root system architecture, formation of lateral organs, and also development of organs (Schaller, Bishopp, & Kieber, 2015). In addition to this, CK also regulate chlorophyll biosynthesis and chloroplast biogenesis (Cortleven & Schmülling, 2015). Gibberellin are also found to be involved in the stimulation of plant growth and development under various abiotic stress conditions (Ahmad, 2010). Iqbal and Ashraf (2013) have described the efficiency of gibberellic acid in the uptake, and ion partitioning within the plant system and thereby enhance growth rate and maintain the plant metabolism under normal and stressed conditions. These compounds have important role also in the seed dormancy, formation of floral organs, and lateral shoot growth (Olszewski, Sun, & Gubler, 2002). Additionally, the crosstalk between gibberellic acid with other phytohormones also mediates the enhanced stress tolerance in the plant system. The functioning of all these phytohormones are directly or indirectly influenced by the type and concentrations of hormones produced by plant microbiome

PGPB can also have the ability to produce ACC deaminase enzyme, which reduces the level of stress hormone ethylene in the plant. However, the increased level of ethylene turns out as stress ethylene which thereby hinders the plant growth and development. The plant-associated bacteria with the potential for 1-aminocyclopropane-1-carboxylate deaminase (ACCD) production (Glick, 2007) can maintain the threshold level of stress ethylene in plants. The enzyme breaks down ACC, an immediate precursor of ethylene in to α-ketobutyrate, and ammonia and result in the decrease of ethylene level (Glick, 2014). ACC deaminase is a multimeric enzyme with a monomeric subunit molecular mass of approximately 35–42 kDa. It is a sulfhydryl enzyme that utilizes pyridoxal 5-phosphate as an essential cofactor. Pyridoxal phosphate is tightly bound to the enzyme in the amount of approximately one molecule per subunit. Also several *d*-amino acids, notably *d*-serine, and *d*-cysteine can act as substrates for ACC deaminase. Previous studies have demonstrated a diverse range of bacteria, including *Pseudomonas* spp., *Paraburkholderia* spp., *Arthrobacter* spp., *Bacillus* spp., and *Pantoea* spp. to have plant growth promoting properties due to the phosphate solubilization, nitrogen fixation, indole acetic acid, and ACC deaminase production (Rohini et al., 2018; Sabu, Aswani, Jishma, et al., 2019). This property of attributing tolerance to abiotic stressors by ACCD activity and other additional mechanisms of PGPB to ameliorate stresses in host plants are referred to as induced systemic tolerance (Yang, Kloepper, & Ryu, 2009).

6.5 Plant microbiome as a tool for plant disease management

Plant diseases are the major challenges to agricultural productivity. Even though various agrochemicals like chemical fertilizers and pesticides have been used to control these plant diseases, the intensive use of these chemicals negatively influence the ecosystem functioning. Here comes the importance of suitable biological control agents with environment friendly applications for the sustainable agricultural production. There are many reports on the identification and biocontrol activity of plant-associated bacteria and fungi which further suggest the beneficial role of these microorganisms in the agriculture (Table 6.1). Considering the plant microbiome and its metabolic interplay with the host plants, these microorganisms have the potential to be explored as efficient biocontrol agents (Berg et al., 2017). This is because all microorganisms associated with plants form a network which in turn decipher the biocontrol property to the associated plants. The term

Table 6.1 List of selected endophytic bacteria and fungi with biocontrol potential.

Microorganisms	Target pathogen	Reference
Paenibacillus polymyxa	*Rhizoctonia solani*	Cho et al. (2007)
Bacillus sp.	*R. solani*	Cho et al. (2007)
Pseudomonas putida	*Phytophthora capsici*	Aravind, Kumar, Eapen, and Ramana (2009)
Pseudomonas aeruginosa	*P. capsici*	Aravind et al. (2009)
B. pumilus	*Fusarium oxysporum, Pythium ultimum, R. solani*, and *P. capsici*	Hong and Park (2016)
B. amyloliquefaciens BmB 1	*Rhizoctonia sp.* and *Pythium sp.*	Jasim et al. (2016)
Serratia sp.	*P. myriotylum, F. oxysporum, R. solani, P. infestans*, and *Sclerotium rolfsii*	Sabu, Aswani, Jishma, et al. (2019)
Bacillus sp.Fcl1	*Corynespora cassiicola, F. oxysporum, Colletotrichum acutatum*, and *P. myriotylum*	Jayakumar, Krishna, Mohan, Nair, and Radhakrishnan (2019)
P. aeruginosa	*P. myriotylum*	Jasim et al. (2014)
B. safensis B21	*Magnaporthe oryzae*	Rong et al. (2020)
Bacillus sp.	*Curvularia lunata*	Xie et al. (2020)
B. stratosphericus LW-03	*F. oxysporum, Botryosphaeria dothidea, Botrytis cinerea, and F. fujikuroi*	Khan et al. (2020)
B. pumilus	*Aspergillus flavus, Aspergillus fumigatus, Penicillium citrinum, F. oxysporum, and Rhizopus stolonifer*	Erjaee, Shekarforoush, Hosseinzadeh, Dehghani, and Winter (2020)
B. velezensis OEE1	*Verticillium dahliae*	Azabou et al. (2020)
Endophytic fungi		
Piriformospora indica	*R. solani*	Nassimi and Taheri (2017)
Cryptosporiopsis quercina	*Pyricularia oryzae*	Li, Strobel, Harper, Lobkovsky, and Clardy (2000)
Penicillium sp.	*Pseudomonas syringae pv. tomato*	Hossain, Sultana, Kubota, and Hyakumachi (2008)
Acremonium sp.	*P. myriotylum*	Anisha and Radhakrishnan (2015)
Aspergillus niger	*P. myriotylum*	Anisha and Radhakrishnan (2017)
Phlebia sp.	*P. myriotylum*	Anisha and Radhakrishnan (2017)

(*Continued*)

Table 6.1 List of selected endophytic bacteria and fungi with biocontrol potential. *Continued*

Microorganisms	Target pathogen	Reference
F. oxysporum	*P. myriotylum*	Anisha and Radhakrishnan (2017)
Rhizopycnis vagum	*C. cassiicola, F. oxysporum, Colletotrichum accutatum, and P. myriotylum*	Anisha et al. (2018)
Paraconiothyrium sp.	*P. myriotylum*	Anisha et al. (2018)
Diaporthe phaseolorum Stdif6	*Cochliobolus miyabeanus, Diaporthe citri, Exserohilum turcicum, Pestalotiopsis theae, Colletotrichum capsici, Alternaria oleracea, and Ceratocystis paradoxa*	Wu et al. (2020)

"biological control" was first used by C. F. Von in 1914 (Cook, 1988) and it include the use of beneficial organisms, genes, or their metabolic products to control the infections caused by plant pathogens and thereby protect the plant from pathogen attack (Tranier et al., 2014; Vinale et al., 2008).

6.5.1 Endophytes as biological control agents

Endophytic microorganisms generally inhabit within their host plant without causing any harm to the host. Therefore these microorganisms are considered an important bioresource for potential candidate strains with a wide range of plant beneficial features (De Silva, Brooks, Lumyong, & Hyde, 2019; Potshangbam, Devi, Sahoo, & Strobel, 2017; Singh, Srivastava, Kumar, Singh, & Pandey, 2020). Microbial biological control agents protect plant from diseases through direct or indirect mechanisms of action. The direct interaction involves the production of siderophores, defense-related enzymes, and antibiosis, while the indirect antagonistic activity involves induced resistance and priming and also through the pathogenesis-related genes (Conrath, Beckers, Langenbach, & Jaskiewicz, 2015; Kumar et al., 2021). The direct and indirect modes of action have been described as follows:

1. *Siderophore production*

   Siderophores are small organic molecules produced by microbes under iron-limiting conditions which function to promote plant growth and yield. They can also have the ability to detect the iron content in different environments and may serve as a potential biosensor for field applications (Saha et al., 2016). Siderophores can also act as a potential biocontrol agent against various phytopathogens. The role of siderophore production as an important mechanism in biocontrol has also been demonstrated in relation to the antifungal activity of *Pseudomonas* spp. against phytopathogenic fungi (Raaijmakers et al., 1995).

2. *Defense enzymes*

   Plant disease management mainly involves the use of chemical fungicides, bactericides, and insecticides. However, there are several defense-related enzymes such as peroxidase, polyphenol oxidase , phenylalanine ammonia lyase, chitinase, and β-1,3-glucanase which are correlated with the disease resistance in plants (Gajanayaka, Prasannath, & De Costa, 2014; Prasannath, 2017). Chitinases are considered a large group of defense-related enzymes which degrade chitin, the cell wall component of many phytopathogens, and thereby protect the plant from pathogenic attack (Jalil, Mishra, & Ansari, 2015). The inhibitory activities of endophytic *Streptomyces* against *Colletotrichum sublineolum* has been described to be due to the chitinase production (Quecine et al., 2008). Similarly, the endophytic *Bacillus cereus* has also been reported for activity against the root rot pathogen *Rhizoctonia solani* through chitinase production. The possible mechanism of action of chitinase produced by *Bacillus* spp. in fungal disease suppression has been well described in many previous reports (Bressan & Figueiredo, 2010). In addition to this, the exogenous application of chitin supplement has also been found to enhance the induced systemic resistance against pathogenic microorganisms. In a previous study, bacterial blight disease caused by *Xanthomonas axonopodis* pv. *malvacearum* (Xam) has been treated with endophytic *Bacillus* EPCO 102 and EPCO 16 and *Pseudomonas fluorescens* Pf1 with or without the addition of chitin. Here, reduced disease incidence could be observed with the chitin supplemented bioformulation. The formulation application has also resulted in the higher production of defense enzymes such as chitinase, peroxidase, PPO, and phenylalanine ammonia-lyase, etc. (Rajendran, Saravanakumar, Raguchander, & Samiyappan, 2006). Also, microbial consortia containing Bacillus *subtilis* (EPC 5), *P. fluorescens* (Pf1), and *Trichoderma viride* (Tv1) were reported o provide higher level of disease protection from *Ganoderma lucidum* (Leys) Karst infection in coconut palm through the induction of defense enzymes, including peroxidase, polyphenol oxidase (PPO), phenylalanine ammonia lyase (PAL), and phenolics (Rajendran, Akila, Karthikeyan, Raguchander, & Samiyappan, 2015).

3. *Antibiosis*

   This mainly involves the production of antimicrobial secondary metabolites which are heterogeneous groups of organic, low-molecular weight compounds produced by microorganisms that help to defend a broad range of plant pathogens. A large number of antibiotics have been reported from actinomycetes with the ability to produce 8700 different antibiotics, bacteria with 2900 antibiotics, and fungi with 4900 types (Köhl, Kolnaar, & Ravensberg, 2019). The production of antimicrobial metabolites has been reported from various bacteria belonging to *Agrobacterium*, *Bacillus*, *Pantoea*,

Pseudomonas, *Serratia*, *Stenotrophomonas*, *Streptomyces*, etc. The production of antimicrobial lipopeptides from *Bacillus* spp. has also been investigated and was found to have remarkable antiphytopathogenic activity (Jasim, Sreelakshmi, Mathew, & Radhakrishnan, 2016; Jasim, Benny, Sabu, Mathew, & Radhakrishnan, 2016; Ongena & Jacques, 2008). *Pseudomonas* have been reported well for the production of antibiotic metabolites like 2,4-diacetylphloroglucinol (DAPG), pyrrolnitrin, pyocyanin, pyouteroin, and phenazine (Jasim, Anisha, et al., 2014; Jayakumar, Perinchery, Jaffer, & Radhakrishnan, 2018; Raaijmakers & Mazzola, 2012). The plant-associated fungi can also act as antagonists through the production of various antimicrobial compounds. For example, the endophytic fungi *Trichoderma*, *Fusarium*, and *Acremonium* have already been reported to produce gliovirin, gliotoxin, viridian, etc. with promising biological application (Anisha & Radhakrishnan, 2015; Ghorbanpour, Omidvari, Abbaszadeh-Dahaji, Omidvar, & Kariman, 2018).

The screening of candidate microorganisms for its antagonistic activity initially starts with in vitro assays by dual culture method. The testing of inhibitory activity of potential antagonists by dual-culture method has several advantages. This fast and reproducible results generated through this method which can easily be quantified by measuring the zone of inhibition and thereby the percentage of inhibition can be calculated. Several reports have also demonstrated dual-culture method as a basic screening for the selection of potential bacterial or fungal antagonistic agents (Anisha, Jishma, Bilzamol, & Radhakrishnan, 2018; Anisha, Sachidanandan, & Radhakrishnan, 2018; Sabu, Aswani, Jishma, et al., 2019). Further metabolite characterization from potential organisms and evaluation of its toxicity effects are basic steps to assess field applicability of selected organisms.

4. *Induced resistance and priming*

Plants themselves are capable of defending diseases caused by pathogenic microorganisms through various physical and chemical mechanisms. Interaction with beneficial microorganisms or their metabolites has been reported to activate various defense mechanisms like induced systemic resistance. Enhancing the resistance is one of the most potential agronomic strategies to prevent biotic losses in crops. Induction of resistance in plants against the pathogens is considered to be executed through the pathogen-associated molecular pattern (PAMP)-triggered immunity (PTI) and effector-triggered immunity (ETI). PTI response involves membrane-embedded receptors like proteins, kinases, and cytoplasmic kinases, while ETI functions when cytoplasmic resistance proteins detect specific pathogen effectors. In the case of fungal pathogens, PAMPs are specifically mentioned as Fungal-Associated Molecular Patterns (FAMPs) that are the broad structural patterns of fungi interacting specifically with the plant system. The FAMP-triggered immunity has also been suggested to play a key role in immunological response of plants toward most of the fungal pathogens. Chitin, chitosan, and chito-oligosaccharides are the well-known FAMPs that interact with both dicot and monocot plants. However, the resistance can be induced locally at the site of infection and can spread via signaling throughout the plant resulting in systemic acquired resistance. This type of induced resistance is considered a direct reaction to the stimulus of necrotizing pathogens (Conrath et al., 2015; Mauch-Mani, Baccelli, Luna, & Flors, 2017; Boller & Felix, 2009). In addition, the siderophores produced by iron-competing bacteria and antibiotics such as DAPG and pyocyanin, biosurfactants, and 2R,3R-butanediol produced from bacteria were also reported to act as elicitors of ISR (Pieterse et al., 2014).

5. *Pathogenesis-related proteins*

Pathogenesis related proteins (PRPs) are defined as a set of proteins associated with host defense induced under pathological conditions. These proteins exhibit high degree of pathogen specificity and are coordinated at the level of transcription. PRPs are well known for their role in acquired resistance which is in correlation with necrotic lesions in plants (Golshani, Fakheri, Behshad, & Vashvaei, 2015). There are also reports on the endophytic bacteria and fungi mediated induction of pathogenesis related activity and thereby increased disease resistance. PRPs can be present throughout the plant parts and a single plant can have the presence of large number of PRPs belonging to different classes with different functions. Interestingly, numerous PRPs have been detected in a wide variety of plants, including rice, wheat, maize, sorghum, barley, tomato, pearl millet, bean, chickpea, soybean, pepper, sunflower, carrot, pepper, grape vine, alfalfa, celery, rubber, and many others. This has been well demonstrated with the endophytic bacteria *Bacillus amyloliquefaciens* which has been illustrated to increase the activities of PRPs during the protection from litchi downy blight disease caused by *Peronophthora litchi* (Cai, Lin, Chen, & Hu, 2008).

6.6 Modulation of plant microbiome through microbial inoculation

The development and application of microbial-based formulations typically starts with an in vitro screening of strains for various plant growth promoting characteristics like phosphate solubilization, nitrogen fixation, ACC deaminase activity and also for its capability for the production of siderophores, plant hormones, and antibiotics. The most promising strains selected will then be tested under the greenhouse condition and further by field study (Xu, Jing, Guo, Li, & Zhang, 2020). Through this bottom-up approach, potential candidate organisms can be identified with multiple plant beneficial features. Sabu, Aswani, Nidheesh et al. (2019) have demonstrated the efficacy of the bacterial strain *Burkholderia* spp. in *Capsicum frutescens* grown under field conditions. The study revealed the effect of selected bacterial species in triggering the early flowering and fruiting with the enhanced protection of plants from insect attack. However, the competitive ability of microbial inoculants is not only the basis of selection criterion for its suitability for bioformulations. The dose of inoculated culture, soil type, texture, successful colonization efficiency of the candidate, and other physiological activities can also determine the competitive ability of these bio-inoculants (Khare & Arora, 2015). Suitable carrier with enhanced shelf-life is also a key factor for the successful delivery of bioformulations. These activities may be tightly regulated by microbiome or its interactions with plant holobiont.

The use of microbial consortia is another approach for the preparation of bioformulations. This is prepared by the combination of two or more microorganisms either with plant growth promoting traits like plant growth enhancement or biocontrol properties or with both the properties. However, some consortia have shown to reduce the plant beneficial features as compared to the single inoculum which indicates the need for smart and knowledge-driven selection of compatible strains in the consortia (Herrera Paredes et al., 2018). The identification of candidate organisms for microbial consortia starts with a selection of potential microorganisms followed by culture-dependent screenings (Armanhi et al., 2018). With the increasing advances in science and technology, new approaches like top-down methods are available for studying microbiome characteristics at a molecular level. Through this approach, prediction of potential candidate organisms with multibeneficial features can be identified in an accurate manner. The method is also feasible with the direct identification of core microbiome from any samples based on single amplicon variants by nucleic acid high throughput sequencing (Callahan et al., 2016). The application of beneficial microorganisms in any form likely to modulate plant microbiome which has to be unraveled by detailed studies.

6.7 Conclusion

Plant microbiota and their interactions within the plants are highly diverse and various factors could determine the microbiome assembly and its functioning. Multibeneficial features of these microorganisms help to improve plant growth and protection and hence have potential application in agriculture. Several approaches like the development of bio-formulations and designing of smart microbial consortia may help to minimize the use of agrochemicals and thereby favour sustainable agricultural productivity. The understanding on plant microbiota, functional mechanisms, and its modulatory effect in resident microbiome can substantially increase the basic knowledge and thereby promote effective application of microbial inoculants for sustainable agricultural practices. Hence, a detailed insight into the concept of plant microbiome modulation for plant disease management is highly important to generate ideas for the development of efficient bioformulations and suitable delivery approaches for field applications.

Acknowledgment

Authors acknowledge Kerala State Council for Science, Technology and Environment—Kerala Biotechnology Commission—Young Investigators Programme in Biotechnology (673/2017/KSCSTE dated October 13, 2017) and JAIVAM project (3972/AD A7/2019 dated August 17, 2019), Kerala state plan fund project and DST-PURSE P II Programme.

References

Abdelrahman, M., Abdel-Motaal, F., El-Sayed, M., Jogaiah, S., Shigyo, M., Ito, S.-i, & Lam-Son Phan, T. (2016). Dissection of *Trichoderma longibrachiatum*-induced defense in onion (*Allium cepa* L.) against *Fusarium oxysporum* f. sp. cepa by target metabolite profiling. *Plant Science*, *246*, 128–138.

Ahmad, P. (2010). Growth and antioxidant responses in mustard (*Brassica juncea* L.) plants subjected to combined effect of gibberellic acid and salinity. *Archives of Agronomy and Soil Science*, *56*, 575–588.

Anisha, C., Jishma, P., Bilzamol, V. S., & Radhakrishnan, E. K. (2018). Effect of ginger endophyte *Rhizopycnis vagum* on rhizome bud formation and protection from phytopathogens. *Biocatalysis and Agricultural Biotechnology*, *14*, 116–119.

Anisha, C., & Radhakrishnan, E. K. (2015). Gliotoxin-producing endophytic *Acremonium* sp. from *Zingiber officinale* found antagonistic to soft rot pathogen *Pythium myriotylum*. *Applied Biochemistry and Biotechnology*, *175*(7), 3458–3467.

Anisha, C., & Radhakrishnan, E. K. (2017). Metabolite analysis of endophytic fungi from cultivars of *Zingiber officinale* Rosc. identifies myriad of bioactive compounds including tyrosol. *3 Biotech*, *7*(2), 146.

Anisha, C., Sachidanandan, P., & Radhakrishnan, E. K. (2018). Endophytic *Paraconiothyrium* sp. from *Zingiber officinale* Rosc. displays broad-spectrum antimicrobial activity by production of danthron. *Current Microbiology*, *75*(3), 343–352.

Aravind, R., Kumar, A., Eapen, S. J., & Ramana, K. V. (2009). Endophytic bacterial flora in root and stem tissues of black pepper (*Piper nigrum* L.) genotype: Isolation, identification and evaluation against *Phytophthora capsici*. *Letters in Applied Microbiology*, *48*(1), 58–64.

Armanhi, J. S. L., de Souza, R. S. C., Damasceno, N. D. B., de Araújo, L. M., Imperial, J., & Arruda, P. (2018). A community-based culture collection for targeting novel plant growth-promoting bacteria from the sugarcane microbiome. *Frontiers in Plant Science*, *8*, 2191.

Arnold, A. E., Mejía, L. C., Kyllo, D., Rojas, E. I., Maynard, Z., Robbins, N., & Herre, E. A. (2003). Fungal endophytes limit pathogen damage in a tropical tree. *Proceedings of the National Academy of Sciences*, *100*(26), 15649–15654.

Asgher, M., Khan, M. I. R., Anjum, N. A., & Khan, N. A. (2015). Minimising toxicity of cadmium in plants-role of plant growth regulators. *Protoplasma*, *252*, 399–413. Available from https://doi.org/10.1007/s00709-014-0710-4.

Azabou, M. C., Gharbi, Y., Medhioub, I., Ennouri, K., Barham, H., Tounsi, S., & Triki, M. A. (2020). The endophytic strain *Bacillus velezensis* OEE1: An efficient biocontrol agent against *Verticillium wilt* of olive and a potential plant growth promoting bacteria. *Biological Control*, *142*, 104168.

Berendsen, R. L., Pieterse, C. M., & Bakker, P. A. (2012). The rhizosphere microbiome and plant health. *Trends in Plant Science*, *17*, 478–486. Available from https://doi.org/10.1016/j.tplants.2012.04.001.

Berg, G., Köberl, M., Rybakova, D., Müller, H., Grosch, R., & Smalla, K. (2017). Plant microbial diversity is suggested as the key to future biocontrol and health trends. *FEMS Microbiology Ecology*, *93*(5).

Boller, T., & Felix, G. (2009). A renaissance of elicitors: Perception of microbe-associated molecular patterns and danger signals by pattern-recognition receptors. *Annual Review of Plant Biology*, *60*, 379–406.

Braun-Kiewnick, A., Jacobsen, B. J., & Sands, D. C. (2000). Biological control of *Pseudomonas syringae* pv. syringae, the causal agent of basal kernel blight of barley, by antagonistic Pantoea agglomerans. *Phytopathology*, *90*(4), 368–375.

Bressan, W., & Figueiredo, J. F. (2010). Chitinolytic *Bacillus* spp. isolates antagonistic to *Fusarium moniliforme* in maize. *Journal of Plant Pathology*, 343–347.

Cai, X. Q., Lin, N., Chen, W., & Hu, F.P. (2008). Control effects on litchi downy blight disease by endophytic bacterial strain TB2 and its pathogenesis-related proteins. In *III international symposium on longan, lychee, and other fruit trees in Sapindaceae family 863* (pp. 631–636).

Callahan, B. J., McMurdie, P. J., Rosen, M. J., Han, A. W., Johnson, A. J. A., & Holmes, S. P. (2016). DADA2: High-resolution sample inference from Illumina amplicon data. *Nature Methods*, *13*(7), 581–583.

Chen, H., Wu, H., Yan, B., Zhao, H., Liu, F., Zhang, H., & Liang, Z. (2018). Core microbiome of medicinal plant *Salvia miltiorrhiza* seed: A rich reservoir of beneficial microbes for secondary metabolism? *International Journal of Molecular Sciences*, *19*(3), 672.

Cho, K. M., Hong, S. Y., Lee, S. M., Kim, Y. H., Kahng, G. G., Lim, Y. P., & Yun, H. D. (2007). Endophytic bacterial communities in ginseng and their antifungal activity against pathogens. *Microbial Ecology*, *54*(2), 341–351.

Choi, O., Kim, J., Kim, J. G., Jeong, Y., Moon, J. S., Park, C. S., & Hwang, I. (2008). Pyrroloquinoline quinone is a plant growth promotion factor produced by *Pseudomonas fluorescens* B16. *Plant Physiology*, *146*(2), 657–668.

Conrath, U., Beckers, G. J., Langenbach, C. J., & Jaskiewicz, M. R. (2015). Priming for enhanced defense. *Annual Review of Phytopathology*, *53*.

Cook, R. J. (1988). Biological control and holistic plant-health care in agriculture. *American Journal of Alternative Agriculture*, *3*(2–3), 51–62.

Cortleven, A., & Schmülling, T. (2015). Regulation of chloroplast development and function by cytokinin. *Journal of Experimental Botany*, *66*(16), 4999–5013.

De Silva, N. I., Brooks, S., Lumyong, S., & Hyde, K. D. (2019). Use of endophytes as biocontrol agents. *Fungal Biology Reviews*, *33*(2), 133–148.

Deppenmeier, U., Hoffmeister, M., & Prust, C. (2002). Biochemistry and biotechnological applications of Gluconobacter strains. *Applied Microbiology and Biotechnology*, *60*(3), 233–242.

Edwards, J., Johnson, C., Santos-Medellín, C., Lurie, E., Podishetty, N. K., Bhatnagar, S., & Sundaresan, V. (2015). Structure, variation, and assembly of the root-associated microbiomes of rice. *Proceedings of the National Academy of Sciences*, *112*(8), E911–E920.

Egamberdiyeva, D., & Höflich, G. (2004). Effect of plant growth-promoting bacteria on growth and nutrient uptake of cotton and pea in a semi-arid region of Uzbekistan. *Journal of Arid Environments*, *56*(2), 293–301.

Erjaee, Z., Shekarforoush, S. S., Hosseinzadeh, S., Dehghani, A., & Winter, D. (2020). Identification of antifungal intracellular proteins of endophytic *Bacillus pumilus* by LC–MS/MS analysis. *International Journal of Peptide Research and Therapeutics, 26*, 1–7.

Farrar, K., Bryant, D., & Cope-Selby, N. (2014). Understanding and engineering beneficial plant–microbe interactions: Plant growth promotion in energy crops. *Plant Biotechnology Journal, 12*, 1193–1206.

Freilich, S., Zarecki, R., Eilam, O., Segal, E. S., Henry, C. S., Kupiec, M., & Ruppin, E. (2011). Competitive and cooperative metabolic interactions in bacterial communities. *Nature Communications, 2*(1), 1–7.

Gajanayaka, G. M. D. R., Prasannath, K., & De Costa, D. M. (2014). Variation of chitinase and β-1,3-glucanase activities in tomato and chilli tissues grown under different crop management practices and agroecological regions. *Proceedings of the Peradeniya University International Research Sessions, 18*, 519.

Ghorbanpour, M., Omidvari, M., Abbaszadeh-Dahaji, P., Omidvar, R., & Kariman, K. (2018). Mechanisms underlying the protective effects of beneficial fungi against plant diseases. *Biological Control, 117*, 147–157.

Glick, B. R. (2007). Promotion of plant growth by bacterial ACC deaminase. *Critical Reviews in Plant Sciences, 26*, 227–242. Available from https://doi.org/10.1080/07352680701572966.

Glick, B. R. (2014). Bacteria with ACC deaminase can promote plant growth and help to feed the world. *Microbiological Research, 169*, 30–39. Available from https://doi.org/10.1016/j.micres.2013.09.009.

Golshani, F., Fakheri, B. A., Behshad, E., & Vashvaei, R. M. (2015). PRs proteins and their mechanism in plants. *Biological Forum, 7*(1), 477.

Gyaneshwar, P., Kumar, G. N., Parekh, L., & Poole, P. (2002). *Role of soil microorganisms in improving P nutrition of plants. Food security in nutrient-stressed environments: Exploiting plants' genetic capabilities* (pp. 133–143). Springer.

Han, S. H., Kim, C. H., Lee, J. H., Park, J. Y., Cho, S. M., Park, S. K., & Kim, Y. C. (2008). Inactivation of PQQ genes of Enterobacter intermedium 60–2G reduces antifungal activity and induction of systemic resistance. *FEMS Microbiology Letters, 282* (1), 140–146.

Hardoim, P. R., van Overbeek, L. S., Berg, G., Pirttilä, A. M., Compant, S., Campisano, A., … Sessitsch, A. (2015). The hidden world within plants: ecological and evolutionary considerations for defining functioning of microbial endophytes. *Microbiology and Molecular Biology Reviews: MMBR, 79*(3), 293–320. Available from https://doi.org/10.1128/MMBR.00050-14.

Herrera Paredes, S., Gao, T., Law, T. F., Finkel, O. M., Mucyn, T., Teixeira, P. J. P. L., … Gabriel, Castrillo (2018). Design of synthetic bacterial communities for predictable plant phenotypes. *PLoS Biology, 16*, e2003962. Available from https://doi.org/10.1111/j.1461-0248.2007.01139.x.

Hong, C. E., & Park, J. M. (2016). Endophytic bacteria as biocontrol agents against plant pathogens: Current state-of-the-art. *Plant Biotechnology Reports, 10*(6), 353–357.

Hossain, M. M., Sultana, F., Kubota, M., & Hyakumachi, M. (2008). Differential inducible defense mechanisms against bacterial speck pathogen in *Arabidopsis thaliana* by plant-growth-promoting-fungus *Penicillium* sp. GP16–2 and its cell free filtrate. *Plant and Soil, 304*(1–2), 227–239.

Igiehon, N. O., & Babalola, O. O. (2017). Biofertilizers and sustainable agriculture: Exploring arbuscular mycorrhizal fungi. *Applied Microbiology and Biotechnology*, *101*, 4871–4881. Available from https://doi.org/10.1007/s00253-017-8344-z.

Iqbal, M., & Ashraf, M. (2013). Gibberellic acid mediated induction of salt tolerance in wheat plants: Growth, ionic partitioning, photosynthesis, yield and hormonal homeostasis. *Environmental and Experimental Botany*, *86*, 76–85. Available from https://doi.org/10.1016/j.envexpbot.2010.06.002.

Jalil, S. U., Mishra, M. A. N. E. E. S. H., & Ansari, M. I. (2015). Current view on chitinase for plant defence. *Trends Bioscience*, *8*(24), 6733–6743.

Jasim, B., Anisha, C., Rohini, S., Kurian, J. M., Jyothis, M., & Radhakrishnan, E. K. (2014). Phenazine carboxylic acid production and rhizome protective effect of endophytic *Pseudomonas aeruginosa* isolated from *Zingiber officinale*. *World Journal of Microbiology and Biotechnology*, *30*(5), 1649–1654.

Jasim, B., Benny, R., Sabu, R., Mathew, J., & Radhakrishnan, E. K. (2016). Metabolite and mechanistic basis of antifungal property exhibited by Endophytic *Bacillus amyloliquefaciens* BmB 1. *Applied Biochemistry and Biotechnology*, *179*(5), 830–845.

Jasim, B., Jimtha John, C., Shimil, V., Jyothis, M., & Radhakrishnan, E. K. (2014). Studies on the factors modulating indole-3-acetic acid production in endophytic bacterial isolates from *Piper nigrum* and molecular analysis of ipdc gene. *Journal of Applied Microbiology*, *117*(3), 786–799.

Jasim, B., Sreelakshmi, K. S., Mathew, J., & Radhakrishnan, E. K. (2016). Surfactin, iturin, and fengycin biosynthesis by endophytic *Bacillus* sp. from *Bacopa monnieri*. *Microbial Ecology*, *72*(1), 106–119.

Jayakumar, A., Krishna, A., Mohan, M., Nair, I. C., & Radhakrishnan, E. K. (2019). Plant growth enhancement, disease resistance, and elemental modulatory effects of plant probiotic endophytic *Bacillus* sp. Fcl1. *Probiotics and Antimicrobial Proteins*, *11*(2), 526–534.

Jayakumar, A., Perinchery, A., Jaffer, F. M., & Radhakrishnan, E. K. (2018). Differential modulation of phytoelemental composition by selected *Pseudomonas* spp. *3 Biotech*, *8*(9), 377.

Jimtha, J. C., Smitha, P. V., Anisha, C., Deepthi, T., Meekha, G., Radhakrishnan, E. K., & Remakanthan, A. (2014). Isolation of endophytic bacteria from embryogenic suspension culture of banana and assessment of their plant growth promoting properties. *Plant Cell, Tissue and Organ Culture (PCTOC)*, *118*(1), 57–66.

Jones, P., Garcia, B., Furches, A., Tuskan, G., & Jacobson, D. (2019). Plant host-associated mechanisms for microbial selection. *Frontiers in Plant Science*, *10*, 862.

Khan, M. S., Gao, J., Zhang, M., Chen, X., Du, Y., Yang, F., & Zhang, X. (2020). Isolation and characterization of plant growth-promoting endophytic bacteria *Bacillus stratosphericus* LW-03 from *Lilium wardii*. *3 Biotech*, *10*(7), 1–15.

Khare, E., & Arora, N. K. (2015). *Effects of soil environment on field efficacy of microbial inoculants. Plant microbes symbiosis: Applied facets* (pp. 353–381). New Delhi: Springer.

Köhl, J., Kolnaar, R., & Ravensberg, W. J. (2019). Mode of action of microbial biological control agents against plant diseases: Relevance beyond efficacy. *Frontiers in Plant Science*, *10*, 845.

Kumar, A., Droby, S., Singh, V. K., Singh, S. K., & White, J. F. (2020). Entry, colonization, and distribution of endophytic microorganisms in plants. In A. Kumar, & E. K. Radhakrishnan (Eds.), *Microbial endophytes* (pp. 1–33). Woodhead Publishing. Available from https://doi.org/10.1016/B978-0-12-819654-0.00001-6.

Kumar, A., Singh, R., Yadav, A., Giri, D. D., Singh, P. K., & Pandey, K. D. (2016). Isolation and characterization of bacterial endophytes of *Curcuma longa* L. *3 Biotech*, *6*(1), 60.

Kumar, A., Zhimo, Y., Biasi, A., Salim, S., Feygenberg, O., Wisniewski, M., & Droby, S. (2021). Endophytic microbiome in the carposphere and its importance in fruit physiology and pathology. In D. Spadaro, S. Droby, & M. L. Gullino (Eds.), *Postharvest pathology. Plant pathology in the 21st century* (Vol. 11, pp. 73–88). Cham: Springer. Available from https://doi.org/10.1007/978-3-030-56530-5_5.

Li, J. Y., Strobel, G., Harper, J., Lobkovsky, E., & Clardy, J. (2000). Cryptocin, a potent tetramic acid antimycotic from the endophytic fungus *Cryptosporiopsis* cf. quercina. *Organic Letters*, *2*(6), 767–770.

Mahmud, K., Makaju, S., Ibrahim, R., & Missaoui, A. (2020). Current progress in nitrogen fixing plants and microbiome research. *Plants*, *9*(1), 97.

Margulis, L., & Fester, R. (Eds.), (1991). *Symbiosis as a source of evolutionary innovation: Speciation and morphogenesis*. MIT Press.

Mauch-Mani, B., Baccelli, I., Luna, E., & Flors, V. (2017). Defense priming: An adaptive part of induced resistance. *Annual Review of Plant Biology*, *68*, 485–512.

Meena, V. S., Meena, S. K., Verma, J. P., Kumar, A., Aeron, A., Mishra, P. K., ... Dotaniya, M. L. (2017). Plant beneficial rhizospheric microorganism (PBRM) strategies to improve nutrients use efficiency: A review. *Ecological Engineering*, *107*, 8–32.

Mendes, R., Garbeva, P., & Raaijmakers, J. M. (2016). The rhizosphere microbiome: Significance of plant beneficial, plant pathogenic, and human pathogenic microorganisms. *FEMS Microbiology Reviews*, *37*, 634–663.

Mitter, B., Pfaffenbichler, N., & Sessitsch, A. (2016). Plant–microbe partnerships in 2020. *Microbial Biotechnology*, *9*(5), 635–640.

Nassimi, Z., & Taheri, P. (2017). Endophytic fungus *Piriformospora indica* induced systemic resistance against rice sheath blight via affecting hydrogen peroxide and antioxidants. *Biocontrol Science and Technology*, *27*(2), 252–267.

Olszewski, N., Sun, T. P., & Gubler, F. (2002). Gibberellin signaling, biosynthesis, catabolism, and response pathways. *The Plant Cell*, *14*, 561–580.

Ongena, M., & Jacques, P. (2008). *Bacillus lipopeptides*: Versatile weapons for plant disease biocontrol. *Trends in Microbiology*, *16*(3), 115–125.

Pfeiffer, S., Mitter, B., Oswald, A., Schloter-Hai, B., Schloter, M., Declerck, S., & Sessitsch, A. (2017). Rhizosphere microbiomes of potato cultivated in the High Andes show stable and dynamic core microbiomes with different responses to plant development. *FEMS Microbiology Ecology*, *93*(2), fiw242.

Pieterse, C. M., Zamioudis, C., Berendsen, R. L., Weller, D. M., Van Wees, S. C., & Bakker, P. A. (2014). Induced systemic resistance by beneficial microbes. *Annual Review of Phytopathology*, *52*, 347–375.

Potshangbam, M., Devi, S. I., Sahoo, D., & Strobel, G. A. (2017). Functional characterization of endophytic fungal community associated with *Oryza sativa* L. and *Zea mays* L. *Frontiers in Microbiology*, *8*, 325.

Prasannath, K. (2017). *Plant defense-related enzymes against pathogens: A review*.

Quecine, M. C., Araujo, W. L., Marcon, J., Gai, C. S., Azevedo, J. L. D., & Pizzirani-Kleiner, A. A. (2008). Chitinolytic activity of endophytic Streptomyces and potential for biocontrol. *Letters in Applied Microbiology*, *47*(6), 486–491.

Raaijmakers, J. M., Leeman, M., Van Oorschot, M. M., Van der Sluis, I., Schippers, B., & Bakker, P. A. H. M. (1995). Dose-response relationships in biological control of *Fusarium wilt* of radish by *Pseudomonas* spp. *Phytopathology*, *85*(10), 1075–1080.

Raaijmakers, J. M., & Mazzola, M. (2012). Diversity and natural functions of antibiotics produced by beneficial and plant pathogenic bacteria. *Annual Review of Phytopathology*, *50*, 403–424.

Rajendran, L., Akila, R., Karthikeyan, G., Raguchander, T., & Samiyappan, R. (2015). Defense related enzyme induction in coconut by endophytic bacteria (EPC 5). *Acta Phytopathologica et Entomologica Hungarica*, *50*(1), 29–43.

Rajendran, L., Saravanakumar, D., Raguchander, T., & Samiyappan, R. (2006). Endophytic bacterial induction of defence enzymes against bacterial blight of cotton. *Phytopathologia Mediterranea*, *45*(3), 203–214.

Richardson, A. E., & Simpson, R. J. (2011). Soil microorganisms mediating phosphorus availability update on microbial phosphorus. *Plant Physiology*, *156*(3), 989–996.

Rohini, S., Aswani, R., Kannan, M., Sylas, V. P., & Radhakrishnan, E. K. (2018). Culturable endophytic bacteria of ginger rhizome and their remarkable multi-trait plant growth-promoting features. *Current Microbiology*, *75*(4), 505–511.

Rong, S., Xu, H., Li, L., Chen, R., Gao, X., & Xu, Z. (2020). Antifungal activity of endophytic *Bacillus safensis* B21 and its potential application as a biopesticide to control rice blast. *Pesticide Biochemistry and Physiology*, *162*, 69–77.

Ronner, E., Franke, A. C., Vanlauwe, B., Dianda, M., Edeh, E., Ukem, B., & Giller, K. E. (2016). Understanding variability in soybean yield and response to P-fertilizer and rhizobium inoculants on farmers' fields in northern Nigeria. *Field Crops Research*, *186*, 133–145.

Sabu, R., Aswani, R., Jishma, P., Jasim, B., Mathew, J., & Radhakrishnan, E. K. (2019). Plant growth promoting endophytic *Serratia* sp. ZoB14 protecting ginger from fungal pathogens. *Proceedings of the National Academy of Sciences, India Section B: Biological Sciences*, *89*(1), 213–220.

Sabu, R., Aswani, R., Nidheesh, K. S., Ray, J. G., Remakanthan, A., & Radhakrishnan, E. K. (2019). Beneficial changes in *Capsicum frutescens* due to priming by plant probiotic *Burkholderia* spp. *Probiotics and Antimicrobial Proteins*, *11*(2), 519–525.

Saha, M., Sarkar, S., Sarkar, B., Sharma, B. K., Bhattacharjee, S., & Tribedi, P. (2016). Microbial siderophores and their potential applications: A review. *Environmental Science and Pollution Research*, *23*(5), 3984–3999.

Saharan, B. S., & Vibha, N. (2011). Assessment of plant growth promoting attributes of cotton (*Gossypium hirsutum*) rhizosphere isolates and their potential as bio-inoculants. *Journal of Environmental Research and Development*, *5*(3), 575–583.

Schaller, G. E., Bishopp, A., & Kieber, J. J. (2015). The yin-yang of hormones: Cytokinin and auxin interactions in plant development. *The Plant Cell*, *27*(1), 44–63.

Schmülling, T. (2002). New insights into the functions of cytokinins in plant development. *Journal of Plant Growth Regulation*, *21*(1).

Shen, J., Yuan, L., Zhang, J., Li, H., Bai, Z., Chen, X., & Zhang, F. (2011). Phosphorus dynamics: From soil to plant. *Plant Physiology*, *156*(3), 997–1005.

Shi, S., Tian, L., Nasir, F., Li, X., Li, W., Tran, L. S. P., & Tian, C. (2018). Impact of domestication on the evolution of rhizomicrobiome of rice in response to the presence of *Magnaporthe oryzae*. *Plant Physiology and Biochemistry: PPB/Societe Francaise de Physiologie Vegetale*, *132*, 156–165.

Singh, M., Srivastava, M., Kumar, A., Singh, A. K., & Pandey, K. D. (2020). Endophytic bacteria in plant disease management. In A. Kumar, & V. K. Singh (Eds.), *Microbial endophytes* (pp. 61–89). Woodhead Publishing. Available from https://doi.org/10.1016/B978-0-12-818734-0.00004-8.

Sudisha, J., Mostafa, A., Phan, T. L.-S., & Shin-Ichi, I. (2018). Different mechanisms of Trichoderma virens-mediated resistance in tomato against *Fusarium wilt* involve the jasmonic and salicylic acid pathways. *Molecular Plant Pathology*, *19*(4), 870–882, pmid: 28605157.

Theodorou, M. E., & Plaxton, W. C. (1993). Metabolic adaptations of plant respiration to nutritional phosphate deprivation. *Plant Physiology*, *101*(2), 339–344.

Tian, L., Zhou, X., Ma, L., Xu, S., Nasir, F., & Tian, C. (2017). Root-associated bacterial diversities of *Oryza rufipogon* and *Oryza sativa* and their influencing environmental factors. *Archives of Microbiology*, *199*, 563–571.

Toju, H., Peay, K. G., Yamamichi, M., Narisawa, K., Hiruma, K., Naito, K., & Yoshida, K. (2018). Core microbiomes for sustainable agroecosystems. *Nature Plants*, *4*(5), 247–257.

Tranier, M. S., Pognant-Gros, J., Quiroz, R. D. L. C., González, C. N. A., Mateille, T., & Roussos, S. (2014). Commercial biological control agents targeted against plant-parasitic root-knot nematodes. *Brazilian Archives of Biology and Technology*, *57*(6), 831–841.

Truyens, S., Weyens, N., Cuypers, A., & Vangronsveld, J. (2015). Bacterial seed endophytes: Genera, vertical transmission and interaction with plants. *Environmental Microbiology Reports*, *7*, 40–50.

Vance, C. P., Uhde-Stone, C., & Allan, D. L. (2003). Phosphorus acquisition and use: Critical adaptations by plants for securing a nonrenewable resource. *New Phytologist*, *157*(3), 423–447.

Vinale, F., Sivasithamparam, K., Ghisalberti, E. L., Marra, R., Woo, S. L., & Lorito, M. (2008). Trichoderma–plant–pathogen interactions. *Soil Biology and Biochemistry*, *40* (1), 1–10.

Wei, Z., Yang, T., Friman, V. P., Xu, Y., Shen, Q., & Jousset, A. (2015). Trophic network architecture of root-associated bacterial communities determines pathogen invasion and plant health. *Nature Communications*, *6*(1), 1–9.

Wu, H., Yan, Z., Deng, Y., Wu, Z., Xu, X., Li, X., & Luo, H. (2020). Endophytic fungi from the root tubers of medicinal plant *Stephania dielsiana* and their antimicrobial activity. *Acta Ecologica Sinica*, *40*.

Xie, S., Liu, J., Gu, S., Chen, X., Jiang, H., & Ding, T. (2020). Antifungal activity of volatile compounds produced by endophytic *Bacillus subtilis* DZSY21 against *Curvularia lunata*. *Annals of Microbiology*, *70*(1), 1–10.

Xu, S. J., Jing, Z. Q., Guo, Z. J., Li, Q. Q., & Zhang, X.R. (2020) *Growth-promoting and disease-suppressing effects of Paenibacillus polymyxa strain YCP16–23 on pepper (Capsicum annuum) plants*.

Yang, J., Kloepper, J. W., & Ryu, C. M. (2009). Rhizosphere bacteria help plants tolerate abiotic stress. *Trends in Plant Science*, *14*, 1–4. Available from https://doi.org/10.1016/j.tplants.2008.10.004.

Yeoh, Y. K., Dennis, P. G., Paungfoo-Lonhienne, C., Weber, L., Brackin, R., Ragan, M. A., ... Hugenholtz, P. (2017). Evolutionary conservation of a core root microbiome across plant phyla along a tropical soil chronosequence. *Nature Communications*, *8*, 215.

Zhang, X., Zhang, R., Gao, J., Wang, X., Fan, F., Ma, X., ... Deng, Y. (2017). Thirty-one years of rice-rice-green manure rotations shape the rhizosphere microbial community and enrich beneficial bacteria. *Soil Biology & Biochemistry*, *104*, 208–217.

Zhao, Y., Xiong, Z., Wu, G., Bai, W., Zhu, Z., Gao, Y., ... i, H. (2018). Fungal endophytic communities of two wild rosa varieties with different powdery mildew susceptibilities. *Frontiers in Microbiology*, *9*, 2462.

CHAPTER

Improved designing and development of endophytic bioformulations for plant diseases

7

Prasanna Rajan[1], Reedhu Raj[2], Sijo Mathew[1], Elizabeth Cherian[3] and A. Remakanthan[4]

[1]*Department of Botany, Government College Kottayam, Kottayam, India*
[2]*Department of Botany, Government Victoria College, Palakkad, India*
[3]*Department of Botany, CMS College Kottayam, Kottayam, India*
[4]*Department of Botany, University College, Thiruvananthapuram, Thiruvananthapuram, India*

7.1 Introduction

The year 2020 witnessed the rise of human population to 7.8 billion, and it is predicted that the size of population will cross the line of 8 billion around the period of 2025–30 (Scherbov et al., 2011; UN, 2019). Major part of this population growth is associated with developing countries than developed countries. Considering India, the largest democratic country of the world, with 17.69% of the world's population is in the second position, just behind China (18.46%). Such increase in population will definitely lead to the shortage of resources, particularly food and water, so the situation prompted to make more food from limited area and available water resources (Schneider, 2011) to overcome starvation, malnutrition, etc.

In India the introduction of green revolution strategies helped a lot to overcome such crises that existed before the 1970s by raising more food crops and thereby producing sufficient food products. At present, the country became self-sufficient in food production and agricultural practices in the country have a contribution of about 18% of GDP in 2020–21. The use of novel plant varieties produced by plant breeding techniques and the use of chemical fertilizers, pesticides, weedicides, etc. led to this achievement. By the introduction of green revolution strategies, there occurred tremendous increase in the production and use of various chemical fertilizers, pesticides, and weedicides worldwide and particularly in India.

Biocontrol Mechanisms of Endophytic Microorganisms. DOI: https://doi.org/10.1016/B978-0-323-88478-5.00003-1

7.1.1 Pesticides a burning issue

To avoid crop loss due to pathogen load, farmers started uncontrolled use of chemical pesticides from green revolution period onward, which resulted in a number of impacts on biotic and abiotic components of environment. The major issues such as (1) the emergence of phytopathogens with pesticide resistance, (2) decrease in soil fertility, (3) loss of beneficial microbes from agriculture fields, (4) disappearance of natural predators of insect pests form agriculture fields due to the breakage of food chain, and (5) the presence of pesticides in various food products—mild-to-heavy doses—ultimately led various health issues to human being.

7.1.2 Think green to save future

Definitely we need to think about a new way to keep our nature and protect it for future generations. So, it became essential to find and develop an eco-friendly alternative for mitigating harmful chemical fertilizers, weedicides, and pesticides. One of such alternate thought leads to the concept of "bioformulations" for protecting our crop plants from pathogenic organisms, nutrient deficiency, environmental stresses, and other living organisms from detrimental effects of chemical pesticides. "Bioformulations" are nonchemical formulations with beneficial microorganisms that are involved in nutrient mobilization and plant protection properties (Arora, Khare, & Maheshwari, 2010). Microbes with plant protection properties protect plants from adverse environmental conditions and pathogenic microorganisms. Such microbes, plant-associated microbial community (PAMC), are naturally associated with plant parts like root, stem, and leaves. The major categories of PAMCs are fungi and bacteria, may be associated as epiphytes or endophytes and may be symbionts or saprophytes nutritionally. Among various PAMCs, those associated as endophytes and seen in the rhizospheric region are most diverse and important.

The concept of developing such microbes as microbial biocontrol agents (MBCAs), instead of chemical pesticides, for plant disease control has a number of advantages. The major advantages of MBCAs are as follows: (1) they are eco-friendly, (2) existence of direct relation between plant and microbial community, and (3) possibility for modifying microbes under cultural conditions. But we need to address some challenges—(1) low survival rate of MBCAs under different environmental conditions of the agricultural field, (2) lack of consistent performance of MBCAs, and (3) most of the MBCAs are host-specific so fail to perform well with different crop plants.

7.1.3 Are endophytes a promising candidate?

Among various reasons for the failure of bioformulations, especially in the case of formulations used externally, the most important one is instability of biocontrol properties of bioformulations under different environmental conditions.

At present, active researches are going on the world around to develop endophytes as promising MBCAs. Since they are localized within plant body, changes in external environmental factors will not affect much the endophytic MBCAs. In plants the presence of endophytes is mapped in root, stem, leaves, seeds, etc. (Hallmann et al., 1997; Patriquin & Dobbereiner, 1978), and there exists a close relationship between host plants and endophytic microbes (Tewari, Shrivas, Hariprasad, & Sharma, 2019). In addition, they have rapid colonizing ability within host plant and are able to reach target site easily that enables to avoid repeated field applications (Devi & Momota, 2015).

This chapter aims at the elucidation of various mechanisms employed by the endophytes in the protection of plants against phytopathogens and the various strategies for their modification and commercialization. The chapter also focuses on the different formulation procedures developed and tries to unravel the recent success stories in the trial to transfer MBCAs from laboratory to field.

7.2 Mechanism deployed by endophytes in plant protection

Plant microbiome includes plant growth–promoting rhizospheric bacteria (PGPR) as well as the endophytes that live asymptomatically within the tissues of host. These interact with the host plants and perform various plant beneficial activities to improve their growth and productivity. In addition to these, they also prevent the growth of phytopathogens, and this remarkable capability can be exploited in the use of these microorganisms as effective biocontrol agents. Endophytes offer an advantage over the PGPR in that they are always in direct contact with the host, hence able to interact more efficiently with the host under diverse environmental conditions (Coutinho, Licastro, Mendonça-Previato, Cámara, & Venturi, 2015). The colonizing aptitude of the microbe as well as the resource provision within the host contributes to the distribution of endophytes within a plant (Tewari et al., 2019). There are still gaps in our acquaintance regarding what turns a rhizospheric bacteria into an endophyte and recent molecular characterization of endophytic genes and their comparative genomic studies may unravel novel genes that can contribute to plant–endophyte interactions (Pinski, Betekhtin, Hupert-Kocurek, Mur, & Hasterok, 2019; Santoyo, Moreno-Hagelsieb, del Carmen Orozco-Mosqueda, & Glick, 2016).

Endophytic bacteria and fungi enhance plant health by a plethora of mechanisms. Fadiji and Babalola (2020) categorized these into direct and indirect mechanisms. In this section, we shall discuss the various mechanisms adopted by these organisms in protecting the host plants, thus facilitating their use as biocontrol agents.

7.2.1 Antibiosis

The production of secondary metabolites with antibacterial or antifungal activity by endophytic microorganisms exhibits an immense role in the suppression of the growth of phytopathogenic microorganisms. Hence, such compounds can be referred to as plant defensive compounds (Ray et al., 2019) These are capable of killing or fettering the growth of microorganisms even at low concentrations. Plants host a wide variety of endophytes, association of which boosts the synthesis of such compounds which confers the host with better environmental adaptation and defense enhancement. The 5,8-dimethyl quinolone is an example of a metabolite that is common to almost all the endophytes and hampers the growth of pathogenic microorganisms by forming an irrevocable complex with nucleophilic amino acids (Berdy, 2005).

The diversity of secondary metabolites produced by endophytes is huge. The production of these plant defensive compounds especially alkaloids, flavonoids, isoflavonoids, and diterpenes by the endophytes shoots up during stress conditions (Shwab & Keller, 2008). *Ralstonia*, *Rhizobium*, *Bacillus*, *Acinetobacter*, *Pseudomonas*, *Pantoea*, *Paenibacillus*, *Burkholderia*, *Achromobacter*, *Azospirillum*, *Microbacterium*, *Methylobacterium*, *Variovorax*, and *Enterobacter* are few bacterial endophytes that confer resistance to host in such a situation (Card, Johnson, Teasdale, & Caradus, 2016). Fungal endophytes produce almost all classes of secondary metabolites of which terpenoids and polyketides being most common and flavonoids and lignans being rarest (Mousa & Raizada, 2013). Genetic studies revealed the clustering of genes for the production of antimicrobial secondary metabolites on chromosomes, and this may have contributed to the horizontal transfer of these genes during evolution (Mousa & Raizada, 2013). Some of the major antimicrobial compounds produced by endophytic bacteria, fungi, and actinomycetes are discussed by Fadiji and Babalola (2020).

7.2.2 Production of hydrolyzing enzymes

Endophytic fungi and bacteria successfully colonize the host plants by invading the cells through wounds or natural openings and remain in a mutualistic association with their phyto partner. The invasion of some bacterial and fungal endophytes is accomplished by breaking the plant cell wall components by the release of lytic enzymes capable of hydrolyzing cellulose, pectin, hemicellulose, chitin, and proteins (Pinski et al., 2019). Khan, Shahzad, Al-Harrasi, and Lee (2017) have given a detailed account on the extracellular enzyme production by fungal and bacterial endophytes. Expansins, which are limited to bacteria from genera *Bacillus*, *Xanthomonas*, *Xylella*, *Ralstonia*, and *Erwinia* are capable of creating extensions or creeps in plant cell wall instead of breaks (Kerff et al., 2008). Many of these enzymes are involved in controlling phytopathogens and are also found to digest the fungal cell wall (Fadiji & Babalola, 2020). Gao, Dai, and Liu (2010) have cited works where reduced biocontrol activity is shown when one or more of the lytic enzyme–producing genes in endophytic bacteria are mutated.

However, this property cannot be solely used for the control of phytopathogens and can enhance the antagonistic activity of endophytes when combined with other mechanisms (Fadiji & Babalola, 2020; Gao et al., 2010). Moreover, a majority of the endophytic bacteria do not possess the genes for cell wall–hydrolyzing enzymes in their genome (Pinski et al., 2019).

7.2.3 Production of phytohormones

The many phytohormones of the plants offer distinguishing roles in their growth and development. Diverse studies report the production of phytohormones especially IAA (Indole-3-acetic acid) by the endophytes that accomplish many physiological functions such as increasing the root length and surface area, expanding the plant cell wall, thereby increasing the secretion of exudates and results in the growth of beneficial microorganisms. An interesting fact is that the type of pathway employed by the microorganism for the production of phytohormone in the plants greatly influences the selection of associated microbe. The beneficial microbes normally synthesize IAA by Indole-3-pyruvate pathway, whereas pathogenic ones opt Indole-3-acetamide pathway (Hardoim, van Overbeek, & van Elsas, 2008).

A plethora of research findings suggest the importance of IAA produced by endophytes and its growth-promoting capabilities. A recent study by Mehmood et al. (2019) on the IAA production by *Aspergillus awamori*, an endophytic fungus isolated from *Withania somnifera*, showed growth-enhancing activities in *Zea mays*. The study also concluded that IAA was involved in the initial cross-talk between the two partners and acts as a signaling molecule between them. Another interesting work reports the production of IAA by the bacterial endophyte *Staphylococcus* sp. (Jayakumar, Krishna, Nair, & Radhakrishnan, 2020; Jayakumar, Nair, & Radhakrishnan, 2020; Jayakumar, Padmakumar, Nair, & Radhakrishnan, 2020) and *Bacillus subtilis* Dcl1 (Jayakumar, Krishna, Mohan, Nair, & Radhakrishnan, 2019; Jayakumar, Krishna, et al., 2020; Jayakumar, Nair, et al., 2020; Jayakumar, Padmakumar, et al., 2020) isolated from the rhizome of *Curcuma longa* under both drought stressed and normal conditions. The microbes also showed synergistic plant growth–promoting effect in *Vigna unguiculata* along with supplemented silicate sources, under drought conditions.

Similarly, Rohini, Aswani, Kannan, Sylas, and Radhakrishnan (2018) isolated 96 endophytic bacteria from the rhizome of *Zingiber officinalis* of which 16 endophytes were with multiple growth-promoting properties like IAA production, phosphate solubilization, and ACC deaminase (1-aminocyclopropane-1-carboxylate deaminase) activity. Moreover, the isolates showed enhanced growth in *V. unguiculata* var Lola which makes the work extremely remarkable and significant in the agricultural field.

There are only a few bacterial endophytes that produce Gibberellic acids, whereas a number of fungal endophytes have been reported to produce GAs (Gibberellic acid). Ali, Charles, and Glick (2017) reviewed numerous studies that exhibit improvement in growth of host plants such as increase in shoot length, shoot fresh and dry biomass, leaf area, chlorophyll content, and photosynthetic area when inoculated with the endophyte partner. The paper also highlights the role of endophytic GAs in overcoming salinity stress.

Cytokinins are a group of hormones involved in plant growth, development, and physiology. These are produced not only by plants but also by other prokaryotic and eukaryotic organisms, both beneficial and pathogenic ones. Akhtar, Mekureyaw, Pandey, and Roitsch (2020) discuss in detail the role of cytokinin in the interaction of plants with bacterial and fungal pathogens, insect–pests, and finally, the crosstalk of cytokinin with other plant hormones in aiding defense mechanisms of plants. Although cytokinin-producing endophytic members are moderately less in number, they play a significant role in improving the growth of plants under drought conditions (Ali et al., 2017).

Methylobacterium oryzae, a stress tolerant beneficial plant endophyte, was inoculated into lentils and exposed to drought conditions during early vegetative and reproductive phase. Inoculation improved the growth, physiological parameters and showed enhanced levels of cytokinins in drought stressed lentils (Jorge et al., 2019).

Increased levels of ethylene indicated conditions of plant stress; plants harboring endophytes that produce ACC deaminase are of great help in combating the stress conditions as it is an immediate precursor of ethylene (Glick, 2014). Inoculation with endophytes capable of producing ACC deaminase and persisting in the crop plants can be of immense application in combating stress condition (Liu et al., 2017). del Carmen Orozco-Mosqueda, Glick, and Santoyo (2020) review in detail the recent research progress on ACC deaminase produced by endophytes and rhizospheric bacteria, especially *Bacillus*, in improving plant growth under stress conditions. Jayakumar, Krishna, et al. (2020), Jayakumar, Nair, et al. (2020), and Jayakumar, Padmakumar, et al. (2020) isolated drought tolerant strains of bacterial endophytes from the xerophyte *Ananas comosus* and demonstrated the capability of these endophytes in IAA, ACC deaminase production and nitrogen fixation under drought conditions. The study also showed improvement in growth parameters of *Vigna radiata* when primed with the selected endophytes. Thus the synergistic effect of phytohormones produced by MBCAs helps plant enhance its growth parameters, physiological conditions, and defense system against phytopathogens.

7.2.4 Phosphate solubilization

Phosphate forms the second major nutrient required for the growth and development of plants. Hence, its unavailability in the soil can be considered a major limiting factor for plant growth and it demands the application of phosphate fertilizers. Endophytes are capable of secreting organic acids and acid phosphatases which can convert phosphate complexes into soluble orthophosphates that can be readily absorbed by the plants, a noticeable character that can be equated with that of PGPR. These organic acids chelate the cation attached to phosphate

(Kpomblekou-A & Tabatabai, 1994). In return, PGPR gain the root borne exudates necessary for its growth (Khan, Zaidi, Ahemad, Oves, & Wani, 2010).

Mehta, Sharma, Putatunda, and Walia (2019) conclude phosphate solubilizing endophytic fungi as more vigorous colonizers than nonendophytic microbes. The paper also discusses in detail the role of different endophytic fungi and mycorrhiza in mineral solubilization, their mode of action in growth promotion, and increasing crop productivity. Such endophytes capable of phosphate solubilization can be employed as live microbial biofertilizer, a better substitute for chemical fertilizer (Otieno et al., 2015).

7.2.5 Siderophore production

The endophytes are capable of secreting siderophores, which are small molecules capable of chelating Fe and are produced under iron limiting conditions by a majority of the anaerobic microorganisms (Khokhar, 2012). The potential of siderophores is the capability to chelate iron and make it available to the plants, thereby permitting the phytopathogens in the neighborhood to starve and deplete in number. This property can be exploited in the application of such endophytes as effective biocontrol agents. The level of iron in the soil, the form of available iron, pH, presence of other trace elements in the soil, and adequate supply of carbon, nitrogen, as well as phosphorous contribute to the production of siderophores by the endophytes (Duffy & Défago, 1999).

Siderophores produced by *Pseudomonas* were found to control *Fusarium oxysporum* that causes wilt diseases in potatoes (Schippers, Bakker, & Bakker, 1987). Maheshwari, Bhutani, and Suneja (2019) isolated siderophore-producing endophytes from both the roots and nodules of *Pisum sativum* and *Cicer arietinum*, which also showed growth-promoting activities. Siderophore-producing endophytic fungi were isolated from a medicinal orchid, *Cymbidium aloifolium*, by Chowdappa, Jagannath, Konappa, Udayashankar, and Jogaiah (2020). The solvent-extracted siderophores were found to be active against *Ralstonia solanacearum*, causing bacterial wilt in groundnutand *Xanthomonas oryzae*pv. *oryzae* that causes bacterial blight disease in rice. Interestingly, a modified microtiter plate was designed by Arora and Verma (2017) which allows simultaneous analysis of siderophore production by several bacteria both qualitatively and quantitatively.

7.2.6 ACC utilization

Various stress conditions such as salinity, drought, flooding, soil toxicity with heavy metals and organics, and various bacterial and fungal pathogens trigger the synthesis of ethylene in plants, thereby limiting its growth. The endophytes expressing ACC deaminase enzyme have been shown to protect plants from different abiotic and biotic stresses bytrapping the precursor of ethylene (ACC) and converting it to ammonia and 2-oxobutanol by the enzyme ACC deaminase. Ali et al. (2017) reviewed several studies that demonstrated the activity of ACC deaminase under stress conditions.

One such instance is that of *Brachybacterium paraconglomeratum*, an ACC deaminase–producing salt tolerant bacterial endophyte, isolated from the surface-sterilized roots of a medicinal plant, *Chlorophytum borivilianum*, that is shown to promote plant growth under stressful conditions. Biochemical analysis of these plants under salinity stress conditions showed the presence of ACC, proline, malondialdehyde (MDA), and abscisic acid (ABA) in higher amounts. Plants treated with *B. paraconglomeratum* showed decreased level of ACC, MDA, ABA and increase in total biomass, IAA, and chlorophyll content (Barnawal et al., 2016).

ACC deaminase is also reported to help efficient colonization of the associated endophytic microbes as they help to reduce the amount of ethylene within the plant (Hardoim et al., 2018).

7.2.7 Competition with pathogens

PGPR as well as endophytes can inhabit the host plants and exist in harmony with the host cells. These organisms compete with the organisms in the vicinity of the host that may be potential phytopathogens. Endophytes exhibit nutrient and colonizing site competition, stimulate the host plant defense, and produce antimicrobial agents to overcome the phytopathogen (Ting, Mah, & Tee, 2010). Trivedi, Leach, Tringe, Sa, and Singh (2020) reviewed the various works in association with plant–microbe interactions and its role in modulating growth properties of plants as well as plant health, by inhibiting phytopathogens.

7.2.8 Increased lignin biosynthesis

Lignin biosynthesis in response to attack by phytopathogens prevents the spread of pathogenic bacteria and reduces the infiltration of fungal toxins and enzymes into the plant cells (Pinski et al., 2019). Mejía et al. (2014) reported the stimulation of lignin biosynthesis by endophytes in the leaf tissues of *Theobroma cacao*. Inoculation with *Bacillus amyloliquefaciens* pb1 in roots of cotton showed a surge in lignin biosynthesis, which can be linked to enhanced expression of peroxidase enzymes (Irizarry & White, 2018).

7.2.9 Induction of plant resistance

Induced systemic resistance (ISR) is conferred by some beneficial microorganisms that enhance the resistance of the plant, thereby exhibiting various defense responses to wider group of pathogens. ISR is mediated by jasmonic acid or ethylene and does not end up in the production of pathogenesis-related (PR) proteins. Systemic acquired resistance, mediated by salicylic acid, is induced by the infection of certain pathogenic microorganisms leading to the production of PR proteins. PR proteins are responsible for lysing the invading cells and reinforce the cell wall boundaries, thus protecting the host plant from pathogenic attack (Fadiji et al., 2020).

7.2.10 Stimulation of plant secondary metabolite production

Plant secondary metabolites are of immense importance in the field of defense from phytopathogens, especially alkaloids. Endophytes are known to stimulate the production of secondary metabolites by the host plants; hence, coculturing can be adopted which can many folds increase the secretion of secondary metabolites by the plant cells. Several works by Abdelwahab et al. (2018), Ding, Wang, Guo, and Wang (2018), Li et al. (2019), and Xu et al. (2018) demonstrate the enhancement of bioactive secondary metabolite production on coculturing with endophytes.

7.2.11 Promoting plant growth and physiology

Since endophytes can easily inhabit and remain in harmony with the host plant, these can easily be inoculated and established into the host plant, thereby promoting the growth characters such as increased growth, better tolerance to stress, thus, capable of withstanding any soil conditions. Improved growth characters of host plants can be attributed to the phytohormones secreted by the endophytes, rendering the plants capable of tolerating a variety of biotic and abiotic stresses (Eid, Salim, Hassan, Ismail, & Fouda, 2019; Gao et al., 2010). Rho et al. (2018) analyzed the biomass accumulation of 42 host plant species from different literatures by 94 endophytic strains under three abiotic stress conditions: drought, nitrogen deficiency, and excessive salinity. Interestingly, increased biomass was observed in all the three stress conditions among the wide range of plant species with little evidence on plant–endophyte specificity.

7.2.12 Hyperparasitism and predation

Endophytes show hyperparasitism and predatory characters in nutrient-deficient conditions. Hyperparasitism is the condition where endophytes attack the hyphae of pathogens by penetrating and lysing them. *Trichoderma* forms an excellent case of mycoparasitism by producing a variety of enzymes that can digest fungal cell wall (Druzhinina et al., 2018; Gao et al., 2010). *Trichoderma* also possess a distinguishing property of parasitizing related fungi which distinguishes it from other mycoparasitic fungi (Druzhinina et al., 2018).

Thus it can be concluded that different microbes in a single host utilize different mechanisms for controlling phytopathogens. The combination of these mechanisms employed by the associated microorganisms makes them fruitful candidate for the development of effective MBCAs. The isolated microbiome can later be improved using various techniques so as to improve its efficiency, shelf life, stress tolerance capability, and mode of action.

7.3 Techniques for improvement of MBCAs

Numerous antagonistic microorganisms capable of acting as effective MBCAs in laboratory or controlled conditions are known. They form the key component of sustainable disease control measures and can be integrated into crop production. Many of them fail to perform well in field conditions due to many biotic and abiotic factors that adversely affect their performance. Various measures can be utilized for the improvement of one or many of the characters of the microbial agent or for improving the conditions of their growth, which can be discussed under the following headings.

7.3.1 Molecular methods for the improvement of microbial biocontrol agents

Genetic improvement of microorganisms to improve their biocontrol efficacy requires a clear understanding of the genes responsible for their biocontrol activity (Raguchander, Saravanakumar, & Balasubramanian, 2011). Research during the last years enabled the scientist to better understand the biochemical mechanism and the genes involved in the biocontrol activity. The biocontrol activity of microorganisms may not be dependent on a single gene, instead, it will be reliant on the regulation and interaction of many genes (Raguchander et al., 2011).

Strategies like targeted gene sequence, differential display techniques, and whole-genome sequencing can be employed for a better understanding of the mechanism of gene action. Since targeted gene sequences focus on only one or few genes, it demonstrates only a small fraction of genes involved in biocontrol properties (Raguchander et al., 2011). Studies on differential expression of genes under varied exposure conditions like the presence or absence of pathogen can reveal the overexpression or repression of novel genes. Further characterization and property studies of these genes can contribute to our understanding of the mechanism of gene action (Raguchander et al., 2011). The whole-genome sequencing of biocontrol agents will also be a promising strategy. The whole-genome sequencing of *Pseudomonas fluorescens* Pf-5 revealed that their genome size is 7.07 Mb of which 6% is for the synthesis of secondary metabolites like antibiotics and siderophores (Loper, Kobayashi, & Paulsen, 2007).

The genetic improvement of biocontrol agents for enhancing the biocontrol activity can be achieved through methods like protoplast fusion, physical and chemical mutation, and genetic recombination, which are discussed next.

7.3.1.1 Protoplast fusion

Protoplast fusion is comparatively an easy method in strain improvement. In the case of filamentous fungi, it is an effective tool for bringing genetic recombination and developing superior hybrid strains. It can be used to combine the advantages of two distinct strains like disease resistance, enhanced growth, nitrogen fixation, improved production of secondary metabolites, and enhanced resistance against stress conditions (Ram et al., 2018).

Lakhani, Vakharia, Makhlouf, Eissa, and Hassan (2016) conclude protoplast fusion as an effective tool for the production of superior hybrid strains of *Trichoderma* and for enhancing its antagonistic activity against a wide variety of soil borne pathogens. Electro-fusedcells of *Trichoderma harzianum* and *Trichoderma longibrachiatum* showed enhanced biocontrol efficacy and tolerance to fungicides copper sulfate and carbendazim (Mrinalini & Lalithakumari, 1998). The interspecific protoplast fusion between *T. harzianum* and *Trichoderma viride* produced fusants with enhanced extracellular enzyme production and biocontrol activity of *Trichoderma* spp. against grapevine pathogens, *Macrophomina phaseolina*, *Pythium ultimum*, and *Sclerotium rolfsii*. They showed higher production of β-1,3-glucanase, chitinase, and protease (Hassan, 2014).

Fusants that resulted from the protoplast fusion of *B. amyloliquefaciens* subsp. *plantarum* SA5 and *Lysinibacillus sphaericus* Amira strain were tested for their chitinase and nematicidal activities and were found to show better activity against the Root-knot Nematode *Meloidogyne incognita* (Abdel-Salam, Ameen, Soliman, Elkelany, & Asar, 2018). Self-fused *Streptomyces griseus* showed increased bioactivity against *F. oxysporum* f. sp. *lycopersici* and increased chitinase production (Anitha & Rebeeth, 2016). Gaziea, Shereen, Laila, and Eman (2020) attempted combining *Bacillus thuringiensis* I 977 and *Pseudomonas aeruginosa*, and the fusants exhibited increased growth parameters in grapevines and improved biocontrol against Root-knot Nematode *M. incognita*.

Protoplast fusion can also be used as a method for constructing new strains for bioremediation. Fusants obtained by the fusion of *T. viride* and *Trichoderma Koningii* were tested for their ability to remove Zn ion using FTIR (Fourier-transform infrared spectroscopy) technique and concluded that Zn tolerance was higher in fusants when compared with parental strains (Mazrou et al., 2020).

7.3.1.2 Genetic recombination

Several genes are involved in imparting the biocontrol property of endophytes which can be modified to improve the efficacy of biocontrol agents. Sarethy and Saharan (2021) focuses on the importance of omics-based technologies in the development and commercialization of MBCAs. Root colonization is a prerequisite for the successful establishment of endophytes on the host plant. Disruption of global transcription regulator AbrB gene, which negatively regulates root colonization and biofilm formation in *Bacillus amyloliquifaciens* SQR9, enhanced its colonization efficiency leading to its enhanced biocontrol efficacy against cucumber and watermelon wilt disease (Weng, Wang, Li, Shen, & Zhang, 2013).

Genetically engineered *Bacillus velezensis* SQR9 (formerly known as *B. amyloliquifaciens* SQR9) with xylose inducible degQ gene performs well in biofilm formation, antibiotic expression, colonization activity, and biocontrol activity (Xu et al., 2019).

Streptomyces lydicus A01 can produce natamycin that can bind the ergosterol of the fungal cell membrane and inhibits the growth of *Botrytis cinerea*, while *T. harzianum* P1 has high chitinase activity to decompose its cell wall. The conjugal transformant (CT) of *S. lydicus* A01 with the *chit42* gene from *T. harzianum* has improved biocontrol efficacy against B. cinerea, as it can produce both natamycin (antibiotic) and chitinase (Wu et al., 2013). Similarly, the CT of *S. lydicus* A01 with the *chit33* gene from *T. harzianum* has improved biocontrol efficacy against several plant fungal pathogens, including *Fusarium* spp. (Xu et al., 2019). *Clonostachys rosea* is a biocontrol fungus in which endochitinase-encoding gene *Chi67−1* showed an upregulation when induced by sclerotia and showed higher efficiency against *Sclerotinia* rot of soybean (Sun, Sun, Zhou, & Li, 2017).

Streptobacillus lydicus A01 wild type, which has no detectable glucanase activity on transforming with β-1,3-1,4-glucanase gene from *Paenibacillus polymyxa* had natamycin and chitinase as that of the wild type and high glucanase activity. The transformed strain showed substantially high biocontrol efficacy than the wild type (Li et al., 2015).

P. fluorescens strains HC1−07 and HC9−07 produce the cyclic lipopeptide (CLP) and phenazine-1-carboxylic acid (PCA), respectively. Phenazines are a group of colored secondary metabolite which is active against a wide range of plant pathogens. CLPs have antimicrobial, cytotoxic, and surfactant properties. The seven-gene operon for the synthesis of PCA from *Pseudomonas synxantha* 2−79 was introduced into *P. fluorescens* HC1−07 and the recombinant strain showed increased biocontrol activity against *Gaeumannomyces graminis* var. *tritici*, causal agent of take-all disease of wheat (Yang, Mavrodi, Mavrodi, Thomashow, & Weller, 2017).

The cell wall−degrading enzymes secreted by *Trichoderma* spp. as a tool in their defense against phytopathogen are predominantly glycosylated proteins. The key enzyme in their posttranslational *O*-glycosylation process is dolichyl phosphate mannose (DPM) synthase. The overexpression of DPM1 gene from *Saccharomyces cerevisiae* on *Trichoderma atroviride* showed doubled DPM synthase activity and elevated cellulolytic activity. The transformants showed improved antifungal activity against the plant pathogen *P. ultimum* (Zembek et al., 2011).

7.3.1.3 Mutagenesis

Random mutagenesis through chemical and physical mutagen is an important method of genetic manipulation strategy for improving the antagonistic activity of biocontrol agents. The method forms an excellent tool for the development of *Trichoderma* with capability of enhanced secretory enzymes (Singh, Maurya, & Upadhyay, 2016). Haggag and Mohamed (2002) demonstrated the enhanced antifungal metabolite production by three species of *Trichoderma* on exposure to gamma radiations, thus showing effective biocontrol against *Sclerotium cepivorum*, the causative organism of onion white rot disease. They concluded that gamma irradiation was capable of improving the activity of exoenzymes as well as increasing the production of antibiotics and phenolics in *Trichoderma*. Gamma-induced mutants of *T. harzianum* exhibited stronger growth inhibition and colonization against *M. phaseolina* and *Rhizoctonia solani* (Abbasi, Safaie, Shams-Bakhsh, & Shahbazi, 2016). Similarly gamma-induced mutants of *Trichoderma viridae* also showed improved biocontrol activity against soil-borne pathogen *F. oxysporum* f. sp. *ciceri*, *S. rolfsii*, and *Rhizoctonia bataticola* (Wagh, Ingle, Dandale, & Mane, 2015).

B. thuringiensis NM101−19 was subjected to various doses of gamma radiation so as to improve its chitinase activity (Gomaa & El-Mahdy, 2018). Similarly, the use of physical, chemical, and site-directed mutagenesis for genetically improving the chitinolytic activity of *Streptomyces griseorubens* E44G resulted in 1.39-fold increase on comparing with the wild type (Hafez et al., 2020). The use of in vitro mutagenesis along with genetic transformation using antifungal genes can produce both mutant and transgenic lines (Bermúdez-Caraballoso, Cruz-Martín, & Concepción-Hernández, 2020).

7.3.2 Combined application of MBCAs

The combined application of MBCAs, with different strategies of action, can improve the biocontrol efficacy mainly by increasing their consistency and stability in action. Compared with the use of a single MBCA, the combined use of MBCAs can have increased, decreased, or similar biocontrol efficiency depending on the type of interaction between them (Xu, Jeffries, Pautasso, & Jeger, 2011).

Xu et al. (2011) compared the different studies done on the combined use of BCAs to curb plant diseases and concluded that in such cases antagonistic interactions are more likely to occur than synergistic interactions. Many works using the combined application of MBCAs showing improvement in the desirable characteristics have been reported from time to time.

De Jaeger, De la Providencia, Rouhier, and Declerck (2011) formulated *T. harzianum* with *Glomus* sp. in alginate beads, which can be of application in sustainable agriculture. Improved efficiency and consistency in action against rice blast pathogen was observed when *T. harzianum* strain Th 3 and *P. fluorescens* strain RRb11 were applied in a combined manner, where the rice plants showed an improvement in growth parameters also (Jambhulkar et al., 2018). Experiments by Chemeltorit, Mutaqin, and Widodo (2017) demonstrated that combining *Trichoderma hamatum* THSW 13 and *P. aeruginosa* BJ10−86 showed the inhibition of *Phytophthora capsici* by 73.2% under in vitro conditions, which was similar to mefenoxam treatment. This suggests the possibility of replacing fungicide with biocontrol agents after field trials. Marian, Morita, Koyama, Suga, and Shimizu (2019) presented a similar effect when a combination of *Mitsuaria* sp. and nonpathogenic *Ralstonia* sp. improved the biocontrol efficacy against tomato bacterial wilt caused by *Ralstonia pseudosolanacearum*.

Another experiment validates that the efficacy of this approach was demonstrated in chick pea when a mixture of *B. subtilis* and *T. harzianum* could suppress the Fusarium wilt and improved the growth parameters (Zaim, Bekkar, & Belabid, 2018).

Arbuscular mycorrhizal fungi, entomopathogenic *Pseudomonas* bacteria, and entomopathogenic nematodes were combined and found to have biocontrol activity against *Diabrotica virgifera*, the western corn root worm (Jaffuel et al., 2019).

Allaga et al. (2020) combined beneficial bacteria and fungi to form a bioinoculant for the biocontrol of plant pathogens, phosphorous solubilization, stem degradation, humification, and nitrogen fixation. *Trichoderma asperellum* (biocontrol component), *Trichoderma atrobrunneum* (phosphorous solubilization and production of cellulose degrading enzymes), *Streptomyces albus* (humus producing component), and *Azotobacter vinelandii* (nitrogen fixation) were used for the preparation of bioinoculant and were proved to be with immense application potential.

7.3.3 Enhancing stress tolerance capability of MBCAs

The successful application of a bioformulation to the field depends on the capability of MBCAs in overcoming the biotic and abiotic stress conditions prevalent there. When microorganisms are exposed to mild stress, induction of general and specific proteins, as well as physiological changes to withstand lethal challenges of the same stress conditions, is activated (Ang et al., 1991). Cross protection is another mechanism where adaptation to one stress can render the cells resistant to other stresses.

Screening and selection of pertinent stress-tolerant strains will be of much utility in the establishment of the same in the field conditions. Poosapati, Ravulapalli, Tippirishetty, Vishwanathaswamy, and Chunduri (2014) isolated thermotolerant strains of *Trichoderma* from various agroclimatic zones of India. From 250 isolates of *Trichoderma*, *Trichoderma asperillum* TaDOR673 was able to tolerate the highest temperature of 52°C for a longer duration and showed an efficient biocontrol activity of 79.7% against collar rot disease of groundnut caused by *S. rolfsii*; other isolates, namely, *T. harzianum* TaDOR671, *T. asperillum* TaDOR79, *T. asperillum* TaDOR293, and *T. asperillum* TaDOR79, were also able to survive in high temperature. On exposure to NaCl concentrations, these strains were able to survive in higher saline concentrations. Praveen Kumar, Mir Hassan Ahmed, Desai, Leo Daniel Amalraj, and Rasul (2014) attempted the screening and isolation of abiotic stress-tolerant and plant growth–promoting strains of *Pseudomonas* and *Bacillus* spp. with improved biocontrol activity.

The microbiota of plants from hostile environment can be of immense help in alleviating abiotic stress in crop plants. A large group of culturable endophytes were isolated from the roots of halophyte *Limoniastrum monopetalum* and found to have improved biocontrol activity against phytopathogens and displayed protection of crop plants against abiotic stresses (Slama et al., 2019).

Stress adaptation is another technique adopted to improve the stress tolerance of MBCAs. Exposing the biocontrol agents to mild shock can increase their tolerance on subsequent exposure to stress conditions. Daranas, Badosa, Francés, Montesinos, and Bonaterra (2018) subjected *Lactobacillus plantarum* strains PM411 and TC 92 to hyperosmotic and acidic conditions during growth. The combined stress-adapted cells of *L. plantarum* strain PM411 cells showed improved biocontrol activity with more consistent action on different plant hosts in the greenhouse and under field conditions. Mild heat-shock treatment of biocontrol yeast, *Rhodotorula mucilaginosa*, not only increased its tolerance to high temperature, oxidative stress, salt stress, and low pH but also exhibited improved efficacy against *Penicillium expansum* infection of apples (Cheng et al., 2016). Pretreating *Candida oleophila* with mild salt stress (1 M NaCl for 30 minutes) increased its tolerance to stress by 6 M NaCl and low pH; pretreated cells showed more efficient biocontrol efficiency against the infection of *P. expansum* and *B. cinerea* in kiwifruits (Wang et al., 2018). Sui and Liu (2014) established the increased thermotolerance and biocontrol efficacy of antagonistic yeast *Pichia guilliermondii* on treatment with 10% (w/v) glucose.

Another interesting mechanism of enhancing stress tolerance is the shift in the morphology of MBCAs in response to stress conditions. *Pichia kudriavzevii*, antagonistic yeast used as a postharvest biocontrol agent, reversibly shifted from yeast-like morphology to biofilm morphology, which increased the oxidative stress tolerance, biocontrol efficacy, and antioxidant activity (Chi et al., 2015). Li et al. (2016) presented the shift from single-cell morphology to pseudohyphal morphology by changing agar concentrations in *Candida diversa*, resulting in increased tolerance to heat and oxidative stress and improved biocontrol efficacy against *B. cinerea*.

Recently, Kucuk and Gezer (2020) studied the stress tolerance of mutant isolates of *Trichoderma* species produced by exposure to UV (Ultraviolet) radiation and demonstrated the tolerance of some of these mutant strains to abiotic factors like high-temperature drought and saline conditions, which suggests their possible application as biocontrol agents.

7.3.4 Addition of organic amendments

Organic amendments are exogenously added materials that include crop residues, organic wastes, peat, compost, and biochar that can indirectly clampdown the phytopathogens. These provide the soil microbiota with varied food base which modifies the microbial community of the soil (Bonanomi, Lorito, Vinale, & Woo, 2018) and their interactions may influence the expression of biological control traits, thus exhibiting antagonistic effect against soil-borne pathogens (Nelson, 2004).

Hartmann, Frey, Mayer, Mäder, and Widmer (2015) demonstrated that the application and quality of organic components added to the soil as the focal points of soil microflora diversity. Microbial biodiversity of the soil, decomposition of organic matter, and plant growth were integrated and analyzed to conclude that eukarya played a major role compared to bacterial members in organic carbon recycling and supporting plant growth (Bonanomi et al., 2016). Bonanomi et al. (2018) had reviewed in detail the various feeding preferences of organic amendments by soil microbes and plants. This can be of prodigious assistance in the designing of well-characterized organic amendments that can be integrated with MBCAs to address the problem of soil-borne pathogens.

The addition of organic amendments can steer and strengthen complex microbe–eukaryote association (Suleiman et al., 2019). Various studies can be plotted to illustrate the effect of organic amendments on the improvement of beneficial characters of MBCAs. Yu, Wang, Yin, Wang, and Zheng (2008) reported the enhancement in antagonistic activity of *Cryptococcus laurentii* against the postharvest blue mold rot caused by *P. expansum* in pear fruit. Kang (2011) demonstrated the improvement in biocontrol activity of *Pseudomonas* sp. NJ134 against *F. oxysporum*, the causative organism of Fusarium wilt after the amendment of culture with mannitol. Enhanced disease suppression was observed in *Phytophthora* blight of pepper by *Pseudomonas* sp. when used in combination with olive oil (Özyilmaz & Benlioglu, 2013).

Wu et al. (2015) demonstrated that the addition of pectin as organic amendment decreased the occurrence of tobacco bacterial wilt by inducing colonization and antibiotic secretion by *B. amyloliquefaciens* SQY 162 in the rhizosphere of tobacco. Moreover, pectin amendment was found to increase phosphorous solubilizing bacteria in soil. Marine actinomycetes supplemented with chitin offered effective biocontrol against *R. solani* under greenhouse (Yandigeri, Malviya, Solanki, Shrivastava, & Sivakumar, 2015). Gramisci, Lutz, Lopes, and Sangorrín (2018) demonstrated the improved antagonistic activity of *Pichia membranifaciens* and *Vishniacozymavictoriae* against the postharvest disease of pear fruits by the addition of several amino acids and $CaCl_2$. When exposed to 5% or 10% (w/v) maltose or lactose, biocontrol yeast *C. oleophila* exhibited enhanced enzyme activity, viability, and ATP levels, thus improving their thermotolerance and biocontrol efficiency (Zheng et al., 2019).

Thus integration of properly characterized and selected organic amendments and MBCAs can address the problem of phytopathogens sustainably.

The following session deals with the various formulation procedures, which is of prime importance in the delivery of MBCAs to the field as well as its commercialization.

7.4 Formulation procedure

The formulation of MBCAs forms a decisive step concerning its application and commercialization. Additionally, it affects the performance of the MBCAs and improves the shelf life, thus contributing to their overall efficacy as a substitute for chemical pesticides (Fravel, 2005). The majority of the formulation procedures are company secrets and confidential. Many pieces of literature are available regarding the optimization of these procedures (Marian & Shimizu, 2019).

The majority of the MBCAs are commercially presented in the form of wettable powders, liquids, or granular formulations; wettable powder form being the most widely used due to its numerous advantages like easy handling, lower cost of storage and transportation, and lower risk of contamination (Marian & Shimizu, 2019). Optimization of formulation procedure must be done to improve the general growth conditions of the microbe and preconditioning steps must be followed so that the MBCAs are made capable of enduring the formulation events, dehydration, and storage conditions (Melin, Schnürer, & Håkansson, 2011).

In general, a formulation must stabilize the microorganism during all the stages of its production and application, protect the organism from harmful environmental conditions, must aid in the application of the product, and enhance the activity of the organism at the application site. Dry, powdered, or moist biomass can be formulated into dust, pellets, gels supplemented with inert carriers or food bases (Lewis, 1991). The choice of formulation procedure and stabilization technique varies according to the envisioned application. In this section, we shall discuss the various formulation procedures opted for the preparation of MBCAs.

7.4.1 Drying methods

Drying is one of the earliest methods adopted for the preparation of microbial bioformulations. A major shortcoming of this method is the humidity of the environment that negatively affects the viability of the microorganism. Storage under vacuum is adopted as an alternative method to this. Treating them with an osmoprotectant before drying, proper optimization of the drying process, and the use of additives or nutrients help in overcoming the disadvantages of drying as well as increase the efficiency of the process without compromising on the viability (Fages, 1992; Umashankar, Chandralekha, Dandavate, Tavanandi, & Raghavarao, 2019). The following are some of the drying methods commonly adopted:

1. Spray drying and freez–drying

   A predetermined quantity of both the microbial cells and carrier material is dissolved in distilled water to obtain a slurry before the drying process. In freeze-drying the slurry was frozen at −20°C followed by primary drying at −51°C under the pressure of <12 kPa and secondary drying at 26°C, taking almost 20 hours to complete the whole process. For spray drying the slurry was continuously stirred and fed into the sprayer to obtain uniform spray-dried powder samples (Umashankar et al., 2019).
2. Fluidized bed drying

   Carbó, Torres, Usall, Fons, and Teixidó (2017) report the fluidized bed drying technique using *Candidasake* CPA-1 cells, which is effective against postharvest diseases of fruits. Cells obtained after centrifugation were mixed with a carrier to form a dough which was introduced into the tubes of a fluidized bed dryer, subsequently, the dry extruded particles were rehydrated using phosphate buffer.

3. Low-temperature low humidity drying (LTLHD)
   Umashankar et al. (2019) briefly described the technique of LTLHD by using cultures of *P. aeruginosa* and carrier materials (whey protein, corn starch, and trehalose). The predetermined liquid suspension of *P. aeruginosa* cells was mixed with carrier materials to form a dough which was fed into a granulator to obtain uniformly sized pellets. These granules were fed into fluidized bed dryer equipped with a dehumidifier and air recirculation system. The drying was carried out in batch mode, maintaining temperature (50°C ± 2°C) and relative humidity (15°C ± 1°C) to obtain powdered samples (Umashankar et al., 2019). Compared with the conventional drying techniques, LTLHD offers to dry "heat-sensitive microorganisms" as well as reduces the time taken for drying (Marian & Shimizu, 2019).
4. Fluid bed spray drying (FBSD)
   FBSD is a recent technology developed as a better substitute for freeze–drying, spray drying, or fluidized bed drying. Gotor Vila (2017) described in detail the technique of FBSD and compared its performance with traditional drying methods. In the previous experiment, cultured cells of *B. amyloliquefaciens* CPA-8 suspensions were sprayed using a peristaltic pump, into a fluid bed spray dryer loaded with powdered carrier material, to obtain FBSD products.

7.4.2 Encapsulation methods

Formulation of MBCA by encapsulation method uses entrapping of potential microorganisms that can act as biocontrol agents within a system that is capable of improving the competence of MBCA by shielding them from biotic and abiotic stresses and maintaining a beneficial microenvironment or a micro-niche. This results in extended shelf life and extended periods of activity with the reduced number of field applications (Cassidy, Lee, & Trevors, 1996; Vemmer & Patel, 2013). Moreover, the nature of encapsulating material allows the slow release of microorganisms into the field by its degradation, guaranteeing sufficient time for its establishment. Paula, Carolino, Paula, and Samuels (2011) reported the addition of low levels of low-dose pesticides along with the biocontrol agent, thus improving the overall performance of biocontrol agents and reducing the environmental risk of the pesticide. Vassilev et al. (2020) proposed replacing expensive polymers with low-cost natural additives and underlined the beneficial effects of these carrier additives on the microorganisms.

Various encapsulation procedures are followed for the preparation of MBCAs of which ionic gelation using alginate is being extensively used (McLoughlin, 1994). A detailed review of the various encapsulation methods for MBCAs and the polymers used as encapsulating agents has been done by Vemmer and Patel (2013). A brief account of the various encapsulation method is given next.

1. A warm polymer solution with living cells is allowed to drip into a cold collecting or solidifying solution. Polymers such as carrageenan, gelatin, agar, or agarose can be used as they form a gel when temperature decreases. Droplet emulsification using thermal gelation, ionic gelation, and interfacial emulsion polymerization yields smaller particles with a higher size distribution compared to dripping methods (Vemmer & Patel, 2013).
2. Bead formation occurs when a solution of polyuronic acids such as sodium alginate or sodium pectinate with microbial bioactive agents is dipped into a solution of cross-linking agents containing divalent cations. The formulation of biocontrol agents using alginate has been successfully demonstrated for weed-killing fungi such as *Alternaria*, *Fusarium*, and *Phyllosticta* spp.; antagonistic fungi such as *Trichoderma*, *Gliocladium*, *Talaromyces*, and *Penicillium* spp.; bacteria such as *Pseudomonas* and *Bacillus* spp.; and insect-killing nematodes such as *Steinernema* and *Heterorhabditis* spp. by Connick (1988).

   Hollow beads can be formed when the cross-linking solution is supplemented with a viscosity-enhancing polymer (e.g., carboxy methyl cellulose) and then dipped into alginate solution (Vemmer & Patel, 2013).
3. Spray drying involves dispersing a polymer solution into a hot air stream. This method was confirmed by Tamez-Guerra et al. (1996) for the preparation of *B. thuringiensis* formulations and by Muñoz-Celaya et al. (2012) for *T. harzianum* conidia.
4. The complex coacervation method embraces dipping a solution of polyelectrolyte holding living cells into another solution of polyelectrolyte with counter charge. This results in the formation of a slightly hydrated colloid.
5. Polyvinyl alcohol with living cells or enzymes and glycerol is extruded onto a surface to produce lens-shaped droplets (LentiKats), which is another recent procedure (Schlieker & Vorlop, 2006).
6. Sol–gel technology provides a new possibility for the development of polymers using inorganic materials such as glasses or ceramics. The bioapplications of sol–gel technology were reviewed in detail by Avnir, Coradin, Lev, and Livage (2006).
7. Vemmer and Patel (2013) designed two coating techniques that utilize polyelectrolytes with altering charges for the preparation of bioformulations. Ionic polymer coating involves the suspension of polyanionic beads in polycationic solution, which can ultimately give rise to alternate multiple coatings. Layer-by-layer coating, used in nanotechnology, forms multilayer capsules by repeated exposure of colloids to polyelectrolytes of alternating charges.

Combination of one or more of the previous methods is adopted for the formulation of biocontrol agents. López-Cruz, Ragazzo-Sánchez, and Calderón-Santoyo

(2020) formulated a hydratable powder of microencapsulated *Meyerozyma guilliermondii* produced by spray drying to control anthracnose disease on mango fruit. The formulation utilized sodium alginate and soy protein isolate as encapsulating agents.

Metarhizium brunneum strain BIPESCO5 mycelium was encapsulated in calcium alginate, and an enhanced drying survival of 31.5% was observed. Also, the encapsulated ones showed 3.8- to 7-fold increase in endophytism when compared to nonencapsulated ones (Krell, Jakobs-Schoenwandt, Vidal, & Patel, 2018). Humbert, Przyklenk, Vemmer, and Patel (2017) compared the effect of calcium chloride and calcium gluconate while encapsulating the biocontrol fungi *M. brunneum* and *S. cerevisiae* and demonstrated that application of calcium gluconate as crosslinker will enhance drying survival and shelf life.

The selection of appropriate formulation procedure depends on the microbial population used, conditions for its optimal activity, and the one that does not compromise on the biocontrol efficacy, microbial population, and storage stability is normally selected (Wong, Saidi, Vadamalai, Teh, & Zulperi, 2019).

7.5 Future prospects

Agriculture forms an important sector of developing as well as developed countries as it satisfies the basic requirement of our life. The widespread adoption of green revolution technologies has cemented the way for satisfying the increasing demand for food by increasing population; but the extensive use of chemical fertilizers, pesticides, and insecticides has posted a query on the health of our biosphere. Adoption of sustainable agriculture methods by eliminating the use of agrochemicals has become the need of the hour. The development and use of MBCAs will be a promising, sustainable, and greener alternative to these chemical additives.

The success of MBCAs depends on the isolation and culturing of plant-associated microbiota, identifying the possible role in the life cycle of their host, evaluating the compatibility of successful MBCAs with other crop plants and other microorganisms, designing appropriate formulation recipes, shelf-life analysis, and finally providing the technical support for the commercialization and application of these formulations. A major shortcoming of MBCAs is the poor performance in the field and inconsistency in their action which can be contributed to the antagonistic activity of existing microbes, adaptability issues, and other incompatibility issues.

Recent studies have unraveled the ability of endophytes to synthesize nanoparticles that has wide range of biomedical applications. Green synthesis of nanoparticles is advantageous as it is eco-friendly and enjoys the ability to tolerate and bioaccumulate the metal of interest. Thus discovering, developing, and commercializing MBCAs that can be integrated into crop production require an in-depth understanding of the plant–microbe interactions. The current approaches like "microbiome engineering" and breeding of "microbe optimized crops" offer great hope in the development of a sustainable method to address the increasing food demand.

References

Abbasi, S., Safaie, N., Shams-Bakhsh, M., & Shahbazi, S. (2016). Biocontrol activities of gamma induced mutants of *Trichoderma harzianum* against some soilborne fungal pathogens and their DNA fingerprinting. *Iranian Journal of Biotechnology*, *14*(4), 260.

Abdel-Salam, M. S., Ameen, H. H., Soliman, G. M., Elkelany, U. S., & Asar, A. M. (2018). Improving the nematicidal potential of *Bacillus amyloliquefaciens* and *Lysinibacillus sphaericus* against the root-knot nematode *Meloidogyne incognita* using protoplast fusion technique. *Egyptian Journal of Biological Pest Control*, *28*(1), 1–6.

Abdelwahab, M. F., Kurtán, T., Mándi, A., Müller, W. E., Fouad, M. A., Kamel, M. S., ... Proksch, P. (2018). Induced secondary metabolites from the endophytic fungus *Aspergillus versicolor* through bacterial co-culture and OSMAC approaches. *Tetrahedron Letters*, *59*(27), 2647–2652.

Akhtar, S. S., Mekureyaw, M. F., Pandey, C., & Roitsch, T. (2020). Role of cytokinins for interactions of plants with microbial pathogens and pest insects. *Frontiers in Plant Science*, *10*, 1777.

Ali, S., Charles, T. C., & Glick, B. R. (2017). *Endophytic phytohormones and their role in plant growth promotion. Functional importance of the plant microbiome* (pp. 89–105). Cham: Springer.

Allaga, H., Bóka, B., Poór, P., Nagy, V. D., Szűcs, A., Stankovics, I., ... Körmöczi, P. (2020). A composite bioinoculant based on the combined application of beneficial bacteria and fungi. *Agronomy*, *10*(2), 220.

Ang, D., Ziegelhoffer, T., Maddock, A., Zeilstra-Ryalls, J., Georgopoulos, C., Fayet, O., ... Zylicz, M. (1991). *The biological role of the universally conserved* E. coli *heat shock proteins. Heat shock* (pp. 45–53). Springer.

Anitha, A., & Rebeeth, M. (2016). Self-fusion of streptomyces griseus enhances chitinase production and biocontrol activity against *Fusarium oxysporum* F. Sp. Lycopersici. *Biosciences Biotechnology Research Asia*, *6*(1), 175–180.

Arora, N. K., Khare, E., & Maheshwari, D. K. (2010). Plant growth promoting rhizobacteria: Constraints in bioformulations, commercialization and future strategies. In D. K. Maheshwari (Ed.), *Plant growth and health promoting bacteria* (pp. 97–116). Berlin: Springer-Verlag.

Arora, N. K., & Verma, M. (2017). Modified microplate method for rapid and efficient estimation of siderophore produced by bacteria. *3 Biotech*, *7*(6), 1–9.

Avnir, D., Coradin, T., Lev, O., & Livage, J. (2006). Recent bio-applications of sol–gel materials. *Journal of Materials Chemistry*, *16*(11), 1013–1030.

Barnawal, D., Bharti, N., Tripathi, A., Pandey, S. S., Chanotiya, C. S., & Kalra, A. (2016). ACC-deaminase-producing endophyte *Brachybacterium paraconglomeratum* strain SMR20 ameliorates Chlorophytum salinity stress via altering phytohormone generation. *Journal of Plant Growth Regulation*, *35*(2), 553–564.

Berdy, J. C. (2005). Bioactive microbial metabolites. *The Journal of Antibiotics*, *58*(1), 1–26.

Bermúdez-Caraballoso, I., Cruz-Martín, M., & Concepción-Hernández, M. (2020). *Biotechnological tools for the development of Foc TR4-resistant or-tolerant Musa spp. Cultivars. Agricultural, forestry and bioindustry biotechnology and biodiscovery* (pp. 403–431). Cham: Springer.

Bonanomi, G., De Filippis, F., Cesarano, G., La Storia, A., Ercolini, D., & Scala, F. (2016). Organic farming induces changes in soil microbiota that affect agro-ecosystem functions. *Soil Biology and Biochemistry*, *103*, 327–336.

Bonanomi, G., Lorito, M., Vinale, F., & Woo, S. L. (2018). Organic amendments, beneficial microbes, and soil microbiota: Toward a unified framework for disease suppression. *Annual Review of Phytopathology*, *56*, 1–20.

Carbó, A., Torres, R., Usall, J., Fons, E., & Teixidó, N. (2017). Dry formulations of the biocontrol agent Candida sake CPA-1 using fluidised bed drying to control the main postharvest diseases on fruits. *Journal of the Science of Food and Agriculture*, *97*(11), 3691–3698.

Card, S., Johnson, L., Teasdale, S., & Caradus, J. (2016). Deciphering endophyte behaviour: The link between endophyte biology and efficacious biological control agents. *FEMS Microbiology Ecology*, *92*(8), 1–19. Available from https://doi.org/10.1093/femsec/fiw114.

Cassidy, M. B., Lee, H., & Trevors, J. T. (1996). Environmental applications of immobilized microbial cells: A review. *Journal of Industrial Microbiology and Biotechnology*, *16*(2), 79–101.

Chemeltorit, P. P., Mutaqin, K. H., & Widodo, W. (2017). Combining *Trichoderma hamatum* THSW13 and *Pseudomonas aeruginosa* BJ10–86: A synergistic chili pepper seed treatment for *Phytophthora capsici* infested soil. *European Journal of Plant Pathology*, *147*(1), 157–166.

Cheng, Z., Chi, M., Li, G., Chen, H., Sui, Y., Sun, H., ... Liu, J. (2016). Heat shock improves stress tolerance and biocontrol performance of *Rhodotorula mucilaginosa*. *Biological Control*, *95*, 49–56.

Chi, M., Li, G., Liu, Y., Liu, G., Li, M., Zhang, X., ... Liu, J. (2015). Increase in antioxidant enzyme activity, stress tolerance and biocontrol efficacy of *Pichia kudriavzevii* with the transition from a yeast-like to biofilm morphology. *Biological Control*, *90*, 113–119.

Chowdappa, S., Jagannath, S., Konappa, N., Udayashankar, A. C., & Jogaiah, S. (2020). Detection and characterization of antibacterial siderophores secreted by endophytic fungi from *Cymbidium aloifolium*. *Biomolecules*, *10*(10), 1412.

Connick, W. J., Jr (1988). *Formulation of living biological control agents with alginate. Pesticide formulations, ACS symposium series* (pp. 241–250). American Chemical Society.

Coutinho, B. G., Licastro, D., Mendonça-Previato, L., Cámara, M., & Venturi, V. (2015). Plant-influenced gene expression in the rice endophyte *Burkholderia kururiensis* M130. *Molecular Plant-Microbe Interactions*, *28*(1), 10–21.

Daranas, N., Badosa, E., Francés, J., Montesinos, E., & Bonaterra, A. (2018). Enhancing water stress tolerance improves fitness in biological control strains of *Lactobacillus plantarum* in plant environments. *PLoS One*, *13*(1), e0190931.

De Jaeger, N., De la Providencia, I. E., Rouhier, H., & Declerck, S. (2011). Co-entrapment of *Trichoderma harzianum* and *Glomus* sp. within alginate beads: Impact on the arbuscular mycorrhizal fungi life cycle. *Journal of Applied Microbiology*, *111*(1), 125–135.

del Carmen Orozco-Mosqueda, M., Glick, B. R., & Santoyo, G. (2020). ACC deaminase in plant growth-promoting bacteria (PGPB): An efficient mechanism to counter salt stress in crops. *Microbiological Research*, *235*, 126439.

Devi, S. I., & Momota, P. (2015). *Plant-endophyte interaction and its unrelenting contribution towards plant health. Plant microbes symbiosis: Applied facets* (pp. 147–162). New Delhi: Springer.

Ding, C. H., Wang, Q. B., Guo, S., & Wang, Z. Y. (2018). The improvement of bioactive secondary metabolites accumulation in Rumex gmelini Turcz through co-culture with endophytic fungi. *Brazilian Journal of Microbiology*, *49*(2), 362–369.

Druzhinina, I. S., Chenthamara, K., Zhang, J., Atanasova, L., Yang, D., Miao, Y., ... Kubicek, C. P. (2018). Massive lateral transfer of genes encoding plant cell wall-degrading enzymes to the mycoparasitic fungus Trichoderma from its plant-associated hosts. *PLoS Genetics*, *14*(4), e1007322.

Duffy, B. K., & Défago, G. (1999). Environmental factors modulating antibiotic and siderophore biosynthesis by *Pseudomonas fluorescens* biocontrol strains. *Applied and Environmental Microbiology*, *65*(6), 2429–2438.

Eid, A. M., Salim, S. S., Hassan, S. E. D., Ismail, M. A., & Fouda, A. (2019). *Role of endophytes in plant health and abiotic stress management. Microbiome in plant health and disease* (pp. 119–144). Singapore: Springer.

Fadiji, A. E., & Babalola, O. O. (2020). Elucidating mechanisms of endophytes used in plant protection and other bioactivities with multifunctional prospects. *Frontiers in Bioengineering and Biotechnology*, *8*, 467.

Fages, J. (1992). An industrial view of *Azospirillum* inoculants: Formulation and application technology. *Symbiosis*, *13*, 15–26.

Fravel, D. R. (2005). Commercialization and implementation of biocontrol. *Annual Review of Phytopathology*, *43*, 337–359.

Gao, F. K., Dai, C. C., & Liu, X. Z. (2010). Mechanisms of fungal endophytes in plant protection against pathogens. *African Journal of Microbiology Research*, *4*(13), 1346–1351.

Gaziea, S. M., Shereen, M. A., Laila, H. F., & Eman, E. H. S. (2020). Efficiency of biological control of root-knot nematodes in infected grapevines seedling by genetic improved bacteria. *Plant Archives*, *20*(1), 951–961.

Glick, B. R. (2014). Bacteria with ACC deaminase can promote plant growth and help to feed the world. *Microbiological Research*, *169*(1), 30–39.

Gomaa, E. Z., & El-Mahdy, O. M. (2018). Improvement of chitinase production by *Bacillus thuringiensis* NM101–19 for antifungal biocontrol through physical mutation. *Microbiology (Reading, England)*, *87*(4), 472–485.

Gotor Vila, A. M. (2017). *New advances in the control of brown rot in stone fruit using the biocontrol agent Bacillus amyloliquefaciens CPA-8* (Doctoral dissertation). Universitat de Lleida.

Gramisci, B. R., Lutz, M. C., Lopes, C. A., & Sangorrín, M. P. (2018). Enhancing the efficacy of yeast biocontrol agents against postharvest pathogens through nutrient profiling and the use of other additives. *Biological Control*, *121*, 151–158.

Hafez, E. E., Rashad, Y. M., Abdulkhair, W. M., Al-Askar, A. A., Ghoneem, K. M., Baka, Z. A., & Shabana, Y. M. (2020). Improving the chitinolytic activity of *Streptomyces griseorubens* E44G by mutagenesis. *Journal of Microbiology, Biotechnology and Food Sciences*, *10*(1), 1156–1160.

Haggag, W. M., & Mohamed, H. A. A. (2002). Enhancement of antifungal metabolite production from gamma-ray induced mutants of some *Trichoderma* species for control onion white disease. *Plant Pathology Bulletin*, *11*, 45–56.

Hardoim, P. R., van Overbeek, L. S., & van Elsas, J. D. (2008). Properties of bacterial endophytes and their proposed role in plant growth. *Trends in Microbiology*, *16*(10), 463–471.

Hartmann, M., Frey, B., Mayer, J., Mäder, P., & Widmer, F. (2015). Distinct soil microbial diversity under long-term organic and conventional farming. *ISME Journal*, *9*(5), 1177–1194.

Hassan, M. M. (2014). Influence of protoplast fusion between two *Trichoderma* spp. on extracellular enzymes production and antagonistic activity. *Biotechnology & Biotechnological Equipment*, *28*(6), 1014–1023.

Humbert, P., Przyklenk, M., Vemmer, M., & Patel, A. V. (2017). Calcium gluconate as cross-linker improves survival and shelf life of encapsulated and dried *Metarhizium brunneum* and *Saccharomyces cerevisiae* for the application as biological control agents. *Journal of Microencapsulation*, *34*(1), 47–56.

Irizarry, I., & White, J. F. (2018). *Bacillus amyloliquefaciens* alters gene expression, ROS production and lignin synthesis in cotton seedling roots. *Journal of Applied Microbiology*, *124*(6), 1589–1603.

Jaffuel, G., Imperiali, N., Shelby, K., Campos-Herrera, R., Geisert, R., Maurhofer, M., ... Hibbard, B. E. (2019). Protecting maize from rootworm damage with the combined application of arbuscular mycorrhizal fungi, Pseudomonas bacteria and entomopathogenic nematodes. *Scientific Reports*, *9*(1), 1–12.

Jambhulkar, P. P., Sharma, P., Manokaran, R., Lakshman, D. K., Rokadia, P., & Jambhulkar, N. (2018). Assessing synergism of combined applications of *Trichoderma harzianum* and *Pseudomonas fluorescens* to control blast and bacterial leaf blight of rice. *European Journal of Plant Pathology*, *152*(3), 747–757.

Jayakumar, A., Krishna, A., Mohan, M., Nair, I. C., & Radhakrishnan, E. K. (2019). Plant growth enhancement, disease resistance, and elemental modulatory effects of plant probiotic endophytic *Bacillus* sp. Fcl1. *Probiotics and Antimicrobial Proteins*, *11*(2), 526–534.

Jayakumar, A., Krishna, A., Nair, I. C., & Radhakrishnan, E. K. (2020). Drought-tolerant and plant growth-promoting endophytic *Staphylococcus* sp. having synergistic effect with silicate supplementation. *Archives of Microbiology*, *202*, 1899–1906.

Jayakumar, A., Nair, I. C., & Radhakrishnan, E. K. (2020). Environmental adaptations of an extremely plant beneficial *Bacillus subtilis* Dcl1 identified through the genomic and metabolomic analysis. *Microbial Ecology*, *81*, 1–16.

Jayakumar, A., Padmakumar, P., Nair, I. C., & Radhakrishnan, E. K. (2020). Drought tolerant bacterial endophytes with potential plant probiotic effects from *Ananas comosus*. *Biologia (Lahore, Pakistan)*, *75*, 1769–1778.

Jorge, G. L., Kisiala, A., Morrison, E., Aoki, M., Nogueira, A. P. O., & Emery, R. N. (2019). Endosymbiotic *Methylobacterium oryzae* mitigates the impact of limited water availability in lentil (*Lens culinaris* Medik.) by increasing plant cytokinin levels. *Environmental and Experimental Botany*, *162*, 525–540.

Kang, B. R. (2011). Mannitol amendment as a carbon source in a bean-based formulation enhances biocontrol efficacy of a 2,4-diacetylphloroglucinol-producing *Pseudomonas* sp. nj134 against tomato Fusarium wilt. *The Plant Pathology Journal*, *27*(4), 390–395.

Kerff, F., Amoroso, A., Herman, R., Sauvage, E., Petrella, S., Filée, P., ... Cosgrove, D. J. (2008). Crystal structure and activity of *Bacillus subtilis* YoaJ (EXLX1), a bacterial expansin that promotes root colonization. *Proceedings of the National Academy of Sciences of the United States of America, 105*(44), 16876–16881.

Khan, A. L., Shahzad, R., Al-Harrasi, A., & Lee, I. J. (2017). *Endophytic microbes: A resource for producing extracellular enzymes. Endophytes: Crop productivity and protection* (pp. 95–110). Cham: Springer.

Khan, M. S., Zaidi, A., Ahemad, M., Oves, M., & Wani, P. A. (2010). Plant growth promotion by phosphate solubilizing fungi—current perspective. *Archives of Agronomy and Soil Science, 56*(1), 73–98.

Khokhar, M. (2012). Biological control of plant pathogens using biotechnological aspects: A review. *Journal of Plant Pathology & Microbiology, 03*(05). Available from https://doi.org/10.4172/scientificreports.277.

Kpomblekou-A, K., & Tabatabai, M. A. (1994). Effect of organic acids on release of phosphorus from phosphate rocks1. *Soil Science, 158*(6), 442–453.

Krell, V., Jakobs-Schoenwandt, D., Vidal, S., & Patel, A. V. (2018). Encapsulation of *Metarhizium brunneum* enhances endophytism in tomato plants. *Biological control, 116*, 62–73.

Kucuk, C., & Gezer, T. (2020). In vitro assessment of protease production and stress tolerance of mutant isolates of *Trichoderma* sp. *Journal of Agriculture and Applied Biology, 1*(2), 92–99.

Lakhani, H. N., Vakharia, D. N., Makhlouf, A. H., Eissa, R. A., & Hassan, M. M. (2016). Influence of protoplast fusion in *Trichoderma* spp. on controlling some soil borne diseases. *Journal of Plant Pathology & Microbiology, 7*(370), 2.

Lewis, J. A. (1991). *Formulation and delivery systems of biocontrol agents with emphasis on fungi. The rhizosphere and plant growth* (pp. 279–287). Dordrecht: Springer.

Li, G., Chi, M., Chen, H., Sui, Y., Li, Y., Liu, Y., ... Liu, J. (2016). Stress tolerance and biocontrol performance of the yeast antagonist, *Candida diversa*, change with morphology transition. *Environmental Science and Pollution Research, 23*(3), 2962–2967.

Li, H. T., Zhou, H., Duan, R. T., Li, H. Y., Tang, L. H., Yang, X. Q., ... Ding, Z. T. (2019). Inducing secondary metabolite production by co-culture of the endophytic fungus *Phoma* sp. and the symbiotic fungus *Armillaria* sp. *Journal of Natural Products, 82*(4), 1009–1013.

Li, J., Liu, W., Luo, L., Dong, D., Liu, T., Zhang, T., ... Wu, H. (2015). Expression of *Paenibacillus polymyxa* β-1, 3–1,4-glucanase in *Streptomyces lydicus* A01 improves its biocontrol effect against *Botrytis cinerea*. *Biological Control, 90*, 141–147.

Liu, H., Carvalhais, L. C., Crawford, M., Singh, E., Dennis, P. G., Pieterse, C. M., & Schenk, P. M. (2017). Inner plant values: Diversity, colonization and benefits from endophytic bacteria. *Frontiers in Microbiology, 8*, 2552.

Loper, J. E., Kobayashi, D. Y., & Paulsen, I. T. (2007). The genomic sequence of *Pseudomonas fluorescens* Pf-5: Insights into biological control. *Phytopathology, 97*(2), 233–238.

López-Cruz, R., Ragazzo-Sánchez, J. A., & Calderón-Santoyo, M. (2020). Microencapsulation of *Meyerozyma guilliermondii* by spray drying using sodium alginate and soy protein isolate as wall materials: A biocontrol formulation for anthracnose disease of mango. *Biocontrol Science and Technology, 30*(10), 1116–1132.

Maheshwari, R., Bhutani, N., & Suneja, P. (2019). Screening and characterization of siderophore producing endophytic bacteria from *Cicer arietinum* and *Pisum sativum* plants. *Journal of Applied Biology & Biotechnology*, *7*, 7–14.

Marian, M., Morita, A., Koyama, H., Suga, H., & Shimizu, M. (2019). Enhanced biocontrol of tomato bacterial wilt using the combined application of *Mitsuaria* sp. TWR114 and nonpathogenic *Ralstonia* sp. TCR112. *Journal of General Plant Pathology*, *85*(2), 142–154.

Marian, M., & Shimizu, M. (2019). Improving performance of microbial biocontrol agents against plant diseases. *Journal of General Plant Pathology*, *85*(5), 329–336.

Mazrou, Y. S., Neha, B., Kandoliya, U. K., Srutiben, G., Hardik, L., Gaber, A., ... Hassan, M. M. (2020). Selection and characterization of novel zinc-tolerant Trichoderma strains obtained by protoplast fusion. *Journal of Environmental Biology*, *41*(4), 718–726.

McLoughlin, A. J. (1994). *Controlled release of immobilized cells as a strategy to regulate ecological competence of inocula. Biotechnics/wastewater* (pp. 1–45). Berlin, Heidelberg: Springer.

Mehmood, A., Hussain, A., Irshad, M., Hamayun, M., Iqbal, A., & Khan, N. (2019). In vitro production of IAA by endophytic fungus *Aspergillus awamori* and its growth promoting activities in *Zea mays*. *Symbiosis*, *77*(3), 225–235.

Mehta, P., Sharma, R., Putatunda, C., & Walia, A. (2019). *Endophytic fungi: Role in phosphate solubilization. Advances in endophytic fungal research* (pp. 183–209). Cham: Springer.

Mejía, L. C., Herre, E. A., Sparks, J. P., Winter, K., García, M. N., Van Bael, S. A., ... Maximova, S. N. (2014). Pervasive effects of a dominant foliar endophytic fungus on host genetic and phenotypic expression in a tropical tree. *Frontiers in Microbiology*, *5*, 479.

Melin, P., Schnürer, J., & Håkansson, S. (2011). Formulation and stabilisation of the biocontrol yeast Pichia anomala. *Antonie Van Leeuwenhoek*, *99*(1), 107–112.

Mousa, W. K., & Raizada, M. N. (2013). The diversity of anti-microbial secondary metabolites produced by fungal endophytes: An interdisciplinary perspective. *Frontiers in Microbiology*, *4*, 1–18. Available from https://doi.org/10.3389/fmicb.2013.00065.

Mrinalini, C., & Lalithakumari, D. (1998). Integration of enhanced biocontrol efficacy and fungicide tolerance in *Trichoderma* spp. by electrofusion/Integration einerverbesserten-Wirksamkeit der biologischenBekämpfung und der Flungizidtoleranz in *Trichoderma* spp. durchElektrofusion. *ZeitschriftfürPflanzenkrankheiten und Pflanzenschutz/Journal of Plant Diseases and Protection*, *105*(1), 34–40.

Muñoz-Celaya, A. L., Ortiz-García, M., Vernon-Carter, E. J., Jauregui-Rincón, J., Galindo, E., & Serrano-Carreón, L. (2012). Spray-drying microencapsulation of *Trichoderma harzianumconidias* in carbohydrate polymers matrices. *Carbohydrate Polymers*, *88*(4), 1141–1148.

Nelson, E. B. (2004). Microbial dynamics and interactions in the spermosphere. *Annual Review of Phytopathology*, *42*, 271–309.

Otieno, N., Lally, R. D., Kiwanuka, S., Lloyd, A., Ryan, D., Germaine, K. J., & Dowling, D. N. (2015). Plant growth promotion induced by phosphate solubilizing endophytic Pseudomonas isolates. *Frontiers in Microbiology*, *6*, 745.

Özyilmaz, Ü., & Benlioglu, K. (2013). Enhanced biological control of phytophthora blight of pepper by biosurfactant-producing Pseudomonas. *The Plant Pathology Journal*, *29*(4), 418.

Paula, A. R., Carolino, A. T., Paula, C. O., & Samuels, R. I. (2011). The combination of the entomopathogenic fungus *Metarhizium anisopliae* with the insecticide Imidacloprid increases virulence against the dengue vector *Aedes aegypti* (Diptera: Culicidae). *Parasites & Vectors*, *4*(1), 1–8.

Pinski, A., Betekhtin, A., Hupert-Kocurek, K., Mur, L. A., & Hasterok, R. (2019). Defining the genetic basis of plant–endophytic bacteria interactions. *International Journal of Molecular Sciences*, *20*(8), 1947.

Poosapati, S., Ravulapalli, P. D., Tippirishetty, N., Vishwanathaswamy, D. K., & Chunduri, S. (2014). Selection of high temperature and salinity tolerant Trichoderma isolates with antagonistic activity against *Sclerotium rolfsii*. *SpringerPlus*, *3*(1), 1–11.

Praveen Kumar, G., Mir Hassan Ahmed, S. K., Desai, S., Leo Daniel Amalraj, E., & Rasul, A. (2014). In vitro screening for abiotic stress tolerance in potent biocontrol and plant growth promoting strains of *Pseudomonas* and *Bacillus* spp. *International Journal of Bacteriology*, *2014*.

Raguchander, T., Saravanakumar, D., & Balasubramanian, P. (2011). Molecular approaches to improvement of biocontrol agents of plant diseases. *Journal of Biological Control*, *25*(2), 71–84.

Ram, R. M., Keswani, C., Bisen, K., Tripathi, R., Singh, S. P., & Singh, H. B. (2018). *Biocontrol technology: Eco-friendly approaches for sustainable agriculture*. *Omics technologies and bio-engineering* (pp. 177–190). Academic Press.

Ray, S., Singh, J., Rajput, R. S., Yadav, S., Singh, S., & Singh, H. B. (2019). *A thorough comprehension of host endophytic interaction entailing the biospherical benefits: A metabolomic perspective*. *Endophytes and secondary metabolites* (pp. 657–675). Springer. Available from https://doi.org/10.1007/978-3-319-90484-9_16.

Rho, H., Hsieh, M., Kandel, S. L., Cantillo, J., Doty, S. L., & Kim, S. H. (2018). Do endophytes promote growth of host plants under stress? A *meta*-analysis on plant stress mitigation by endophytes. *Microbial Ecology*, *75*(2), 407–418.

Rohini, S., Aswani, R., Kannan, M., Sylas, V. P., & Radhakrishnan, E. K. (2018). Culturable endophytic bacteria of ginger rhizome and their remarkable multi-trait plant growth-promoting features. *Current Microbiology*, *75*(4), 505–511.

Santoyo, G., Moreno-Hagelsieb, G., del Carmen Orozco-Mosqueda, M., & Glick, B. R. (2016). Plant growth-promoting bacterial endophytes. *Microbiological Research*, *183*, 92–99.

Sarethy, I. P., & Saharan, A. (2021). Genomics, proteomics and transcriptomics in the biological control of plant pathogens: A review. *Indian Phytopathology*, *74*, 1–10.

Schippers, B., Bakker, A. W., & Bakker, P. A. (1987). Interactions of deleterious and beneficial rhizosphere microorganisms and the effect of cropping practices. *Annual Review of Phytopathology*, *25*(1), 339–358.

Schlieker, M., & Vorlop, K. D. (2006). *A novel immobilization method for entrapment: LentiKats®*. *Immobilization of enzymes and cells* (pp. 333–343). Humana Press.

Shwab, E. K., & Keller, N. P. (2008). Regulation of secondary metabolite production in filamentous ascomycetes. *Mycological Research*, *112*(2), 225–230.

Singh, R., Maurya, S., & Upadhyay, R. S. (2016). The improvement of competitive saprophytic capabilities of Trichoderma species through the use of chemical mutagens. *Brazilian Journal of Microbiology*, *47*(1), 10–17.

Slama, B. H., Triki, M. A., ChenariBouket, A., Ben Mefteh, F., Alenezi, F. N., Luptakova, L., & Belbahri, L. (2019). Screening of the high-rhizosphere competent *Limoniastrum monopetalum* culturable endophyte microbiota allows the recovery of multifaceted and versatile biocontrol agents. *Microorganisms*, *7*(8), 249.

Sui, Y., & Liu, J. (2014). Effect of glucose on thermotolerance and biocontrol efficacy of the antagonistic yeast *Pichia guilliermondii*. *Biological Control*, *74*, 59–64.

Suleiman, A. K., Harkes, P., van den Elsen, S., Holterman, M., Korthals, G. W., Helder, J., & Kuramae, E. E. (2019). Organic amendment strengthens interkingdom associations in the soil and rhizosphere of barley (*Hordeum vulgare*). *Science of the Total Environment*, *695*, 133885.

Sun, Z. B., Sun, M. H., Zhou, M., & Li, S. D. (2017). Transformation of the endochitinase gene Chi67–1 in *Clonostachys rosea* 67–1 increases its biocontrol activity against *Sclerotinia sclerotiorum*. *AMB Express*, *7*(1), 1–9.

Tamez-Guerra, P., McGuire, M. R., Medrano-Roldan, H., Galan-Wong, L. J., Shasha, B. S., & Vega, F. E. (1996). Sprayable granule formulations for *Bacillus thuringiensis*. *Journal of Economic Entomology*, *89*(6), 1424–1430.

Tewari, S., Shrivas, V. L., Hariprasad, P., & Sharma, S. (2019). *Harnessing endophytes as biocontrol agents. Plant health under biotic stress* (pp. 189–218). Singapore: Springer.

Ting, A. S. Y., Mah, S. W., & Tee, C. S. (2010). Identification of volatile metabolites from fungal endophytes with biocontrol potential towards *Fusarium oxysporum* F. sp. cubense Race 4. *American Journal of Agricultural and Biological Sciences*, *5*(2), 177–182.

Trivedi, P., Leach, J. E., Tringe, S. G., Sa, T., & Singh, B. K. (2020). Plant–microbiome interactions: From community assembly to plant health. *Nature Reviews Microbiology*, *18*(11), 607–621.

Umashankar, K., Chandralekha, A., Dandavate, T., Tavanandi, H. A., & Raghavarao, K. S. M. S. (2019). A nonconventional method for drying of *Pseudomonas aeruginosa* and its comparison with conventional methods. *Drying Technology*, *37*(7), 839–853.

Vassilev, N., Vassileva, M., Martos, V., Del Moral, L. F. G., Kowalska, J., Tylkowski, B., & Malusá, E. (2020). Formulation of microbial inoculants by encapsulation in natural polysaccharides: Focus on beneficial properties of carrier additives and derivatives. *Frontiers in Plant Science*, *11*, 270. Available from https://www.frontiersin.org/article/10.3389/fpls.2020.00270.

Vemmer, M., & Patel, A. V. (2013). Review of encapsulation methods suitable for microbial biological control agents. *Biological Control*, *67*(3), 380–389. Available from https://doi.org/10.1016/j.biocontrol.2013.09.003.

Wagh, S., Ingle, S. T., Dandale, S., & Mane, S. S. (2015). Improvement in biocontrol ability of *Trichoderma viride* through gamma irradiation. *Trends in Biosciences*, *8*(20), 5622–5626.

Wang, Y., Luo, Y., Sui, Y., Xie, Z., Liu, Y., Jiang, M., & Liu, J. (2018). Exposure of *Candida oleophila* to sublethal salt stress induces an antioxidant response and improves biocontrol efficacy. *Biological Control*, *127*, 109–115.

Weng, J., Wang, Y., Li, J., Shen, Q., & Zhang, R. (2013). Enhanced root colonization and biocontrol activity of *Bacillus amyloliquefaciens* SQR9 by abrB gene disruption. *Applied Microbiology and Biotechnology*, *97*(19), 8823–8830.

Wong, C. K. F., Saidi, N. B., Vadamalai, G., Teh, C. Y., & Zulperi, D. (2019). Effect of bioformulations on the biocontrol efficacy, microbial viability and storage stability of a consortium of biocontrol agents against Fusarium wilt of banana. *Journal of Applied Microbiology*, *127*(2), 544–555.

Wu, K., Fang, Z., Guo, R., Pan, B., Shi, W., Yuan, S., ... Shen, Q. (2015). Pectin enhances bio-control efficacy by inducing colonization and secretion of secondary metabolites by *Bacillus amyloliquefaciens* SQY 162 in the rhizosphere of tobacco. *PLoS One*, *10*(5), e0127418.

Wu, Q., Bai, L., Liu, W., Li, Y., Lu, C., Li, Y., & Chen, J. (2013). Construction of *Streptomyces lydicus* A01 transformant with the chit33 gene from *Trichoderma harzianum* CECT2413 and its biocontrol effect on Fusaria. *Chinese Science Bulletin*, *58*(26), 3266–3273.

Xu, F., Wang, S., Li, Y., Zheng, M., Xi, X., Cao, H., ... Han, C. (2018). Yield enhancement strategies of rare pharmaceutical metabolites from endophytes. *Biotechnology Letters*, *40*(5), 797–807.

Xu, X. M., Jeffries, P., Pautasso, M., & Jeger, M. J. (2011). Combined use of biocontrol agents to manage plant diseases in theory and practice. *Phytopathology*, *101*(9), 1024–1031.

Xu, Z., Xie, J., Zhang, H., Wang, D., Shen, Q., & Zhang, R. (2019). Enhanced control of plant wilt disease by a xylose-inducible degQ gene engineered into *Bacillus velezensis* strain SQR9XYQ. *Phytopathology*, *109*(1), 36–43.

Yandigeri, M. S., Malviya, N., Solanki, M. K., Shrivastava, P., & Sivakumar, G. (2015). Chitinolytic Streptomyces vinaceusdrappus S5MW2 isolated from Chilika lake, India enhances plant growth and biocontrol efficacy through chitin supplementation against *Rhizoctonia solani*. *World Journal of Microbiology and Biotechnology*, *31*(8), 1217–1225.

Yang, M., Mavrodi, D. V., Mavrodi, O. V., Thomashow, L. S., & Weller, D. M. (2017). Construction of a recombinant strain of *Pseudomonas fluorescens* producing both phenazine-1-carboxylic acid and cyclic lipopeptide for the biocontrol of take-all disease of wheat. *European Journal of Plant Pathology*, *149*(3), 683–694.

Yu, T., Wang, L., Yin, Y., Wang, Y., & Zheng, X. (2008). Effect of chitin on the antagonistic activity of *Cryptococcus laurentii* against *Penicillium expansum* in pear fruit. *International Journal of Food Microbiology*, *122*(1–2), 44–48.

Zaim, S., Bekkar, A. A., & Belabid, L. (2018). Efficacy of *Bacillus subtilis* and *Trichoderma harzianum* combination on chickpea Fusarium wilt caused by F. oxysporum f. sp. ciceris. *Archives of Phytopathology and Plant Protection*, *51*(3–4), 217–226.

Zembek, P., Perlińska-Lenart, U., Brunner, K., Reithner, B., Palamarczyk, G., Mach, R. L., & Kruszewska, J. S. (2011). Elevated activity of dolichyl phosphate mannose synthase enhances biocontrol abilities of *Trichoderma atroviride*. *Molecular Plant-Microbe Interactions*, *24*(12), 1522–1529.

Zheng, F., Zhang, W., Sui, Y., Ding, R., Yi, W., Hu, Y., ... Zhu, C. (2019). Sugar protectants improve the thermotolerance and biocontrol efficacy of the biocontrol Yeast, *Candida oleophila*. *Frontiers in Microbiology*, *10*, 187.

Further reading

Cabrefiga, J., Francés, J., Montesinos, E., & Bonaterra, A. (2011). Improvement of fitness and efficacy of a fire blight biocontrol agent via nutritional enhancement combined with osmoadaptation. *Applied and Environmental Microbiology*, *77*(10), 3174–3181.

Cañamás, T. P., Viñas, I., Abadias, M., Usall, J., Torres, R., & Teixidó, N. (2009). Acid tolerance response induced in the biocontrol agent Pantoea agglomerans CPA-2 and effect on its survival ability in acidic environments. *Microbiological Research*, *164*(4), 438–450.

Moslemy, P., Guiot, S. R., & Neufeld, R. J. (2002). Production of size-controlled gellan gum microbeads encapsulating gasoline-degrading bacteria. *Enzyme and Microbial Technology*, *30*(1), 10–18.

Pal, K. K., Tilak, K. V. B. R., Saxcna, A. K., Dey, R., & Singh, C. S. (2001). Suppression of maize root diseases caused by *Macrophomina phaseolina*, *Fusarium moniliforme* and *Fusarium graminearum* by plant growth promoting rhizobacteria. *Microbiological Research*, *156*(3), 209–223.

Strobel, S. A., Allen, K., Roberts, C., Jimenez, D., Scher, H. B., & Jeoh, T. (2018). Industrially-scalable microencapsulation of plant beneficial bacteria in dry cross-linked alginate matrix. *Industrial Biotechnology*, *14*(3), 138–147.

Szczech, M., & Maciorowski, R. (2016). Microencapsulation technique with organic additives for biocontrol agents. *Journal of Horticultural Research*, *24*(1), 111–122.

Tu, L., He, Y., Yang, H., Wu, Z., & Yi, L. (2015). Preparation and characterization of alginate–gelatin microencapsulated *Bacillus subtilis* SL-13 by emulsification/internal gelation. *Journal of Biomaterials Science, Polymer Edition*, *26*(12), 735–749.

Wei, Z., Huang, J., Yang, C., Xu, Y., Shen, Q., & Chen, W. (2015). Screening of suitable carriers for *Bacillus amyloliquefaciens* strain QL-18 to enhance the biocontrol of tomato bacterial wilt. *Crop Protection*, *75*, 96–103.

Wiyono, S., Schulz, D. F., & Wolf, G. A. (2008). Improvement of the formulation and antagonistic activity of *Pseudomonas fluorescens* B5 through selective additives in the pelleting process. *Biological Control*, *46*(3), 348–357.

Zhan, Y., Xu, Q., Yang, M. M., Yang, H. T., Liu, H. X., Wang, Y. P., & Guo, J. H. (2012). Screening of freeze-dried protective agents for the formulation of biocontrol strains, *Bacillus cereus* AR156, *Burkholderia vietnamiensis* B418 and *Pantoea agglomerans* 2Re40. *Letters in Applied Microbiology*, *54*(1), 10–17.

Zohar-Perez, C., Chernin, L., Chet, I., & Nussinovitch, A. (2003). Structure of dried cellular alginate matrix containing fillers provides extra protection for microorganisms against UVC radiation. *Radiation Research*, *160*(2), 198–204.

CHAPTER

Novel trends in endophytic applications for plant disease management

8

Priya Jaiswal[1,2], Sristi Kar[1], Sankalp Misra[1,2], Vijaykant Dixit[1], Shashank Kumar Mishra[1,2] and Puneet Singh Chauhan[1,2]

[1]*Microbial Technology Division, Council of Scientific and Industrial Research-National Botanical Research Institute (CSIR-NBRI), India*

[2]*Academy of Scientific and Innovative Research (AcSIR), India*

8.1 Introduction

Plant growth and development are under a constant threat of pathogen infestation. To evade agricultural losses caused by pathogen infestation in crops bounty of agrochemicals are being used. Despite their effectiveness in reducing diseases in plants, chemical pesticides are not a sustainable source as they become an environmental concern and their residues remain within the plants which reduce the quality of grains, fruits, and vegetables (Singh, Singh et al., 2020). Agrochemicals consist of fertilizers (nitrogen potassium and phosphorus) and pesticides (insecticides, fungicides, and herbicides). Long-term use of chemical fertilizers containing nitrogen and ammonium salts causes a reduction in soil pH and promotes wilt diseases in plants. Fungal pathogens are the most prevalent pathogens that attack plants at the time of sowing. Fungal pathogens pose a threat to food security, for example, toxin produced by *Fusarium graminearum* deoxynivalenol is a potential health risk for humans and animals (Weller, Culbreath, Gianessi, & Godfrey, 2014). Plant pathogens with rigorous use of pesticides over the years have gradually developed resistance mechanisms against them. Microbial pathogens owing to their capability of higher mutation rates and gene transfer mechanisms have a higher evolution potential and develop a resistance mechanism against chemical control agents. For example, (1) a single-site mutation in the gene coding for β-tubulin in *Botrytis cinerea* (the causative agent of gray molds in cultivated vegetables, fruits, and flowers) makes it resistant to benzimidazole group of fungicides, (2) the development of resistance against quinone outside inhibitors (QoI) fungicides (strobilurin fungicides) in powdery mildews in cucumber and wheat pathogen *Mycosphaerella graminicola* (De Silva et al., 2018). Hence, there is a requisite for finding an alternative being more efficient, environment-friendly, and reduce the dependency on synthetic pesticides in agriculture (Santoyo et al., 2012). Microorganisms have several mechanisms to tolerate abiotic and biotic stress.

Biocontrol Mechanisms of Endophytic Microorganisms. DOI: https://doi.org/10.1016/B978-0-323-88478-5.00014-6

Several microbes have the capability to inhibit the growth of opportunistic pathogens and protect plants; hence, they can be a potential bioresource for integrated plant disease management. To control plant pathogens biocontrol can be used as an alternative method with minimal impact on the environment. The method of biological control of plant disease management was first used by C. F. Von in 1914. Biological control is described as the use of useful microbes, their products (metabolites, enzymes, and genes) that help in eliminating the negative effect of plant pathogens and maintain plant health and growth. According to International Biocontrol Manufacturers Association, biocontrol is defined as the use of microbial agents or products that naturally act as an antagonist to crop pests and pathogens. Although biocontrol is an environment-friendly approach, there are several constraints in their production, the development of this technology has been restricted. Several microbial strains (bacteria and fungi) are being reported to possess different mechanisms to combat pathogen infestation. Parasitic *Trichoderma lignorum* and antibiotic-producing genera like *Aspergillus*, *Penicillium*, *Trichoderma*, and *Streptomyces* species are used to control plant pathogens. Siderophore [complex iron (III)] production by *Erwinia carotovora* in soils also inhibits the growth of pathogens by making iron unavailable for pathogens.

8.2 Endophytic microorganisms as biocontrol agents

Plants and animals live in close association with microorganisms. In animals, the gut microbiome plays a significant role in boosting their host health. Similarly, plants are also inhabited by a hidden diversity of microbes (fungi and bacteria). These microbes present within the plant tissues form endophytic diversity. Endophytes are present locally, intracellularly, or systemically within their host without causing any harm to their host plants. Endophytes use their plant host not only as a habitat and source of nutrition but also aid them by enhancing their development, growth, adaption, and stress tolerance (Saikkonen, Faeth, Helander, & Sullivan, 1998; Wani, Ashraf, Mohiuddin, & Riyaz-Ul-Hassan, 2015). Endophytes are ubiquitous in various below-ground and above-ground plant tissues of spermatophytes, ferns, mosses, lycophytes, liverworts, and hornworts from natural forests and agricultural ecosystems (Gupta et al., 2019; Kumar et al., 2016). Endophytes are supposed to have originated from microbial communities present in the rhizosphere and phyllosphere. They obtain their nutrients directly from their hosts and in return, they improve their growth by alleviating stress conditions. They spend at least a part of their life cycle within their host tissues or cells; therefore they are better accustomed to the environment present within the host. Interaction and association between plants and bacteria is a well-established subject. These studies validated that several genera of bacteria have a positive impact on plant growth and health. Likewise, plants also chose their microbiome to obtain benefits from microbial colonizers, including those residing within plant tissues, called endophytes. Any bacteria can be functionally termed an endophyte if it has been isolated by disinfecting the plant tissues or isolated from the interiors of the plant, and their occupancy does not cause any visible damage to the plant (Hardoim et al., 2015; Kumar, Droby, Singh, Singh, & White, 2020). Since bacterial endophytes reside within plant tissues in close contact, therefore, they can readily exert a direct beneficial effect (Santoyo et al., 2016). The rhizosphere is the first site of colonization for bacteria; thus the root ecosystem is assumed to be one of the main sources for endophytic bacterial colonization (Hardoim et al., 2015). Plants are colonized by plant growth–promoting bacterial endophytes which can be exploited to suppress the activity of plant pathogens (Table 8.1). Endophytic diversity varies with different stages of plant

Table 8.1 List of endophytes and their active compounds in association with different host plants.

Endophyte	Plant	Activity	Compound produced	Reference
Bacillus subtilis	*Allamanda cathartica*	Antifungal	Terpene	Nithya and Muthumary (2011)
Bacillus atrophaeus, *Bacillus mojavensis*	*Glycyrrhiza uralensis*	Antifungal	1,2-Bezenedicarboxyl acid, methyl ester, decanedioic acid, bis (2-ehtylhexyl) ester	Mohamad et al. (2018)
Phoma sp.	*Cinnamomum mollissimum*	Antifungal	5-Hydroxyramulsin	Santiago, Fitchett, Munro, Jalil, and Santhanam (2012)
Phomopsis sp.	*Aconitum carmichaeli*	Antifungal	Gavodermside and Clavasterols	Wu, Huang, Miao, and Chen (2013)
Aspergillus sp.	*Bauhinia guianensis*	Antibacterial	Fumigaclavine C and Pseurotin C	Pinheiro et al. (2013)
Streptomyces sp.	*Grevillea pteridifolia*	Antibacterial	Kakadumycin A Echinodermycin	Castillo et al. (2003)
Streptomyces sp. *TP-A0595*	*Allium tuberosum*	Antifungal	6-Prenylindole	Singh and Dubey (2018)
Aeromicrobium ponti	*Vochysia divergens*	Antibacterial	1-Acetyl-*b*-carboline, Indole-3-carbaldehyde, 3-(hydroxyacetyl)-indole, brevianamide F, and cyclo-(L-Pro-L-Phe)	Gos et al. (2017)

growth and development. Hence, prophylactic measures can be brought into action to evict potential plant pathogens by inoculating biocontrol agents (or other biocompounds) at different growth stages. For example—(1) biocontrol endophytes can be applied as biocontrol agents in the agricultural soils during the stage of seed sowing pathogens and (2) they can also be applied during developmental stages until obtaining the plant product (fruit, vegetable, and seed). Potential endophytes with biocontrol properties have been screened via in vivo assays. The parameters for biocontrol efficacy of the endophytes include a percentage of control in the incidence of disease under greenhouse conditions and measurements of physiological plant health, biochemical assays (enzymatic assay), plant hormones production, and biomass production. Examples of endophytic isolates possessing antagonistic activities against plant pathogens have been listed in Table 8.1.

Endophytes with biocontrol potential also hold some of the plant growth–promoting attribute, for example—nitrogen fixation, mobilization of immobilized nutrients-like phosphorus through the production of organic acids, and iron through the production of siderophores (Glick, 2012; Singh, Srivastava, Kumar, Singh, & Pandey, 2020). Biocontrol agents also regulate the production of plant hormones like auxins (indole acetic acid or IAA), cytokinins, and ethylene which maintain the morphology and structure of the plant under abiotic and biotic stresses (Fig. 8.1). IAA production by endophytic bacteria increases the surface area and root length in a plant, thereby providing improved access to the nutrients from the soil. During IAA production, cell walls in bacteria are expanded which

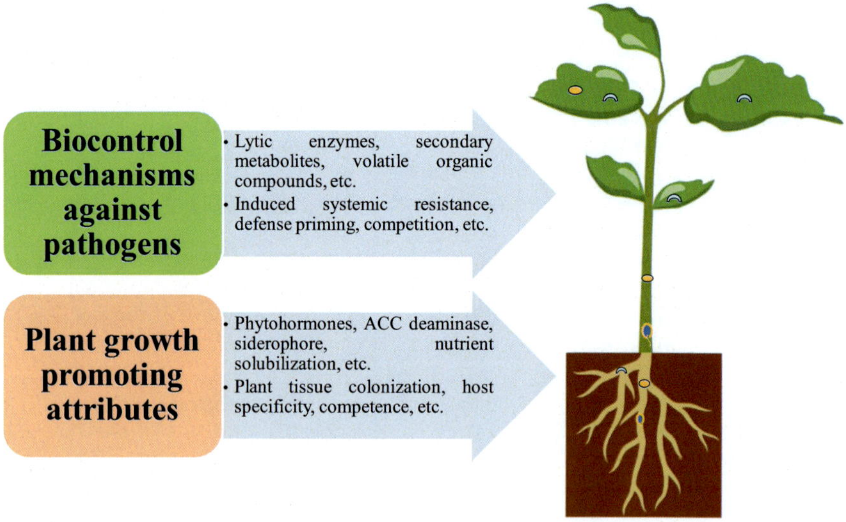

FIGURE 8.1

Illustration depicting endophytic mechanism of biocontrol and plant growth promotion.

enhances the secretion of exudates providing more nutrients for other beneficial bacteria present in the rhizosphere. IAA production by endophytes is recognized as the major effector molecule in plant–microbe interaction, pathogenesis, and phytostimulation (Gao & Tao, 2012; Kumar et al., 2021). Endophytes help in reducing ethylene production under certain stress conditions such as flood, salinity, and pathogens infection (Gamalero & Glick, 2015). Plant growth-promoting bacteria (PGPB) produce 1-aminocyclopropane-1-carboxylate (ACC) which is the immediate precursor of the hormone ethylene in plants. ACC is hydrolyzed by the enzyme ACC deaminase produced by bacteria (Orozco-Mosqueda et al., 2020). Enzyme ACC deaminase degrades ACC molecules into α-ketobutyrate and ammonia. PGPB protect the plants by producing ACC deaminase and the production of ethylene is avoided under biotic and abiotic stress, which, in turn, promotes the growth and facilitating the survival of plants (Santoyo et al., 2020) (Fig. 8.1).

Endophytic bacterial diversity known to produce ACC deaminase has been discovered in genera like *Pseudomonas, Bacillus, Achromobacter, Acinetobacter, Alcaligenes, Agrobacterium, Burkholderia, Enterobacter, Ralstonia, Serratia*, and *Rhizobium* (Kang et al., 2012). PGPB apart from direct action also hold indirect mechanisms of plant growth promotion. Indirect modes include various strategies such as competition for space and nutrients, lytic enzyme production, antibiosis, toxin inhibition, and induction of plant defense mechanisms.

8.3 Biocontrol mechanisms of endophytes

The approach of microbial biotechnology has helped in building up the way that microorganisms assume huge parts in the industry, agribusiness, and medication (Rajamanikyam et al., 2017; Gouda et al., 2016). Having a superior comprehension of the different jobs microorganisms play in the environment will upgrade the manners in which they can be applied in the field of farming in particular for plant development and harvest yield. The universe of endophytes has pulled in light of a legitimate concern for some analysts because of their critical jobs in advancing development and in improving the endurance of plants under extraordinary conditions (Shen et al., 2019). Biologically active metabolites discharged by endophytic microorganisms are helpful in businesses, agribusiness, and the field of medication. Plants perform a significant capacity of deciding the kind of microorganism that can be related to its root exudates (Andreozzi et al., 2019). In this manner the collaboration between endophytic microorganisms and plants incredibly relies upon the capacity of these organisms to utilize the exudates created by the plant roots as their fuel source (Kandel et al., 2017). Endophytes can proficiently upgrade development advancement using different methods of tasks and expanding the opposition of plants to extraordinary conditions (Yadav, 2018). Quite, endophytic microbes have been utilized in the large-scale manufacturing of industrially relevant items like antiinfection agents, proteins, and riboflavin among others (Latz, Jensen, Collinge, & Jørgensen, 2018). The protection from antimicrobial is on the expansion particularly among organic entities that cause disease and this has incredible general well-being suggestions if legitimate consideration is not taken.

8.4 Competition: an eco-friendly reprisal program

Endophytes colonize plant tissues locally or systemically (Latz et al., 2018). They respond by colonizing the hidden spaces making them unavailable for pathogens and sneaking on the available nutrients to carry out their activities thereby creating competition for pathogens (Rodriguez, White, Arnold, & Redman, 2009). Mohandoss and Suryanaraynan (2009) in their study discovered that the application of fungicides caused the destruction of endophytes in mango leaves and facilitates other pathogenic fungi to inhabit the niche. Endophytes employ several other mechanisms in combination with the competition. For example: Lahlali et al. (2014) in their study used endophyte *Heteroconiumchaetospira* which solely on the basis of colonization could not prevent clubroot symptoms in oilseed rape, whereas the symptoms of *Phytophthora* sp. were reduced when a mixture of endophytes was applied. This suggested that other endophytes actively produced some metabolites to prevent the pathogen infestation.

8.5 Antibiosis: strategy for effective biocontrol

To cast out the evolvement of phytopathogenic organisms, the synthesis of various antimicrobial compounds has been extensively studied (Liu, Carvalhais et al., 2017; Liu, Newman, McInroy, Hu, & Kloepper, 2017). The two most common genera explored in this respect are *Pseudomonas* and *Bacillus* studied for their ability to produce antibiotics, such as zymicrolactone A, pyrrolnitrine, ecomycins, 2,4-diacetylphloroglucinol acid, butyroaminectone, oomycin A, 1-carboxamide, phenazine-1-carboxylic acid, pyroluteorine, phenazine-viscosinamide, kyanoaminectone, cepafungins zymicrolactone, aerugina, rhamnolipids, cepacyamide A, pseudomonic acid, and azomycin (Santoyo, Sánchez-Yáñez, & de los Santos-Villalobos, 2019). The genus *Pseudomonas* consists of a number of strains identified for producing a wide variety of antibiotics that are responsible for suppressing agricultural diseases. Various examples have been quoted regarding this concept where microorganisms such as *Pseudomonas fluorescens* are known to produce antibiotics such as 2,4-diacetylphloroglucinol and pyoluteorin that suppress root rot of the tobacco, caused by *Thielaviopsis basicola*. Finally, pyoluteorin and pyrrolnitrin have been shown to effectively suppress watercress disease caused by *Pythium ultimum* and *Rhizoctonia solani*, respectively (Milner, Silo-Suh, Goodman, & Handelsman, 2019; Singh, Singh, & Kumar, 2017).

8.6 Production of lipopeptides: another mechanism for suppressing pathogens

These atoms comprise a cyclic peptide connected to a β-hydroxy or β-amino unsaturated fat chain, ordered into three unique families (iturins, phengicines, and surfactins), in view of their amino corrosive grouping and unsaturated fat length (Falardeau, Wise, Novitsky, & Avis, 2013; Valenzuela-Ruiz et al., 2020). Lipopeptides are incorporated by multicatalyst edifices called nonribosomal peptide synthetase, which are autonomous of courier RNA (Chowdhury, Hartmann, Gao, & Borriss, 2015), which are low-atomic weight compounds with amphiphilic qualities that secure to plants during a few phenological organizes by straightforwardly stifling the development of microorganisms or instigating foundational opposition (Hashem, Tabassum, & Abd_Allah, 2019). As of late, Coutte et al. (2017) announced 263 diverse lipopeptides combined by 11 microbial genera, among which the Bacillus class was the most bountiful maker with 98 distinctive lipopeptides, those were associated with the organic control of a wide scope of phytopathogens (microorganisms, parasites, and oomycetes), causing sicknesses in farming significance crops (Ongena & Jacques, 2008). For instance, various isoforms of phengicines and iturins have been accounted for in sans cell concentrates of fluid societies of *Bacillus subtilis* GA1, with the capacity to hinder *B. cinerea* in apple natural products.

8.7 Production of δ-endotoxins: natural plan for biocontrol

The δ-endotoxins, delivered by *Bacillus thuringiensis* (Bt), are parasporal bodies proteins comprising polypeptide units of various subatomic loads, from 27 to 140 kDa (Villarreal-Delgado et al., 2018). Bt poisons are created during the sporulation stage, the Cry (precious stone) protein is known for its particular poisonous impacts on an objective creature (most have a place with the bug request); moreover, Cyt (cytolytic) proteins have been accounted for with harmful consequences for a wide assortment of creepy crawlies, essentially diptera; nonetheless, its cytotoxicity against mammalian cells has likewise been confirmed (Anaya et al., 2020). The instruments of activity of Cry proteins begin once they are proteolytically prepared through proteases that are found in the midgut of the host, isolating a part of amino acids in the N-terminal locale and at the C-terminal end (contingent upon the idea of the Cry protein), and in this manner delivering dynamic and harmful pieces that cooperate with the receptor proteins present in intestinal cells of the bug, flagging the development of a prepore oligomeric structure and, thus, the lytic pore, which creates an osmotic unevenness, and afterward annihilating the intestinal epithelium and, therefore, causing cell demise (Xu et al., 2014). At present, at any rate strains of Bt mainly 13 in number are utilized in farming, some of them are facultative bacterial endophytes (Regnault-Roger, 2012).

8.8 Lytic enzymes: arsenals of natural origin

Lytic enzymes are associated with the cell divider debasement of phytopathogenic microorganisms, being quite possibly the most detailed natural control instruments primarily against parasitic microbes (Villarreal-Delgado et al., 2018). The contagious cell divider (counting plant microorganisms) comprises glycoproteins, polysaccharides, and different parts that change contingent upon the parasitic species. The polysaccharide division contains up to 80% of the cell divider. These polysaccharides have a deciding primary part in the solidness of the cell divider, through a broad organization of glycosidic securities; in this way the impedance in these securities by lytic proteins can weaken the cell mass of phytopathogenic growths, causing their lysis and cell demise (Jadhav, Shaikh, & Sayyed, 2017). Among the most considered lytic chemicals delivered by natural control specialists are chitinases, cellulases, proteases, and β-1,3-glucanases, which change, puncture, as well as debase the design of the cell divider (Mota, Gomes, Souza Júnior, & Moura, 2017). Mishra and Arora (2012) detailed the job of an extracellular chitinase secluded from *Pseudomonas aeruginosa* in the natural control of *Xanthomonas campestris*, which causes dark decay illness. Besides, strains of *Pseudomonas* secluded from the chickpea rhizosphere have been accounted for to create chitinases and cellulases with opposing action against *R. solani* and *Pythium aphanidermatum* (Sindhu & Dadarwal, 2001). Additionally, *B. subtilis* has been accounted for its organic control limit against *R. solani*, the causative specialist of dark scabies in potatoes, through the creation of chitinases (Saber et al., 2015).

8.9 Siderophore production: indirect mechanism of biocontrol

These metabolites are delivered by organisms in light of a limit of iron in the climate; consequently, some natural control specialists produce these low atomic weight (400–1500 Da) receptor protein structures with a high fondness for iron. Siderophores are auxiliary metabolites that go about as sequestrants of iron, due to their high separation steady by this component (somewhere in the range of 1022 and 1055). Siderophore-creating natural control specialists can utilize iron by two systems: (1) straightforwardly through the Fe^{3+} siderophore complex through the cell layer, or (2) diminished extracellularly to Fe^{2+} buildings (Hider & Kong, 2010). This permits these specialists to control the convergence of iron in their territories through the sequestration of that component (Fe^{3+} siderophore), causing iron not to be accessible for phytopathogenic microorganisms, and confining its development (Kannojia, Choudhary, Srivastava, & Singh, 2019). Right now, a few bacterial strains have been accounted for their capacity to control plant infections through the siderophores creation, restricting the development and colonization of iron-subordinate phytopathogenic microorganisms (Fgaier & Eberl, 2011). Yu et al. (2011) announced that *B. subtilis* CAS15 threatened the development (19%–94%) of 15 contagious phytopathogens having a place with the family *Fusarium, Colletotrichum, Pythium, Magnaporthe*, and *Phytophthora*, through the creation of catecholate-type siderophores (bacillibactin). Then again, de los Santos-Villalobos, Barrera-Galicia, Miranda-Salcedo, & Peña-Cabriales (2012) announced the siderophore-creating limit of *Burkholderia anthina* XXVI, which is engaged with the hindrance of the causal specialist of anthracnose in mango, *Colletotrichum gloeosporioides*, at a base inhibitory centralization of 0.64 μg/mL.

8.10 Induced systemic resistance (ISR): unique reinforcement strategy

Chemical signals (elicitors) created by beneficial microorganisms may induce this mechanism. ISR flagging is subject to jasmonic acid and ethylene (Kannojia et al., 2019). Up until this point, not all the subatomic systems that manage plant valuable microorganism collaborations have been portrayed; nonetheless, the primary courses by which these specialists direct ISR in plants have been recognized: (1) pathogen-associated molecular patterns/microbe-associated molecular patterns, (2) phytohormones, and (3) a few elicitors (volatile organic compounds, siderophores, phytases, miRNAs, among others). ISR has been confirmed in tobacco plants, where PR2 (encodes a β-1,3-glucanase) and PR3 (encodes a chitinase) were initiated in light of volatile compounds delivered by *Bacillus*, giving protection from *R. solani* and *Phytophthora nicotianae*. Notwithstanding pathogenesis-related (PR) qualities, *Bacillus* enacts other assurance systems in plants, which remember primary changes for the cell divider by the collection of lignin, or the creation of optional metabolites like flavonoids, phytoalexins, auxins, and additionally glucosinolates. In this way, ISR has been accounted for in an assortment of harvests (beans, carnations, cucumbers, radishes, tobacco, and tomatoes), fundamentally decreasing the pathogenicity of a few plant pathogen, including growths, microscopic organisms such as fungi and viruses (Kannojia et al., 2019).

8.11 Conclusion

The present chapter attempts to understand the varied mechanisms used by endophytic microorganisms in conferring plant protection against diseases for sustainable agriculture. Endophytic microbes are considered "plant probiotics" that accelerate plant growth by employing different mechanisms. The major advantage of implementing such beneficial microorganisms in agriculture is aimed to reduce the use of different agrochemicals such as chemical fertilizers and pesticides make agriculture more sustainable as well as productive. Many studies are still ongoing toward assessing the ability of endophytes to secrete novel bioactive compounds which will be useful in sustainable agriculture and in enhancing plant growth. These metabolites can also perform pest control and insecticidal activities, together with enhanced plant nutrient uptake under stress conditions such as salinity and drought. In general, novel bioactive compounds produced by endophytes could confer beneficial effects in agriculture and the environment. In the future, bioformulations developed from pure bioactive compounds or a combination of endophytes will be imperative that needs to be explored and worked upon for effective control of phytopathogens.

Acknowledgment

The authors acknowledge the Director, CSIR National Botanical Research Institute for providing facilities and support during the study. This work is supported by the CSIR-Network and in-house projects funded by Council of Scientific and Industrial Research, New Delhi, India.

References

Anaya, P., Onofre, J., Quintero, T., Sánchez, M. C., Gill, J., S S Bravo, A., & Soberón, M. (2020). Oligomerization is a key step for *Bacillus thuringiensis* Cyt1Aa insecticidal activity but not for toxicity against red blood cells. *Insect Biochemistry and Molecular Biology*, *119*, 103317.

Andreozzi, A., Prieto, P., Mercado-Blanco, J., Monaco, S., Zampieri, E., & Romano, S. (2019). Efficient colonization of the endophytes *Herbaspirillum huttiense* RCA24 and *Enterobacter cloacae* RCA25 influences the physiological parameters of Oryza sativa L. cv. *Baldo* rice. *Environmental Microbiology*, *21*, 3489–3504.

Castillo, U., Harper, J. K., Strobel, G. A., Sears, J., Alesi, K., Ford, E., ... Teplo, D. (2003). Kakadumycins, novel antibiotics from *Streptomyces* sp. NRRL 30566: An endophyte of *Grevillea pteridifolia*. *FEMS Microbiology Letters*, *224*, 183–190. Available from https://doi.org/10.1016/S0378-1097(03)00426-9.

Chowdhury, S. P., Hartmann, A., Gao, X., & Borriss, R. (2015). Biocontrol mechanism by root-associated *Bacillus amyloliquefaciens* FZB42—A review. *Frontiers in Microbiology*, *6*, 780.

Coutte, F., Lecouturier, D., Dimitrov, K., Guez, J. S., Delvigne, F., & Dhulster, P. (2017). Microbial lipopeptide production and purification bioprocesses, current progress and future challenges. *Biotechnology Journal*, *12*, 1–10.

de los Santos-Villalobos, S., Barrera-Galicia, G. C., Miranda-Salcedo, M. A., & Peña-Cabriales, J. J. (2012). Burkholderia cepacia XXVI siderophore with biocontrol capacity against *Colletotrichum gloeosporioides*. *World Journal of Microbiology and Biotechnology*, *28*, 2615–2623.

De Silva, N. I., Brooks, S., Lumyong, S., & Hyde, K. D. (2019). Use of endophytes as biocontrol agents. *Fungal Biology Reviews*, *33*, 133–148.

del Carmen Orozco-Mosqueda, M., Glick, B. R., & Santoyo, G. (2020). ACC deaminase in plant growth-promoting bacteria (PGPB): An efficient mechanism to counter salt stress in crops. *Microbiological Research*, *235*, 126439.

Falardeau, J., Wise, C., Novitsky, L., & Avis, T. J. (2013). Ecological and mechanistic insights into the direct and indirect antimicrobial properties of *Bacillus subtilis* lipopeptides on plant pathogens. *Journal of Chemical Ecology*, *39*, 869–878.

Fgaier, H., & Eberl, H. J. (2011). Antagonistic control of microbial pathogens under iron limitations by siderophore producing bacteria in a chemostat setup. *Journal of Theoretical Biology*, *21*, 103–114.

Gamalero, E., & Glick, B. R. (2015). Bacterial modulation of plant ethylene levels. *Plant Physiology*, *169*, 13–22.

Gao, D., & Tao, Y. (2012). Current molecular biologic techniques for characterizing environmental microbial community. *Frontiers of Environmental Science & Engineering*, *6*, 82–97.

Glick, B. R. (2012). Plant growth-promoting bacteria: Mechanisms and applications. *Scientifica*, *2012*, 963401.

Gos, F. M., Savi, D. C., Shaaban, K. A., Thorson, J. S., Aluizio, R., Possiede, Y. M., ... Glienke, C. (2017). Antibacterial activity of endophytic actinomycetes isolated from the medicinal plant *Vochysia divergens* (Pantanal, Brazil). *Frontiers in Microbiology*, *8*, 1642. Available from https://doi.org/10.3389/fmicb.2017.01642.

Gouda, S., Das, G., Sen, S. K., Shin, H. S., & Patra, J. K. (2016). Endophytes: A treasure house of bioactive compounds of medicinal importance. *Frontiers in Microbiology*, *7*, 1538.

Gupta, A., Verma, H., Singh, P. P., Singh, P., Singh, M., Mishra, V., & Kumar, A. (2019). Rhizome endophytes: Roles and applications in sustainable agriculture. In S. Verma, & J. White, Jr (Eds.), *Seed endophytes* (pp. 405–421). Cham: Springer, https://doi.org/10.1007/978-3-030-10504-4_19.

Hardoim, P. R., van Overbeek, L. S., Berg, G., Pirttilä, A. M., Compant, S., Campisano, A., ... Sessitsch, A. (2015). The hidden world within plants: Ecological and evolutionary considerations for defining functioning of microbial endophytes. *Microbiology and Molecular Biology Reviews: MMBR*, *79*, 293–320.

Hashem, A., Tabassum, B., & Abd_Allah, E. F. (2019). *Bacillus subtilis*: A plant-growth promoting rhizobacterium that also impacts biotic stress. *Saudi Journal of Biological Sciences*, *26*, 1291–1297.

Hider, R. C., & Kong, X. (2010). Chemistry and biology of siderophores. *Natural Product Reports*, *27*, 637–657.

Jadhav, H. P., Shaikh, S. S., & Sayyed, R. Z. (2017). *Role of hydrolytic enzymes of rhizoflora in biocontrol of fungal phytopathogens: An overview. Rhizotrophs: Plant growth promotion to bioremediation* (pp. 183–203). Singapore: Springer.

Kandel, S. L., Joubert, P. M., & Doty, S. L. (2017). Bacterial endophyte colonization and distribution within plants. *Microorganisms*, *5*(4), 77.

Kannojia, P., Choudhary., Srivastava, K. K., & Singh, A. K. (2019). *PGPR bioelicitors: Induced systemic resistance (ISR) and proteomic perspective on biocontrol. PGPR amelioration in sustainable agriculture* (pp. 67–84). Woodhead Publishing.

Kang, J. W., Khan, Z., & Doty, S. L. (2012). Biodegradation of trichloroethylene by an endophyte of hybrid poplar. *Applied and Environmental Microbiology*, *78*, 3504–3507.

Kumar, A., Droby, S., Singh, V. K., Singh, S. K., & White, J. F. (2020). Entry, colonization, and distribution of endophytic microorganisms in plants. In A. Kumar, & E. K. Radhakrishnan (Eds.), *Microbial endophytes* (pp. 1–33). Woodhead Publishing. Available from https://doi.org/10.1016/B978–0–12–819654-0.00001–6.

Kumar, A., Singh, R., Yadav, A., Giri, D. D., Singh, P. K., & Pandey, K. D. (2016). Isolation and characterization of bacterial endophytes of *Curcuma longa* L. *3 Biotech*, *6*(1), 60.

Kumar, A., Zhimo, Y., Biasi, A., Salim, S., Feygenberg, O., Wisniewski, M., & Droby, S. (2021). Endophytic microbiome in the carposphere and its importance in fruit physiology and pathology. In D. Spadaro, S. Droby, & M. L. Gullino (Eds.), *Postharvest pathology. Plant pathology in the 21st century* (vol 11, pp. 73–88). Cham: Springer. Available from https://doi.org/10.1007/978–3–030–56530-5_5.

Lahlali, R., McGregor, L., Song, T., Gossen, B., Narisawa, K., & Peng, G. (2014). *Heteroconium chaetospira* induces resistance to clubroot via upregulation of host genes involved in jasmonic acid, ethylene, and auxin biosynthesis. *PLoS One*, *9*. Available from https://doi.org/10.1371/journal.pone.0094144, April.

Latz, M. A. C., Jensen, B., Collinge, D. B., & Jørgensen, H. J. L. (2018). Endophytic fungi as biocontrol agents: Elucidating mechanisms in disease suppression. *Plant Ecology & Diversity*, *11*(5–6), 555–567. Available from https://doi.org/10.1080/17550874.2018.1534146.

Liu, H., Carvalhais., Crawford, L. C., Singh, M., Dennis, E., Pieterse, P. G., ... Schenk, P. M. (2017). Inner plant values: Diversity, colonization and benefits from endophytic bacteria. *Frontiers in Microbiology*, *8*, 1–17.

Liu, K., Newman, M., McInroy, J. A., Hu, C. H., & Kloepper, J. W. (2017). Selection and assessment of plant growth-promoting rhizobacteria for biological control of multiple plant diseases. *Phytopathology*, *107*, 928–936.

Milner, J., Silo-Suh, L., Goodman, R. M., & Handelsman, J. (2019). *Antibiosis and beyond: Genetic diversity, microbial communities, and biological control. Ecological interactions and biological control* (pp. 107–127). CRC Press.

Mishra, S., & Arora, N. K. (2012). Evaluation of rhizospheric *Pseudomonas* and *Bacillus* as biocontrol tool for *Xanthomonas campestris pv. campestris*. *World Journal of Microbiology and Biotechnology*, *28*, 693–702.

Mohamad, O. A. A., Abdalla, O., Li, L., Ma, J., Hatab, S. R., Xu, L., ... Li, W. J. (2018). Evaluation of the antimicrobial activity of endophytic bacterial populations from Chinese traditional medicinal plant licorice and characterization of the bioactive secondary metabolites produced by *Bacillus atrophaeus* against *Verticillium dahliae*. *Frontiers in Microbiology*, *9*, 924. Available from https://doi.org/10.3389/fmicb.2018.00924.

Mohandoss J. & Suryanaraynan T. S., (2009). *Effect of fungicide treatment on foliar fungal endophyte diversity in mango Sydowia-Horn.*

Mota, M. S., Gomes, C. B., Souza Júnior, I. T., & Moura, A. B. (2017). Bacterial selection for biological control of plant disease: Criterion determination and validation. *Brazilian Journal of Microbiology*, *48*, 62–70.

Nithya, K., & Muthumary, J. (2011). Bioactive metabolite produced by *Phomopsis* sp., an endophytic fungus in *Allamanda cathartica* Linn. *Recent Research in Science and Technology*, *3*, 44–48.

Ongena, M., & Jacques, P. (2008). *Bacillus lipopeptides*: Versatile weapons for plant disease biocontrol. *Trends in Microbiology*, *16*, 115–125.

Pinheiro, E. A. A., Carvalho, J. M., Dos Santos, D. C. P., Feitosa, A. D. O., Marinho, P. S. B., Guilhon, G. M. S. P., ... Marinho, A. M. R. M. (2013). Antibacterial activity of alkaloids produced by endophytic fungus *Aspergillus sp*. EJC08 isolated from medical plant *Bauhinia guianensis*. *Natural Product Research*, *27*, 1633–1638. Available from https://doi.org/10.1080/14786419.2012.750316.

Rajamanikyam, M., Vadlapudi, V., & Upadhyayula, S. M. (2017). Endophytic fungi as novel resources of natural therapeutics. *Brazilian Archives of Biology and Technology*, *60*, e17160542.

Regnault-Roger, C. (2012). *Trends for commercialization of biocontrol agent (biopesticide) products. Plant defence: Biological control* (pp. 139–160). Dordrecht: Springer.

Rodriguez, R. J., White, Jf, Arnold, A. E., & Redman, R. S. (2009). Fungal endophytes: Diversity and functional roles. *New Phytologist*, *182*, 182314–182330. Available from https://doi.org/10.1111/j.1469-8137.2009.02773.x.

Saber, W. I., Ghoneem, K. M., Al-Askar, A. A., Rashad, Y. M., Ali, A. A., & Rashad, E. M. (2015). Chitinase production by *Bacillus subtilis* ATCC 11774 and its effect on biocontrol of *Rhizoctonia* diseases of potato. *Acta Biologica Hungarica*, *66*, 436–448.

Saikkonen, K., Faeth, S. H., Helander, M., & Sullivan, T. J. (1998). Fungal endophytes: A continuum of interactions with host plants. *Annual Review of Ecology and Systematics*, *29*, 319–343.

Santiago, C., Fitchett, C., Munro, M. H., Jalil, J., & Santhanam, J. (2012). Cytotoxic and antifungal activities of 5-hydroxyramulosin, a compound produced by an endophytic fungus isolated from *Cinnamomum mollisimum*. *Evidence-Based Complementary and Alternative Medicine: eCAM*, *2012*, 689310.

Santoyo, G., Sánchez-Yáñez, J. M., & de los Santos-Villalobos, S. (2019). *Methods for detecting biocontrol and plant growth-promoting traits in rhizobacteria. Methods in rhizosphere biology research* (pp. 133–149). Singapore: Springer.

Santoyo, G., Orozco-Mosqueda, M. D. C., & Govindappa, M. (2012). Mechanisms of biocontrol and plant growth-promoting activity in soil bacterial species of Bacillus and Pseudomonas: A review. *Biocontrol Science and Technology*, *22*, 855–872.

Santoyo, G., Moreno-Hagelsieb, G., del Carmen Orozco-Mosqueda, M., & Glick, B. R. (2016). Plant growth-promoting bacterial endophytes. *Microbiological Research*, *183*, 92–99.

Shen, F. T., Yen, J. H., Liao, C. S., Chen, W. C., & Chao, Y. T. (2019). Screening of rice endophytic biofertilizers with fungicide tolerance and plant growth-promoting characteristics. *Sustainability*, *11*, 1133.

Sindhu, S. S., & Dadarwal, K. R. (2001). Chitinolytic and cellulolytic *Pseudomonas* sp. antagonistic to fungal pathogens enhances nodulation by *Mesorhizobium* sp. cicer in chickpea. *Microbiological Research*, *156*, 353–358.

Singh, D., Singh, S. K., Modi, A., Singh, P. K., Zhimo, V. Y., & Kumar, A. (2020). Impacts of agrochemicals on soil microbiology and food quality. In M. N. V. Prasad (Ed.), *Agrochemicals detection, treatment and remediation* (pp. 101–116). Butterworth-Heinemann. Available from https://doi.org/10.1016/B978–0–08–103017-2.00004–0.

Singh, M., Srivastava, M., Kumar, A., Singh, A. K., & Pandey, K. D. (2020). Endophytic bacteria in plant disease management. In A. Kumar, & V. K. Singh (Eds.), *Microbial endophytes* (pp. 61–89). Woodhead Publishing. Available from https://doi.org/10.1016/B978–0–12–818734-0.00004–8.

Singh, R., & Dubey, A. K. (2018). Diversity and applications of endophytic actinobacteria of plants in special and other ecological niches. *Frontiers in Microbiology*, *9*, 1767. Available from https://doi.org/10.3389/fmicb.2018.01767.

Singh, V. K., Singh, A. K., & Kumar, A. (2017). Disease management of tomato through PGPB: Current trends and future perspective. *3 Biotech*, *7*(4), 1−10.

Valenzuela-Ruiz, V. V., Gamboa, G. T. G., Rodríguez, E. D. V., Cota, F. I. P., Santoyo, G., & de los Santos Villalobos, S. (2020). Lipopéptidos producidos por agentes de control biológico del género *Bacillus*: Revisión de herramientas analíticas utilizadas para su estudio. *Revista Mexicana de Ciencias Agrícolas*, *11*, 419−432.

Villarreal-Delgado, M. F., Villa Rodríguez, E. D., Cira-Chávez, L. A., Estrada-Alvarado, M. I., Parra-Cota, F. I., & de los Santos-Villalobos, S. (2018). The genus *Bacillus* as a biological control agent and its implications in the agricultural biosecurity. *Mexican Journal of Phytopathology*, *36*, 95−130.

Wani, Z. A., Ashraf, N., Mohiuddin, T., & Riyaz-Ul-Hassan, S. (2015). Plant-endophyte symbiosis, an ecological perspective. *Applied Microbiology and Biotechnology*, *99*, 2955−2965.

Weller, S. C., Culbreath, A. K., Gianessi, L., & Godfrey, L. D. (2014). *The contributions of pesticides to pest management in meeting the global need for food production by 2050*. Ames, IA: Council for Agricultural Science and Technology (CAST).

Wu, S. H., Huang, R., Miao, C. P., & Chen, Y. W. (2013). Two new steroids from an endophytic fungus *Phomopsis sp. Chemistry & Biodiversity*, *10*, 1276−1283. Available from https://doi.org/10.1002/cbdv.201200415.

Xu, X. H., Su, Z. Z., Wang, C., Kubicek, C. P., Feng, X. X., Mao, L. J., & Zhang, C. L. (2014). The rice endophyte *Harpophora oryzae* genome reveals evolution from a pathogen to a mutualistic endophyte. *Scientific Reports*, *4*, 1−9.

Yadav, A., & Yadav, K. (2017). Exploring the potential of endophytes in agriculture: A minireview. *Advances in Plants & Agriculture Research*, *6*(4), 102−106.

Yu, X., Zhang, W., Lang, D., Zhang, X., Cui, G., & Zhang, X. (2019). Interactions between endophytes and plants: Beneficial effect of endophytes to ameliorate biotic and abiotic stresses in plants. *Journal of Plant Biology*, *62*, 1−13.

CHAPTER

Biocontrol applications of microbial metabolites

9

Dibya Jyoti Hazarika[1,2], Merilin Kakoti[1], Ashok Bhattacharyya[3] and Robin Chandra Boro[1]

[1]*Department of Agricultural Biotechnology, Assam Agricultural University, Jorhat, India*

[2]*DBT - North East Centre for Agricultural Biotechnology, Assam Agricultural University, Jorhat, India*

[3]*Department of Plant Pathology, Assam Agricultural University, Jorhat - 785013, Assam, India*

9.1 Introduction

Microbial secondary metabolites are low molecular weight compounds produced through secondary metabolism. Secondary metabolism involves the biosynthetic pathways and their products that are not required for the growth of an organism, rather essential as a defense mechanism to compete in a harsh environment in nature. Diverse categories of small molecules, including antibiotics, toxins, pheromones, pigments, and effectors, are synthesized by microbes, plants, and animals through different secondary metabolism pathways (Demain, 1992). Microbial secondary metabolites are often specific to a particular group of microbes, sometimes produced by specific genera or species. These metabolites possess unusual chemical structures such as β-lactam ring, cyclic peptides with the combination of natural and unnatural (nonprotein) amino acids, large macrolide rings, unusual sugars and oligosaccharides, and other complex structures, which are fairly different from the primary metabolites from which they are synthesized using a series of enzymes.

A large number of secondary metabolites from microbial origin have been discovered from different bacterial, algal, and fungal species. A major fraction of these metabolites play crucial roles in health, nutrition, and agriculture, thus shaping the economy of our society. Several secondary metabolites have emerged as medicines in clinical biology due to their remarkable therapeutic potential. For example, penicillin—the first antibiotic discovered by Alexander Fleming revolutionized clinical biology and opened up a new door for treating infectious diseases. The post-penicillin era has witnessed the discovery of several other natural and synthetic antibiotics, such as streptomycin, tetracycline, and ciprofloxacin.

Biocontrol Mechanisms of Endophytic Microorganisms. DOI: https://doi.org/10.1016/B978-0-323-88478-5.00010-9

Interspecific competition among different microbial species is a crucial aspect of synthesizing secondary metabolites with antibiotic potential (Hibbing, Fuqua, Parsek, & Peterson, 2010). Interspecific competition occurs among different microbes in an ecological niche for nutrition and space. Antibiotic production provides superiority to a particular microbial strain (i.e., the producer) to antagonize the other microbes, thereby gaining enhance access to food and space. This feature empowers a particular species or strain of microbes to be used as a biocontrol agent for agricultural pest management. Biocontrol activity of a microbial strain depends on the type(s) of the antibiotic compound(s) produced by the strain as well as the capability of that strain to synthesize an antibiotic compound in a particular natural environment where it is introduced. In many cases, the biocontrol agent is unable to perform its function in the new environment as the conditions may not be suitable for the production of antibiotics. In such cases, alternate strategies are used to obtain the biocontrol functions of the microbial strains. This chapter reviews the diverse secondary metabolism strategies in microbes, which make them potent biocontrol agents. The chapter also talks about the application potential of microbial metabolites in agricultural biology for controlling important plant diseases. As it is not possible to include all sources of microbial secondary metabolism in a single chapter, our discussions are limited to the biocontrol potential of secondary metabolites from eubacteria and actinomycetes.

9.2 Microbes for biological control

Agricultural crops encounter approximately 70,000 species of pathogenic organisms, out of which only 10% are regarded as serious pests (Pimentel, 1997). Few bacterial species and several fungi are well known for causing important diseases in plants, which result in a substantial yield loss in agricultural crops (Agrios, 2005). To maintain the crop yield qualitatively and quantitatively, the disease management strategies are highly essential. The farmers often use synthetic fungicides and bactericides on a regular basis to control fungal and bacterial diseases in crops. However, excessive use and misuse of synthetic chemicals have adverse effects on human health and the environment, due to their ability to cause acute or chronic toxicity (Samy, Xaviar, Rahman, & Sharifuddin, 1995). In this regard, biological control is emerging as a fruitful alternative measure to synthetic chemicals for controlling plant diseases. The ability of a few microbial species to antagonize the plant pathogenic microorganisms is considered an advantageous phenomenon in the management of plant diseases.

Many microbial species have been reported to show antagonistic potential against plant pathogenic bacteria and fungi. *Bacillus subtilis, Bacillus amyloliquefaciens*, *Burkholderia cepacia*, *Paenibacillus polymyxa*, *Pseudomonas fluorescens*, *Pseudomonas aeruginosa, Serratia liquefaciens, Serratia plymuthica, Serratia proteamaculans*, and many others are important eubacterial species with biocontrol potential against fungal phytopathogens (Table 9.1). Some of these bacterial species produce diverse antifungal metabolites, whereas some others produce chitinolytic enzymes as their weapon against fungi. Members of the genus *Bacillus*, including few important species, such as *B. subtilis, B. amyloliquefaciens*, and *Bacillus pumilus*, are soil-borne; however, these are often reported as endophytes in different plants (Hazarika et al., 2019; Lima et al., 2005; Liu et al., 2009; Sun, Lu, Bie, Lu, & Yang, 2006) and well known for their broad-spectrum biocontrol potential (Hazarika et al., 2019; Liu et al., 2009; Ren et al., 2013). Few actinobacterial species in the genus *Streptomyces*, including *Streptomyces goshikiensis, Streptomyces griseus*, and *Streptomyces hydrogenans*, are well known for their biocontrol potential against different fungal phytopathogens (Faheem et al., 2015; Kaur & Manhas, 2014; Manhas & Kaur, 2016; Nguyen et al., 2012). For instance, *S. hydrogenans* DH16, a soil isolate from Himachal Pradesh, India showed antifungal activity against multiple fungal phytopathogens, namely, *Alternaria alternata, Alternaria brassicicola, Colletotrichum acutatum, Cladosporium herbarum, Exserohilum* sp., *Cercospora* sp., *Fusarium moniliforme*, and *Fusarium oxysporum* with more than 90% disease suppression ability of the culture supernatant against black leaf spot of *Raphanus sativus* caused by *A. brassicicola* (Kaur & Manhas, 2014).

9.3 Antifungal metabolites from microbes

The antifungal properties of microbes can be correlated with their ability to produce selective secondary metabolites and enzymes. Antifungal secondary metabolites can be classified into two major groups: polyketides and nonribosomal peptides (NRPs). There are huge structural dissimilarities between these two groups of metabolites (Fig. 9.1). Apart from these, two other groups of secondary metabolites, terpenoids and shikimic acid derivatives (often found as precursors for polyketide biosynthesis), are also produced by a few eubacterial and actinobacterial species, some of which possess antifungal activity (Gozari, Alborz, El-Seedi, & Jassbi, 2020; Wilson, Patton, Florova, Hale, & Reynolds, 1998). For example, marinocyanins—a group of meroterpenoid from marine actinomycetes strains—show antifungal as well as antibacterial activities (Asolkar et al., 2017). *Lysobacter antibioticus* produces an antifungal compound *p*-aminobenzoic acid, which is a derivative of shikimic acid (Laborda, Zhao, Ling, Hou, & Liu, 2018).

There are fewer reports on the antifungal activities of polyketides from bacterial origin. A few strains of the genus *Burkholderia* synthesize polyketides type of secondary metabolites with antagonistic and growth-suppressive ability against fungi (Flórez et al., 2018; Niehs et al., 2020; Ross, Opel, Scherlach, & Hertweck, 2014). A recent study has identified gladiofungins as an antifungal polyketide produced by *Burkholderia gladioli* (Niehs et al., 2020). The two types of gladiofungins (gladiofungin A and gladiofungin B) are synthesized in *B. gladioli* using

Table 9.1 A comprehensive list of antifungal metabolites from different bacterial species and their modes of action against various fungal pathogens.

Sl. no.	Name of metabolites	Name of the organisms	Target pathogens/diseases	Mechanisms of action/mode of action	References
1	2,4-Diacetylphloroglucinol	*Pseudomonas fluorescens, Pseudomonas* sp.	*Gaeumannomyces graminis var. tritici, Thielaviopsis basicola, Pythium ultimum, Rhizoctonia solani, Fusarium oxysporum f. sp. radicis-lycopersici*	Impairs mitochondrial function	Howell and Stipanovic (1979), Keel et al. (1996), Raaijmakers and Weller (2001), Mazzola, et al. (2004), Almario et al. (2017)
2	Amphisin	*Burkholderia cepacia, Pseudomonas cepacia, P. fluorescens*	*P. ultimum, R. solani, Pyricularia oryzae*	Inhibits membrane transport, impairs mitochondrial function	Homma et al. (1989), Nielsen et al. (2002), Soerensen, Nielsen, Christophersen, Soerensen and Gajhede (2001)
3	Anthranilate	*Pseudomonas aeruginosa*	*Fusarium* spp. *Pythium* spp.	Unclear	Anjaiah et al. (1998)
4	Bacillomycin D	*Bacillus vallismortis, Bacillus amyloliquefaciens, Bacillus subtilis, Bacillus velezensis*	*Fusarium graminearum, Alternaria alternata, Colletotrichum gloeosporioides* (Penz.), *R. solani, Cryphonectria parasitica* and *Phytophthora capsici*	Cell wall and plasma membrane destability	Zhao et al. (2010), Xu et al. (2013), Gong et al. (2014), Jin et al. (2020)
5	Bacilysin	*B. subtilis, B. amyloliquefaciens*	*Candida albicans*	Inhibition of cell wall biosynthesis	Chen et al., 2009, Helfrich and Piel (2016)
6	Fengycin/plipastatin	*B. subtilis, B. amyloliquefaciens*	*Podosphaera fusca, Botrytis cinerea, F. graminearum, Sclerotinia sclerotiorum, F. oxysporum*	Membrane, permeabilization and disruption, plasma membrane destability	Koumoutsi et al. (2004), Ramarathnam et al. (2007), Romero et al. (2007), Aleti, Sessitsch, & Brader (2015), Gong, Li, and Yuan (2015), Cochrane & Vederas (2016)

7	Fusaricidin	*Paenibacillus polymyxa*	*F. oxysporum, Phytophthora sp., Leptosphaeria maculans, Aspergillus sp.*	Damage in membrane structure	Aleti, Sessitsch, and Brader (2015), Beatty and Jensen (2002), Li and Chen (2019), Raza et al. (2009), Lee et al. (2013), Vater, Niu, Dietel and Borriss (2015), Yu and Ye (2016)
8	Hydrogen cyanide	*P. fluorescens, P. aeruginosa*	*T. basicola, R. solani, Meloidogyne javanica*	Inhibits mitochondrial cytochrome oxidase	Voisard, Keel, Haas, and Defago (1989), Siddiqui, Shaukat, Khan, and Ali, (2003), Jayaprakashvel et al. (2010)
9	Iturin	*B. subtilis, B. amyloliquefaciens, Bacillus megaterium*	*B. cinerea, R. solani, Pythium aphanidermatum, P. fusca*	Increasing of membrane permeability	Kefi et al. (2015); Romero et al. (2007); Yu, Sinclair, Hartman, & Bertagnolli (2002); Zouari, Jlaiel, Tounsi, and Trigui (2016)
10	Kurstakin	*Bacillus thuringiensis* subsp. *kurstaki*	*Stachybotrys chartarum*	Unknown	Bechet et al. (2012)
11	Marihysin A	*Bacillus marinus*	*Alternaria solani, F. oxysporum, F. graminearum, Verticillium albo-atrum, Sclerotium* sp., *Penicillium* sp., *R. solani, Colletotrichum* sp.	Unknown	Liu, Zhang, Li, Tao, and Tian (2010)
12	Mojavensin A	*Bacillus mojavensis*	*F. oxysporum f.* sp. *cucumerinum, Valsa mali*	Increasing of membrane permeability	Ma, Wang, Hu, and Wang (2012)

(*Continued*)

Table 9.1 A comprehensive list of antifungal metabolites from different bacterial species and their modes of action against various fungal pathogens. *Continued*

Sl. no.	Name of metabolites	Name of the organisms	Target pathogens/diseases	Mechanisms of action/mode of action	References
13	Mycosubtilin	*B. subtilis*	*F. oxysporum, B. cinerea, P. aphanidermatum, Pichia pastoris, Saccharomyces cerevisiae*	Increasing of membrane permeability, forms pores in the plasma membrane	Besson, Peypoux, and Michel (1978), Leclere et al. (2005)
14	Oomycin A	*P. fluorescens*	*P. ultimum*	Unclear	Howie and Suslow (1991)
15	Paenilamicin	*Paenibacillus larvae*	*S. cerevisiae, P. pastoris, F. oxysporum*	Unknown	Garcia-Gonzalez et al. (2014), Müller et al. (2014),
16	Paenilarvin	*Paenibacillus larvae*	*Mucor hiemalis, Aspergillus clavatus, Penicillium capsulatum, Botryotinia fukeliana*	Increasing of membrane permeability	Sood et al. (2014)
17	Phenazine-1-carboxamide	*P. aeruginosa, Pseudomonas chlororaphis*	*F. oxysporum* f. sp. *radicis-lycopersici*	Oxidative damage	Chin-A-Woeng, Bloemberg, Mulders, Dekkers and Lugtenberg (2000), Mavrodi et al. (2001), Peng et al. (2018), van Rij, Wesselink, Chin-A-Woeng, Bloemberg, and Lugtenberg (2004)
18	Phenazine-1-carboxylic acid	*P. fluorescens, Pseudomonas aureofaciens*	*Gaeumannomyces gramini var. tritici, R. solani*	Oxidative damage	Thomashow, Weller, Bonsall, and Pierson (1990), Pierson and Pierson (1996), Jaaffar, Parejko, Paulitz, Weller, and Thomashow (2017)

19	Pseudane	*P. cepacia*	*P. ultimum, R. solani, P. oryzae*	Inhibits membrane transport	Homma et al. (1989), Nielsen et al. (2002); Nielsen, Christophersen, Anthoni, and Sørensen (1999)
20	Pyocyanin	*P. aeruginosa, P. chlororaphis*	*R. solani, F. graminearum*	Inhibits membrane transport, regulation of pleiotropic drug resistance	Hassan and Fridovich (1980), Baron and Rowe (1981), De Vleesschauwer, Cornelis, and Höfte (2006), Dharni et al. (2012), Houshaymi, Awada, Kedees, and Soayfane (2019), Kerr et al. (1999)
21	Pyoluteorin	*P. fluorescens*	*P. ultimum*	Unclear	Howell and Stipanovic (1980), Hill et al. (1994), Burkhead, Schisler, and Slininger (1994)
22	Pyrrolnitrin	*P. fluorescens, P. cepacia, B. cepacia, Serratia plymuthica*	*Aphanomyces cochlioides, R. solani, Pyrenophora tritici-repens, P. ultimum, P. oryzae*	Impairs electron transport chain, inhibits membrane transport, impairs mitochondrial function	Homma et al. (1989), Hill et al. (1994), Burkhead, Schisler, and Slininger (1994), Nielsen et al. (1999), Nielsen et al. (2002), Levenfors et al. (2004)
23	Soraphen A	*Sorangium cellulosum* (A myxobacterium)	Erysiphe mildew in barley and snow mold in rye. Apple scab and gray mold on grapes	Inhibition of acetyl-CoA carboxylase function	Gerth, Bedorf, Irschik, H€ofle, and Reichenbach (1994), Shen, Volrath, Weatherly, Elich, and Tong (2004)
24	Surfactin	*B. subtilis, B. amyloliquefaciens, Bacillus pumilus, Bacillus licheniformis, Bacillus coagulans*	*B. cinerea, C. albicans, Colletotrichum gossypii, Helminthosporium maydis, Fusarium oxysporum, Gibberella saubinetii, Physalospora piricola, R. solani, S. sclerotiorum*	Pore formation in the cell membrane and induction of apoptosis	Ongena and Jacques (2008), Qi et al. (2010), Beric et al. (2012), Aleti, Sessitsch, and Brader (2015), Kefi et al. (2015), Sumi, Yang, Yeo, and Hahm (2015), Hazarika et al. (2019)

(Continued)

Table 9.1 A comprehensive list of antifungal metabolites from different bacterial species and their modes of action against various fungal pathogens. *Continued*

Sl. no.	Name of metabolites	Name of the organisms	Target pathogens/diseases	Mechanisms of action/mode of action	References
25	Tensin	*P. fluorescens*	*R. solani*	Inhibits membrane transport, impairs mitochondrial function	Nielsen et al. (2000), Henriksen et al. (2000), Nielsen et al. (1999)
26	Viscosinamide	*P. cepacia, P. fluorescens*	*P. ultimum, R. solani, P. oryzae*	Inhibits membrane transport, impairs mitochondrial function	Nielsen et al. (1999), Thrane, Nielsen, Nielsen, Olsson, and Sørensen (2000)
27	Xanthobaccin A	*Lysobacter* sp.	*A. cochlioides*	Not clear (inhibition of cell wall synthesis/ membrane disruption)	Nakayama, Homma, Hashidoko, Mizutani, and Tahara (1999), Islam, Hashidoko, Deora, Ito, and Tahara (2005)
28	Zwittermicin A	*Bacillus cereus*	*Phytophthora medicaginis, Pythium aphanidermatum*	Not clear (may inhibit protein synthesis)	He, Silo-Suh, Handelsman, and Clardy (1994), Kevany, Rasko, and Thomas (2009)

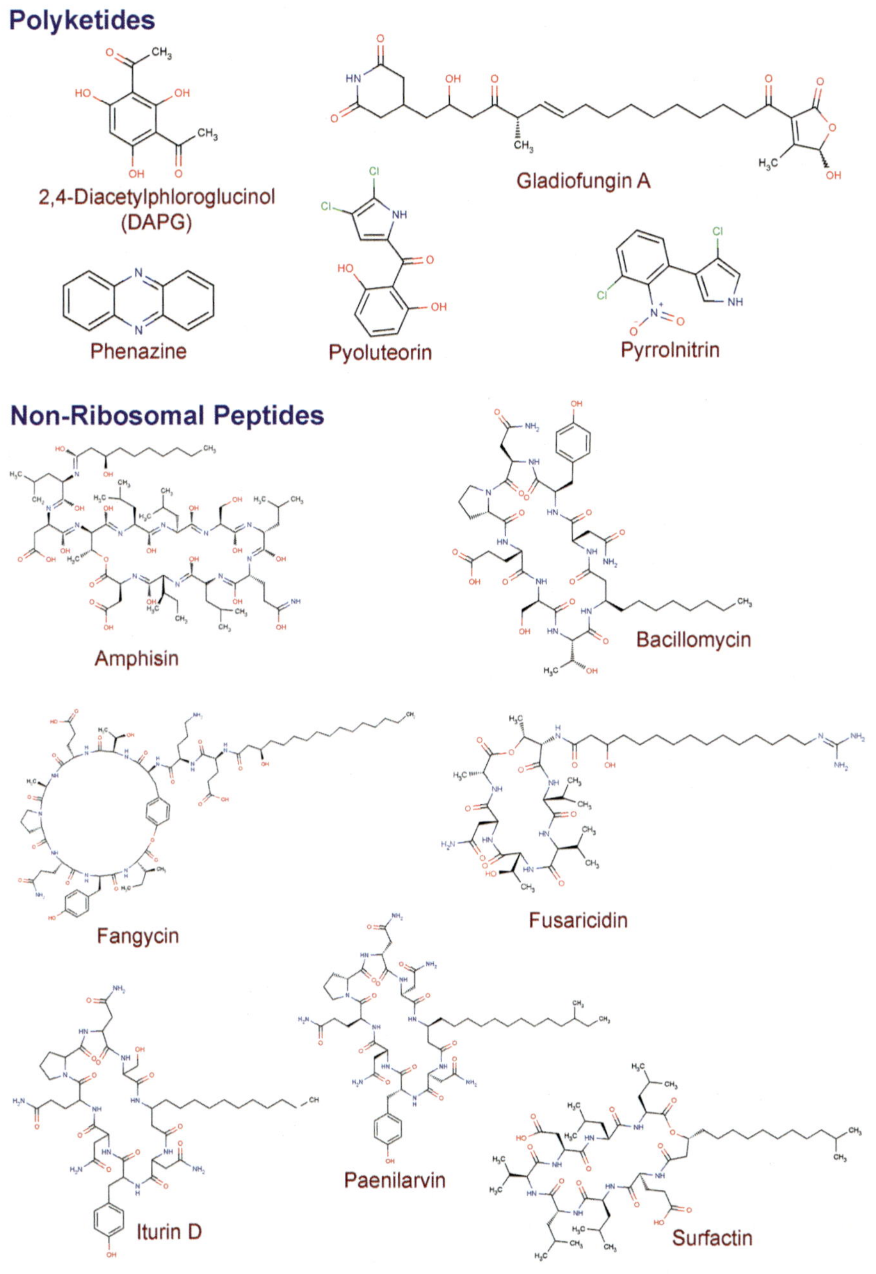

FIGURE 9.1

Chemical structures of few antifungal metabolites produced by different bacterial species.

an orphan gene cluster encoding a trans-AT modular polyketide synthase and accessory proteins. Antifungal activity of gladiofungin A has been demonstrated against *Penicillium notatum*, *Sporobolomyces salmonicolor*, and *Purpureocillium lilacinum* (an entomopathoginic fungus) (Niehs et al., 2020). However, the mode of action of this compound is still unclear. Another study described the production of two antifungal polyketides, pyoluteorin and phenazine, by *P. fluorescens* (Kilani-Feki et al., 2010). These two metabolites are synthesized by the biosynthetic enzymes of pyoluteorin and phenazine gene clusters, respectively (Liu, Dong, Peng, Zhang, & Xu, 2006). These two gene clusters were detected in *P. fluorescens* Pf1TZ and the strain showed strong antifungal activity against several phytopathogenic fungi (Kilani-Feki et al., 2010). 2,4-Diacetylphloroglucinol (DAPG)—an antifungal polyketide—is also produced by a few species of *Pseudomonas* (Almario et al., 2017; Mazzola, Funnell, & Raaijmakers, 2004; Raaijmakers & Weller, 2001). The genetic and physiological investigations suggested that the DAPG acts as a proton ionophore and interferes with the proton gradient across the mitochondrial membrane in *Saccharomyces cerevisiae* (Troppens, Dmitriev, Papkovsky, O'Gara, & Morrissey, 2013). This compound also alters the expression of several fungal cytochrome P450 (*cyp*) genes, targeting the fungal cell membrane formation (Gong et al., 2016). Pyrrolnitrin [3-chloro-4-(2′-nitro-3′-chlorophenyl)pyrrole]—an antibiotic compound first reported from *Pseudomonas pyrrocinia* (Arima, Imanaka, Kousaka, Fukuta, & Tamura, 1964), shows antagonistic activity against yeast, fungi, and Gram-positive bacteria (Arima, Imanaka, Kousaka, Fukuda, & Tamura, 1965). Later, this compound was reported from a large variety of bacterial species, including *B. cepacia*, *Corallococcus exiguous*, *Cystobacter ferrugineus*, *Enterobacter agglomerans*, *Myxococcus fulvus, Pseudomonas* sp., and *Serratia* species (Liu et al., 2007; Van Pée & Ligon, 2000). The antibiotic activity of pyrrolnitrin is mediated by the inhibition of the enzyme glycerol kinase, which leads to the accumulation of glycerol in the cell causing leaky cell membranes (Jespers & De Waard, 1995). Pyrrolnitrin also impairs the electron transport chain and uncouples oxidative phosphorylation in the mitochondria as evidenced in *Neurospora crassa* (Lambowitz & Slayman, 1972). Metabolites from *S. plymuthica* include Heterumalides (NA, B, NE, and X), pyrrolnitrin, and 1-acetyl-7-chloro-1-H-indole. Structural similarities of heterumalides to other compounds suggest that the biosynthesis of these compounds is mediated by a type I polyketide synthase cluster, which is less commonly available in Gram-negative bacteria and detected earlier in few species of the genus *Pseudomonas* (Nowak-Thompson, Gould, & Loper, 1997; Nowak-Thompson, Chaney, Wing, Gould, & Loper, 1999; Rangaswamy, Jiralerspong, Parry, & Bender, 1998). Heterumalide NA, B, and NE were reported to prevent conidia and spore germination of many filamentous fungi as well as Oomycetes at concentrations ranging from 0.4 to 40 μg/mL (Levenfors, Hedman, Thaning, Gerhardson, & Welch, 2004).

There are plenty of reports describing the ability of endophytic and rhizospheric bacteria to produce NRPs type of secondary metabolites with broad-spectrum antifungal activity. NRPs can be further classified into different groups, including the smaller dipeptides, and the larger cyclic peptides and lipopeptides. All NRPs are synthesized by specialized NRP synthetases (NRPSs) and their accessory proteins. NRPS-derived peptide synthesis has several unique features. For example, the noncanonical amino acids can also be utilized for the synthesis of peptides by NRPSs, which enables high structural diversity (Geissler, Heravi, Henkel, & Hausmann, 2019). These large multienzyme complexes are structured in modules that perform iterative functions (Hamdache, Azarken, Lamarti, Aleu, & Collado, 2013).

Few members of the genus *Bacillus* produce multiple lipopeptides that share common structural and functional similarities and dissimilarities. For example, three lipopeptides—surfactin, iturin, and fengycin—share some common features. These lipopeptides work as surface-active agents (surfactants) and possess antimicrobial activities. Several reports have demonstrated antifungal properties of these biosurfactant molecules, which enables their applications as potential biocontrol agents (Alvarez et al., 2012; Hazarika et al., 2019; Kim, Ryu, Kim, & Chi, 2010; Toral, Rodríguez, Béjar, & Sampedro, 2018). It has been reported that fengycin and other lipopeptides target the fungal cell membrane to exhibit antifungal properties (Gong et al., 2015; Tao, Bie, Lv, Zhao, & Lu, 2011). Fusaricidin—a group of cyclic lipopeptides produced by *P. polymyxa*—shows strong antifungal properties and plays a crucial role in controlling *Fusarium* wilt disease of cucumber (Li & Chen, 2019). It was observed from the proteomics data that the compounds affect the energy generation process in the respiratory chain and also inhibit amino acid biosynthesis (Yang et al., 2018). *Paenibacillus ehimensis* IB-X-b produces cyclic lipopeptides of bacillomycin L and fengycin/plipastatin/agrastatin families with strong antifungal potential against phytopathogens (Aktuganov, Jokela, & Kivelä, 2014). Experimental data suggest that bacillomycin L is not only involved in membrane permeabilization but also interacts with the intracellular targets, such as DNA (Zhang, Dong, Shang, Han, & Li, 2013). Bacillomycin D was also reported as a potent antifungal compound from *B. amyloliquefaciens, B. subtilis, B. velezensis*, and few other species (Gong et al., 2014; Xu et al., 2013). A recent study suggested that bacillomycin D showed stronger antifungal activity against *Colletotrichum gloeosporioides* (Penz.) than two other chemical fungicides mancozeb and prochloraz. Bacillomycin D caused damages to the cell wall and plasma membrane of the hyphae and spores of *C. gloeosporioides* (Penz.) and also affected the cytoplasm and organelles inside the fungal hyphae (Jin et al., 2020). *P. fluorescens* is one of the important bacterial species that possess broad-spectrum antifungal activity. Many NRPs are synthesized by *P. fluorescens*, which include viscosinamide (Nielsen, Christophersen, Anthoni, & Sørensen, 1999), tensin (Nielsen, Thrane, Christophersen, Anthoni, & Sørensen, 2000), orphamides (Gross et al., 2007), nunamycin, and nunapeptin (Michelsen, Watrous, & Glaring, 2015). Strains of *Pseudomonas syringae* also synthesize few antifungal peptides grouped as syringomycins, syringostatins, and syringotoxins (De Lucca & Walsh, 2000). Syringomycins—the small lipodepsipeptides are among the most potent antifungal peptides produced by bacteria. Syringomycin E is the most predominant form of this peptide group. This compound was found to be toxic to *Aspergillus flavus, A. fumigatus, A. niger, Fusarium moniliforme*, and *F. oxysporum*, with a 95% mortality at concentrations of 7.8 μg/mL for *A. flavus* and 1.9 μg/mL for the other fungi (De Lucca et al., 1999; Segre et al., 1989). Being an amphiphilic molecule, Syringomycin E interacts with plasma membranes of the target organisms. This compound is responsible to form pores in the plasma membrane, which is the underlying basis for its inhibitory activities (Hutchison, Tester, & Gross, 1995; Takemoto, 1992). Molecular genetic studies with yeast cells suggested that sterols and sphingolipids are important for the antifungal activity of this compound (Takemoto et al., 2003).

Many strains of *Streptomyces* also produce antifungal metabolites belonging to the NRPs group. For example, a new antifungal cyclic peptide Gloeosporiocide has been reported from *Streptomyces morookaense* AM25 isolated from the rhizospheric soil of *Paullinia cupana* in Amazon. Paper disk diffusion assay and microdilution assay revealed strong antifungal activity of this compound against the plant pathogen *C. gloeosporioides* (Vicente Dos Reis et al., 2019). Another example includes valinomycin—a peptide antibiotic reported from *Streptomyces* sp. strain M10. Valinomycin showed strong *in vitro* fungicidal activity against *Botrytis cinerea* and also effectively controlled Botrytis blight development in cucumber plants (Park, Lee, Lee, & Kim, 2008). A study on *Candida albicans* suggested that valinomycin interrupting cell membrane–associated function (Makarasen et al., 2018).

9.4 Antibacterial metabolites from microbes

Several bacterial metabolites belonging to the polyketides and NRPs families show antagonistic activity against bacterial pathogens of plants, animals, and humans (Fig. 9.2). Many strains of *B. subtilis, B. amyloliquefaciens*, and *B. pumilus* produce multiple antimicrobial compounds with remarkable biocontrol potential. In fact, the members of these species possess huge potential to control a broad-spectrum of plant pathogens, including bacteria and fungi. Various antibacterial compounds produced by *B. subtilis* and *B. amyloliquefaciens* include bacaucin, bacillaene, bacilysin, bacitracin, difficidin, gageotetrins, macrolactins, surfactin, and few others (Chakraborty et al., 2020; Hazarika et al., 2019; Patel et al., 1995; Wu et al., 2015; Yuan, Shuangyang, Richard, Erwin, & Kui, 2017). Bacillaene shows broad-spectrum antibacterial activity by inhibiting prokaryotic protein synthesis (Patel et al., 1995). Morphometric studies using scanning electron microscopy and transmission electron microscopy suggested that bacilysin and difficidin act on the cell wall of Gram-negative *Xanthomonas oryzae* pathovers, which resulted in the cell wall rupturing (Wu et al., 2015). Apart from showing antifungal activity, surfactin also possesses strong antibacterial activity

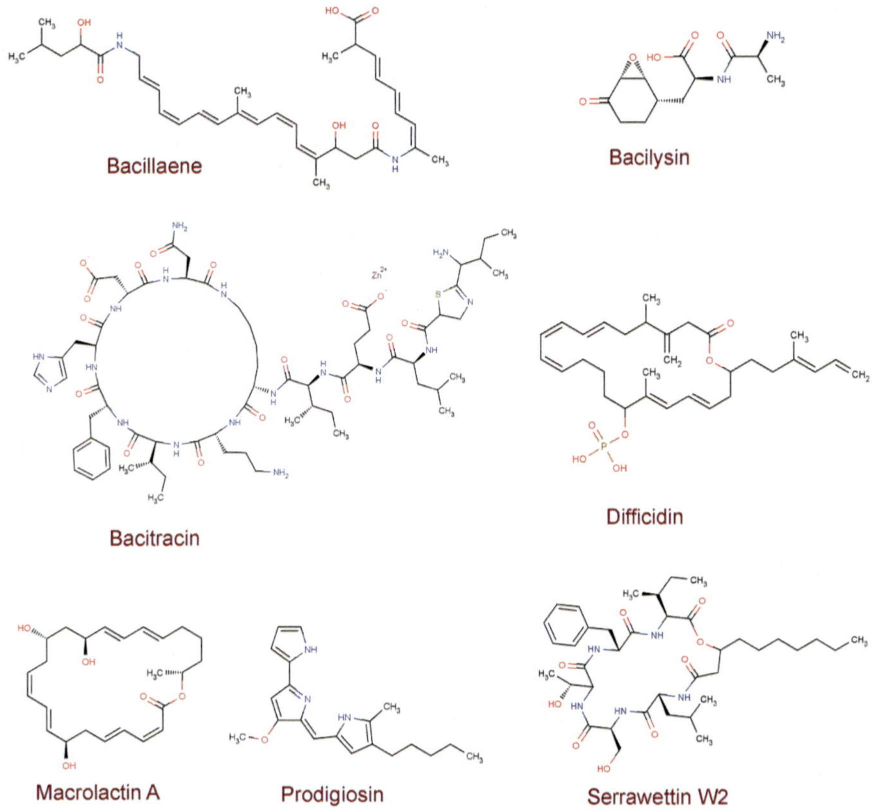

FIGURE 9.2

Chemical structures of few antibacterial metabolites produced by different bacterial species.

due to its ability to increase membrane permeabilization by interacting with, thereby solubilizing phospholipids (Carrillo, Teruel, Aranda, & Ortiz, 2003). Few members of the genus *Serratia*, including *Serratia marcescens, S. nematodiphila*, and *S. surfactantfaciens*, produce a red alkaloid pigment prodigiosin with potent antibacterial activity against various Gram-positive bacteria and few Gram-negative bacteria (Arivizhivendhan et al., 2018; Gondil, Asif, & Bhalla, 2017; Lapenda, Silva, Vicalvi, Sena, & Nascimento, 2015; Su et al., 2016). Prodigiosin causes pore formation in bacterial as well as fungal cell membranes and also shows species-specific inhibitory activity against few fungi, including *F. oxysporum* (Hazarika et al., 2020; Suryawanshi, Patil, Koli, Hallsworth, & Patil, 2017). Studies suggested that prodigiosin interacted with the nucleic acid content of *P. aeruginosa* and inhibited biofilm formation through reactive oxygen species (ROS) mediated damages to the cells (Kimyon, Das, & Ibugo, 2016).

Serrawettins are biosurfactant metabolites produced by *S. marcescens* and few other bacteria of the genus (Matsuyama & Tanikawa, 2011; Su et al., 2016). These NRPs show broad-spectrum antibacterial activity against several pathogenic bacteria (Clements, Ndlovu, & Khan, 2019).

Several endophytic strains of the genus *Streptomyces* exhibit antibacterial activity against plant pathogenic bacteria, including *Ralstonia solanacearum*, *Xanthomonas campestris*, and *X. oryzae* (Hastuti, Lestari, Suwanto, & Saraswati, 2012; Mingma, Pathom-aree, Trakulnaleamsai, Thamchaipenet, & Duangmal, 2014; Tan et al., 2006). Mingma et al. (2014) reported rhizospheric and roots endophytic actinomycetes (belonging to the genera *Amycolatopsis*, *Isoptericola*, *Micromonospora*, *Microbispora*, *Nocardia*, *Nonomuraea*, *Promicromonospora*, and *Pseudonocardia*) from Leguminous plants to show inhibitory properties against soybean pathogen, *X. campestris* pv. *glycine*. These isolates did not show inhibitory activity against three *Rhizobium* strains, suggesting no negative impact on root nodule formation (Mingma et al., 2014). There is much to explore about the bioactive compounds responsible for the antibacterial activities of these actinomycetes. However, most of the well-known antibiotics, such as streptomycin, neomycin, erythromycin, tetracycline, rifamycin, vancomycin, gentamycin, and rifamycin, are naturally produced by different species of *Streptomyces*. These antibiotics can be categorized based on their modes of action as cell wall synthesis inhibitor (e.g., vancomycin), cell membrane targeting, inhibitors of protein synthesis (e.g., erythromycin), etc. Kaufman (2011).

9.5 Insecticidal and nematicidal metabolites from microbes

Bacillus thuringiensis (*Bt*) is the most extensively studied insecticidal bacterial species. Since the first report in 1902 by Ishiwata about an infection of *Bombyx mori*, followed by Berliner's description, *Bt* has become the prime microbial species used in biological control of insect pests (Melo alda, Soccol, & Soccol, 2016). The first formulation of *Bt* biopesticide appeared as the product Sporeine to control various Lepidoptera species that infested different crop species in France (Milner, 1994). Since then, the applicability of *Bt* as biopesticide has extended to a spectrum of susceptible organisms, including the orders of Coleoptera, Culicidae, Hymenoptera, Homoptera, Mallophaga, Simuliidae, and others (Sanchis, 2012). *Bt*-toxins (more often called Cry toxins), the group of insecticidal crystalline proteins, have been found as a major player of the antagonistic activity of *B. thuringiensis* against insects (Lucena et al., 2014); however, complex processes occur simultaneously that culminate in the death of an organism (Melo alda et al., 2016). Thuringiensin—a thermostable secondary metabolite—has been reported from *B. thuringiensis*, which showed insecticidal activity against a wide range of insects belonging to the orders Coleoptera, Diptera,

Hymenoptera, Isoptera, Lepidoptera, and Orthoptera (Liu et al., 2014; Tamez-Guerra et al., 2004; Toledo, Liedo, Williams, & Ibarra, 1999; Tsuchiya et al., 2002). Surfactin from *B. amyloliquefaciens* was reported to show strong aphicidal activity against green peach aphid (*Myzus persicae*) (Yun, Yang, Kim, Kim, & Kim, 2013). Isomers of surfactin consisting of C_{14} [Leu_7], C_{14} [Val_7], and C_{15} [Leu_7] were also reported from *B. subtilis* Y9 with insecticidal activity against this aphid. The isomers with leucine exhibited stronger insecticidal activity than the isomers with valine (Yang et al., 2017). Amines and amides are the naturally available metabolites from endophytes that are lethal to most insects but not mammals. Orfamide A—an insecticidal metabolite—is reported to be produced by different *Pseudomonas* sp. (Flury et al., 2017; Jang et al., 2013). Orfamide A from *Pseudomonas protegens* F6 has been tested for its insecticidal activity against green peach aphid (*M. persicae*), where significant mortality was observed in a dose-dependent manner (Jang et al., 2013).

A number of secondary metabolites from diverse microbial origin have been reported to show remarkable nematicidal activity against soil-borne nematodes. Bacteria exhibiting nematicidal properties include actinomycetes (Yavuzaslanoglu, Yamac, & Nicol, 2011) as well as few eubacterial genera, such as *Burkholderia* (Meyer et al., 2001), *Chromobacterium* (Cronin, Moënne-Loccoz, Dunne, & O'Gara, 1997), *Enterobacter* (Duponnois, Bâ, & Mateille, 1999), *Rhizobium* (Neipp & Becker, 1999), *Serratia* (El-Sherif, Ali, & Barakat, 1994), *Azotobacter*, *Bacillus*, *Clostridium*, *Corynebacterium*, and *Methylobacterium* (Prabhu, Kumar, Subramanian, & Sundaram, 2009). The most extensively studied bacterial species for nematicidal activity is *B. subtilis* (Siddiqui & Shaukat, 2002). Bio Yield (Gustafson LLC, Plano, TX, the Unites States) and BioNem are among the commercially available biological nematicides, and both of these are based on *Bacillus* nematicidal activity (Giannakou, Karpouzas, & Prophetou-Athanasiadou, 2004). Prodigiosin extracted from *S. marcescens* exhibited nematicidal activity against two plant-parasitic nematodes—*Radopholus similis* and *Meloidogyne javanica* at low concentrations [LC_{50} (lethal concentration showing 50% mortality): 83 and 79 μg/mL, respectively] (Rahul et al., 2014).

9.6 Bioformulations for biocontrol activity

The application of beneficial microorganisms as plant growth–promoting microbes (PGPMs) for boosting plant growth and controlling phytopathogens is not a new practice, but still, we are only in the first gear regarding the application of microbe-based products in agriculture. The microbe-based products with different names such as biofertilizers, bioformulations, bioinoculants, biopesticides, and biostimulants are providing a huge impact on agriculture with respect to sustainability and eco-friendliness (Arora & Mishra, 2016). However, sometimes the quality, reliability, and performance of these products are hindering their growth in the market. Hence, new ways and avenues have to be explored to resolve the associated problems and establish their reliability among the end users.

Apart from *Bt*, PGPMs and biocontrol agents are commonly used as cell-based formulations. The requirement for a transition from conventional cell-based bioinoculants to the incorporation of secondary metabolites/additives in bioformulations is clearly demonstrated by several research and trials. Therefore the next-generation bioformulations development aims the use of microbial metabolites/biomolecules as active ingredients for biocontrol of plant diseases. Some of these biomolecules have successfully made their places in bioformulations, being feasible or cost-effective; other such additives are expensive for their inputs despite their requirement at very low concentrations. Blasticidin S, coniothyriomycin, fumaramidmycin, gliovirin, kasugamycin, polyoxin B and D, pyrrolnitrin, rhizocticins strobilurins, and validamycins are the most significant examples of such microbial metabolites used as lead molecules or fungicides (Jayaprakashvel & Mathivanan, 2011). Although *P. syringae* is a good producer of antifungal metabolites, the pathogenic nature of this bacterial species to several crop plants restricts its use in cell-based bioformulations. Risk factors are also associated with the applications of cell-based bioformulations containing opportunistic human pathogens, such as *S. marcescens*, *Chromobacterium* sp., and a few members of Enterobacteriaceae (Batista & Neto, 2017; Lenzi, Marvasi, & Baldi, 2020; Sharma, Bisaria, & Sharma, 2019). In such cases, bacterial cell-free metabolites extracted from the cultures could be beneficial for direct applications in the field. However, downstream processing of few metabolites from microbial cultures is not cost-effective from the farmers' point of view. Recent advances in metabolic engineering have opened up new possibilities for enhanced production (using native strains) or heterologous production (using engineered strains) of such metabolites on a commercial scale making the downstream processing cost-effective (Srivastava, Srivastava, Ramteke, & Mishra, 2019).

9.6.1 Strategies for discovering microbial metabolites

The development of high-throughput omics techniques has largely impacted the discovery of new bioactive secondary metabolites by expanding the understanding of complex mechanisms of secondary metabolites biosynthesis through the organization and regulation of biosynthetic gene clusters (BGCs). The substantial development of sequencing technologies over the last couple of decades combined with efficient bioinformatics tools generated a huge genomic repository with enormous genetic information regarding biosynthesis and metabolism (Palazzotto & Weber, 2018). Genome mining tools have been developed to identify BGCs inside the genomes of specific microorganisms. Dedicated genomc mining tools, such as anti-SMASH (antibiotics and Secondary Metabolites Analysis SHell, accessible at http://antismash.secondarymetabolites.org), PRISM (PRediction Informatics for Secondary Metabolomes, accessible at http://magarveylab.ca/prism), GARLIC (Global Alignment for natuRaL-products chemInformatiCs), GRAPE (Generalized

Retrobiosynthetic Assembly Prediction Engine), and rBAN (Retro-biosynthetic Analysis of Nonribosomal peptides) are available for predicting BGCs in plants, animals, bacterial, and fungal genomes (Blin et al., 2019; Dejong et al., 2016; Medema et al., 2011; Ricart et al., 2019; Skinnider et al., 2015). Genomic analysis revealed that many silent BGCs are available throughout the genomes of different bacterial and fungal species, which can be activated by specific induction factors and interkingdom and intrakingdom interactions among microorganisms (Brakhage & Schroeckh, 2011; Netzker et al., 2015). Microarray and RNA sequencing–based transcriptomics data have suggested the expression of those BGCs in response to particular external factors (Amos et al., 2017; Schroeckh et al., 2009).

Metabolomics, used for the global analysis of low molecular weight compounds, offers platforms for metabolic comparison of various biological samples avoiding redundancy in the chemical structures. Mass spectrometric techniques, such as selected reaction monitoring and multiple reaction monitoring, are extremely beneficial for targeted analysis of metabolites and their structural characterization (Carry et al., 2018; Domingo-Almenara et al., 2018; Wong, Abuhusain, McDonald, & Don, 2012). Liquid Chromatography coupled with tandem Mass Spectrometry (LC–MS/MS) is routinely used for the detection and identification of metabolites from bacteria and fungi. Multivariate statistical analyses such as principal component analysis (PCA) and Partial Least-Squares Discriminant Analysis (PLS-DA) along with the development of the MS/MS database Global Natural Product Social molecular networking platform have upgraded the analysis of MS data minimizing the data redundancy (Crüsemann et al., 2017; Hou et al., 2012). Moreover, advancements of nuclear magnetic resonance (NMR) spectroscopic techniques are also aiding the progress in the discovery of novel bioactive metabolites. The 2D-NMR techniques, such as Correlation Spectroscopy (COSY) and Nuclear Overhauser Effect Spectroscopy (NOESY), have simplified the metabolite screening process from the complex environmental samples (Emwas et al., 2019; Reily & Lindon, 2005).

9.7 Different approaches to enhance the synthesis of microbial secondary metabolites

Microbes in their natural environment will synthesize molecules and express genes that are required for acquiring food and for their multiplication. These are primary metabolites that are basic for an organism's growth and multiplication. Secondary metabolites are synthesized only under certain conditions, such as stress or competition from other organisms. Most of the metabolic pathways are silent under the natural condition and/or *in vitro* condition. Activation of such pathways in a particular organism will require the knowledge of the pathways and the physiology of that organism. Different approaches can be planned for the activation of such silent pathways by using modified cultivation techniques and using molecular techniques (Pettit, 2011; Wakefield, Hassan, Jaspars, Ebel, & Rateb, 2017).

Microbes are cultured in media, having all the essential components for their growth and multiplication. Changes made in the media composition and pH of the media can induce changes in the secondary metabolite production. The addition of a small concentration of dimethyl sulfoxide as a media component increases secondary metabolite production probably by affecting the biochemical pathways (Chen, Wang, Li, Waters, & Davies, 2000).

Microbes encounter and adapt to different biotic and abiotic factors. Microbial cells have the ability to modify their growth pattern, cell morphology, and metabolic pathways that are required to mitigate the external threat. These factors can be used for induction and production of higher amounts of secondary metabolite. Biotic factors include the attack by other microbial pathogens, whereby the microbial pathogen produces molecules to attack and the recipient microbe produces molecules to deter the attacker. Co-cultivation of interspecific microbes induces gene expression and alters their metabolic pathway. Due to changes in metabolite production, there are morphological changes in the microbial cell involved in the interaction (Seyedsayamdost, Traxler, Clardy, & Kolter, 2012; Traxler, Watrous, Alexandrov, Dorrestein, & Kolter, 2013). Olaf et al., (2017) reported increased production of 2,5-bis(1-methylethyl)-pyrazine during the interaction between *Burkholderia* and *Paenibacillus*, which retards the growth of *Burkholderia*. Also, there is an upregulation of the antibiotic resistance gene in *Paenibacillus*. Co-cultivation of the marine fungal isolate *A. fumigatus* Mr2012 with *Streptomyces leeuwenhoekii* strain C34 induced the production of two new derivative molecules, luteoride D and pseurotin G, along with the production of terezine D and 11-O-methylpseurotin A (Wakefield et al., 2017). There was increased production of Chaxapeptin and also induction of pentalenic acid production from the bacterial strain during co-cultivation of *A. fumigatus* Mr2012 with *S. leeuwenhoekii* strain C58, but at the same time, expression of many of the fungal metabolites was suppressed (Wakefield et al., 2017).

By using genetic engineering techniques, silent pathways can be activated or heterologous expression can be done for higher secondary metabolite production. Primary metabolites produced during glycolysis, tricarboxylic acid cycle, and pentose phosphate pathway (PPP) are used by the cell for the synthesis of many primary and secondary metabolites. Therefore by engineering the pathways, the synthesis of primary metabolites can be modulated and thereby the synthesis of secondary metabolite. Xue-Mei, Yong-Keun, Hag, and Soon-Kwang (2017) engineered the PPP for producing higher amounts of NADPH (reduced Nicotinamide Adenine Dinucleotide Phosphate) in *Streptomyces lividans* TK24. The increased concentration of NADPH along with the other intermediates triggered the synthesis of secondary metabolite undecylprodigiosin. WblA protein is involved in disulfide stress response and downregulation of antibiotic production (Jin-Su, Lee, Kim, Lee, & Kim, 2012). The deletion of wblA homolog genes resulted in the overexpression of moenomycin, doxorubicin, violapyrone B, pikromycin, and tylosin biosynthetic genes in *Streptomyces* species (Gust, Challis, Fowler, Kieser, & Chater, 2003; Huang et al., 2016; Min-Woo, Hee-Ju, Si-Sun, & Eung-Soo, 2014; Noh, Kim, Lee, Lee, & Kim, 2010; Paget, Kang, Roe, & Buttner, 1998). Dimitris, Guangde, Yousong, and Hendrik (2018) deleted wblA in *Streptomyces albus*, and the mutant strain overproduced paulomycins A/B with increased biomass. Nah et al. (2013) constructed a triple mutant in *Streptomyces coelicolor* ΔpfkΔwblAΔSCO1712, with increased biosynthesis of actinorhodin and other type II polyketides.

CRISPR-Cas9 (Clustered Regularly Interspaced Short Palindromic Repeats - CRISPR associated nuclease) technology is being used to genetically engineer different organisms. Nielsen, Isbrandt, and Rasmussen (2017) used CRISPR-Cas9 technology to identify novel gene for PKS (polyketide synthase) and NRPS in *T. atroroseus*, involved in secondary metabolite synthesis. CRISPR technique was used to improve the expression of genes involved in epothilone synthesis in *Myxococcus xanthus* (Ran et al., 2018). The use of CRISPR-mediated transcriptional activation (CRISPRa) in *Aspergillus nidulans* increased the production of microperfuranone, by targeting NRP synthetase-like (NRPS-like) gene *micA* (Indra et al., 2020). Nanotechnological approaches are also used to improve secondary metabolite production. The actinobacterium *S. lividans* was found to be associated with electrospun organic nanofibers added in the media, whereby two- to sixfold enhancement in the production of two antibiotics, namely, actinorhodin and undecylprodigiosin, was seen in a calcium-dependent manner (Moffa et al., 2017).

9.8 Conclusion

The use of microbial metabolites produced by antagonistic microorganisms for biological control is a sustainable alternative for chemical pesticides. Biodegradability and targeted mechanisms of action of these biomolecules make them the key choice for plant disease management in an eco-friendly manner (Jayaprakashvel & Mathivanan, 2011). High throughput screening techniques are transforming the discovery of novel microbial biochemicals for biocontrol activities (Zhang et al., 2007). Genome mining platforms have further simplified the discovery of secondary metabolite biosynthetic pathways in different microorganisms thereby discovering many novel compounds. Microbial metabolic engineering is also contributing to the generation of redesigned biosynthetic pathways in synthetic biology for more advanced and complex engineering of microbial metabolism (McArthur & Fong, 2010).

Microbial metabolite–based bioformulations are now building new possibilities for commercial agro-industries, and many of them have started their journey as bioproducts worldwide. However, the end users (farmers) and the path makers are still apprehensive of their application in agriculture. It is high time that we should break the restraints and use those metabolite-based bioformulations for biocontrol of plant diseases.

References

Agrios, G. N. (2005). *Plant pathology* (5th Ed.). London, UK: Elsevier Academic Press.

Aktuganov, G., Jokela, J., Kivelä, H., Khalikova, E., Melentjev, A., Galimzianova, N., … Korpela, T. (2014). Isolation and identification of cyclic lipopeptides from *Paenibacillus ehimensis*, strain IB-X-b. *Journal of Chromatography B: Analytical Technologies in the Biomedical and Life Sciences, 973*, 9–16. Available from https://doi.org/10.1016/j.jchromb.2014.09.042.

Aleti, G., Sessitsch, A., & Brader, G. (2015). Genome mining: prediction of lipopeptides and polyketides from *Bacillus* and related Firmicutes. *Computational and Structural Biotechnology Journal, 13*, 192–203. Available from https://doi.org/10.1016/j.csbj.2015.03.003.

Almario, J., Bruto, M., Vacheron, J., Prigent-Combaret, C., Moënne-Loccoz, Y., & Muller, D. (2017). Distribution of 2,4-diacetylphloroglucinol biosynthetic genes among the *Pseudomonas* spp. reveals unexpected polyphyletism. *Frontiers in Microbiology, 8*.

Alvarez, F., Castro, M., Príncipe, A., Borioli, G., Fischer, S., Mori, G., & Jofre, E. (2012). The plant-associated *Bacillus amyloliquefaciens* strains MEP218 and ARP23 capable of producing the cyclic lipopeptides iturin or surfactin and fengycin are effective in biocontrol of sclerotinia stem rot disease. *Journal of Applied Microbiology, 112*, 159–174. Available from https://doi.org/10.1111/j.1365-2672.2011.05182.x.

Amos, G. C. A., Awakawa, T., Tuttle, R. N., Letzel, A. C., Kim, M. C., Kudo, Y., … Jensen, P. R. (2017). Comparative transcriptomics as a guide to natural product discovery and biosynthetic gene cluster functionality. *Proceedings of the National Academy of Sciences of the United States of America, 114*(52), E11121–E11130. Available from https://doi.org/10.1073/pnas.1714381115.

Anjaiah, V., Koedam, N., Nowak-Thompson, B., Loper, J. E., H€ofte, M., Tambong, J. T., & Cornelis, P. (1998). Involvement of phenazines and anthranilate in the antagonism of *Pseudomonas aeruginosa* PNA1 and Tn5-derivatives towards *Fusarium* sp. and *Pythium* sp. *Molecular Plant Microbe Interactions, 11*, 847–854.

Arima, K., Imanaka, H., Kousaka, M., Fukuda, A., & Tamura, G. (1965). Studies on pyrrolnitrin, a new antibiotic. I. Isolation and properties of pyrrolnitrin. *Journal of Antibiotics, 18*(5), 201–204.

Arima, K., Imanaka, H., Kousaka, M., Fukuta, A., & Tamura, G. (1964). Pyrrolnitrin, a new antibiotic substance, produced by *Pseudomonas*. *Agricultural and Biological Chemistry, 28*(8), 575–576. Available from https://doi.org/10.1080/00021369.1964.10858275.

Arivizhivendhan, K. V., Mahesh, M., Boopathy, R., Swarnalatha, S., Regina Mary, R., & Sekaran, G. (2018). Antioxidant and antimicrobial activity of bioactive prodigiosin produces from *Serratia marcescens* using agricultural waste as a substrate. *Journal of Food Science and Technology, 55*(7), 2661–2670. Available from https://doi.org/10.1007/s13197-018-3188-9.

Arora, N. K., & Mishra, J. (2016). Prospecting the roles of metabolites and additives in future bioformulations for sustainable agriculture. *Applied Soil Ecology, 107*, 405–407. Available from https://doi.org/10.1016/j.apsoil.2016.05.020.

Asolkar, R. N., Singh, A., Jensen, P. R., Aalbersberg, W., Carté, B. K., Feussner, K. D., … Fenical, W. (2017). Marinocyanins, cytotoxic bromo-phenazinone meroterpenoids from a marine bacterium from the streptomycete clade MAR4. *Tetrahedron, 73*(16), 2234–2241. Available from https://doi.org/10.1016/j.tet.2017.03.003.

Baron, S. S., & Rowe, J. J. (1981). Antibiotic action of pyocyanin. *Antimicrobial Agents and Chemotherapy*, *20*, 814–820.

Batista, J. H., & Neto, J. Fd. S. (2017). *Chromobacterium violaceum* pathogenicity: Updates and insights from genome sequencing of novel Chromobacterium species. *Frontiers in Microbiology*, *8*. Available from https://doi.org/10.3389/fmicb.2017.02213.

Beatty, P. H., & Jensen, S. E. (2002). *Paenibacillus polymyxa* produces fusaricidin-type antifungal antibiotics active against *Leptosphaeria maculans*, the causative agent of blackleg disease of canola. *Canadian Journal of Microbiology*, *48*(2), 159–169.

Bechet, M., Caradec, T., Hussein, W., Abderrahmani, A., Chollet, M., Leclere, V., ... Jacques, P. (2012). Structure, biosynthesis, and properties of kurstakins, nonribosomal lipopeptides from *Bacillus* spp. *Applied Microbiology and Biotechnology*, *95*(3), 593–600. Available from https://doi.org/10.1007/s00253-012-4181-2.

Beric, T., Kojic, M., Stankovic, S., Topisirovic, L., Degrassi, G., Myers, M., ... Fira, D. (2012). Antimicrobial activity of *Bacillus* sp natural isolates and their potential use in the biocontrol of phytopathogenic bacteria. *Food Technology and Biotechnology*, *50*(1), 25–31.

Besson, F., Peypoux, F., & Michel, G. (1978). Action of mycosubtilin and of bacillomycin L on *Micrococcus luteus* cells and protoplasts: influence of the polarity of the antibiotics upon their action on the bacterial cytoplasmic membrane. *FEBS Letters*, *90*(1), 36–40.

Blin, K., Shaw, S., Steinke, K., Villebro, R., Ziemert, N., Lee, S. Y., ... Weber, T. (2019). AntiSMASH 5.0: Updates to the secondary metabolite genome mining pipeline. *Nucleic Acids Research*, *47*(1), W81–W87. Available from https://doi.org/10.1093/nar/gkz310.

Brakhage, A. A., & Schroeckh, V. (2011). Fungal secondary metabolites—Strategies to activate silent gene clusters. *Fungal Genetics and Biology*, *48*(1), 15–22. Available from https://doi.org/10.1016/j.fgb.2010.04.004.

Burkhead, K. D., Schisler, D. A., & Slininger, P. J. (1994). Pyrrolnitrin production by biocontrol agent *Pseudomonas cepacia* B37w in culture and in colonized wounds of potatoes. *Applied Environmental Microbiology*, *60*, 2031–2039.

Carrillo, C., Teruel, J. A., Aranda, F. J., & Ortiz, A. (2003). Molecular mechanism of membrane permeabilization by the peptide antibiotic surfactin. *Biochimica et Biophysica Acta – Biomembranes*, *1611*(1–2), 91–97. Available from https://doi.org/10.1016/S0005-2736(03)00029-4.

Carry, E., Zhao, D., Mogno, I., Faith, J., Ho, L., Villani, T., ... Wu, Q. (2018). Targeted analysis of microbial-generated phenolic acid metabolites derived from grape flavanols by gas chromatography-triple quadrupole mass spectrometry. *Journal of Pharmaceutical and Biomedical Analysis*, *159*, 374–383. Available from https://doi.org/10.1016/j.jpba.2018.06.034.

Chakraborty, M., Mahmud, N. U., Gupta, D. R., Tareq, F. S., Shin, H. J., & Islam, T. (2020). Inhibitory effects of linear lipopeptides from a Marine *Bacillus subtilis* on the wheat blast fungus *Magnaporthe oryzae* Triticum. *Frontiers in Microbiology*, *11*. Available from https://doi.org/10.3389/fmicb.2020.00665.

Chen, G., Wang, G. Y. S., Li, X., Waters, B., & Davies, J. (2000). Enhanced production of microbial metabolites in the presence of dimethyl sulfoxide. *Journal of Antibiotics*, *53*(10), 1145–1153. Available from https://doi.org/10.7164/antibiotics.53.1145.

Chen, X. H., Scholz, R., Borriss, M., Junge, H., Mogel, G., Kunz, S., & Borriss, R. (2009). Difficidin and bacilysin produced by plant-associated Bacillus amyloliquefaciens are efficient in controlling fire blight disease. *Journal of Biotechnology*, *140*(1–2), 38–44. Available from https://doi.org/10.1016/j.jbiotec.2008.10.015.

Chin-A-Woeng, T. F., Bloemberg, G. V., Mulders, I. H., Dekkers, L. C., & Lugtenberg, B. J. (2000). Root colonization by phenazine-1-carboxamide-producing bacterium *Pseudomonas chlororaphis* PCL1391 is essential for biocontrol of tomato foot and root rot. *Molecular plant-microbe interactions*, *13*(12), 1340–1345.

Clements, T., Ndlovu, T., & Khan, W. (2019). Broad-spectrum antimicrobial activity of secondary metabolites produced by *Serratia marcescens* strains. *Microbiological Research*, *229*. Available from https://doi.org/10.1016/j.micres.2019.126329.

Cochrane, S. A., & Vederas, J. C. (2016). Lipopeptides from *Bacillus* and *Paenibacillus* spp.: a gold mine of antibiotic candidates. *Medicinal Research Reviews*, *36*(1), 4–31. Available from https://doi.org/10.1002/med.21321.

Cronin, D., Moënne-Loccoz, Y., Dunne, C., & O'Gara, F. (1997). Inhibition of egg hatch of the potato cyst nematode *Globodera rostochiensis* by chitinase-producing bacteria. *European Journal of Plant Pathology*, *103*(5), 433–440. Available from https://doi.org/10.1023/A:1008662729757.

Crüsemann, M., O'Neill, E. C., Larson, C. B., Melnik, A. V., Floros, D. J., da Silva, R. R., ... Moore, B. S. (2017). Prioritizing natural product diversity in a collection of 146 bacterial strains based on growth and extraction protocols. *Journal of Natural Products*, *80*(3), 588–597. Available from https://doi.org/10.1021/acs.jnatprod.6b00722.

De Vleesschauwer, D., Cornelis, P., & Höfte, M. (2006). Redox-active pyocyanin secreted by *Pseudomonas aeruginosa* 7NSK2 triggers systemic resistance to *Magnaporthe grisea* but enhances *Rhizoctonia solani* susceptibility in rice. *Molecular Plant-Microbe Interactions*, *19*(12), 1406–1419.

Dejong, C. A., Chen, G. M., Li, H., Johnston, C. W., Edwards, M. R., Rees, P. N., ... Magarvey, N. A. (2016). Polyketide and nonribosomal peptide retro-biosynthesis and global gene cluster matching. *Nature Chemical Biology*, *12*(12), 1007–1014. Available from https://doi.org/10.1038/nchembio.2188.

Demain, A. L. (1992). Microbial secondary metabolism: A new theoretical frontier for academia, a new opportunity for industry. In *Ciba Foundation symposium* (Vol. 171, pp. 3–23). Netherlands.

De Lucca, A. J., Jacks, T. J., Takemoto, J., Vinyard, B., Peter, J., Navarro, E., & Walsh, T. J. (1999). Fungal lethality, binding, and cytotoxicity of syringomycin-E. *Antimicrobial Agents and Chemotherapy*, *43*(2), 371–373. Available from https://doi.org/10.1128/aac.43.2.371.

De Lucca, A. J., & Walsh, T. J. (2000). Péptidos antifúngicos: Origen, actividad y potencial terapéutico. *Revista Iberoamericana de Micologia*, *17*(4), 116–120.

Dharni, S., Alam, M., Kalani, K., Abdul-Khaliq, A. K., Samad, A., Srivastava, S. K., & Patra, D. D. (2012). Production, purification, and characterization of antifungal metabolite from *Pseudomonas aeruginosa* SD12, a new strain obtained from tannery waste polluted soil. *Journal of microbiology and biotechnology*, *22*(5), 674–683.

Dimitris, K., Guangde, J., Yousong, D., & Hendrik, L. (2018). Rational engineering of *Streptomyces albus* J1074 for the overexpression of secondary metabolite gene clusters. *Microbial Cell Factories*, *17*. Available from https://doi.org/10.1186/s12934-018-0874-2.

Domingo-Almenara, X., Montenegro-Burke, J. R., Ivanisevic, J., Thomas, A., Sidibe, J., Teav, T., ... Nordström, A. (2018). XCMS-MRM and METLIN-MRM: A cloud library and public resource for targeted analysis of small molecules. *Nature Methods*, *15*(9), 681–684. Available from https://doi.org/10.1038/s41592-018-0110-3.

Duponnois, R., Bâ, A. M., & Mateille, T. (1999). Beneficial effects of *Enterobacter cloacae* and *Pseudomonas mendocina* for biocontrol of *Meloidogyne incognita* with the endospore-forming bacterium *Pasteuria penetrans*. *Nematology*, *1*(1), 95–101. Available from https://doi.org/10.1163/156854199507901.

Emwas, A. H., Roy, R., McKay, R. T., Tenori, L., Saccenti, E., Gowda, G. A., ... Wishart, D. S. (2019). NMR spectroscopy for metabolomics research. *Metabolites*, *9*(7). Available from https://doi.org/10.3390/metabo9070123.

El-Sherif, M. A., Ali, A. H., & Barakat, M. I. (1994). Suppressive bacteria associated with plant parasitic nematodes in Egyptian agriculture. *Nematological Research (Japanese Journal of Nematology)*, *24*, 55–59. Available from https://doi.org/10.3725/jjn1993.24.2_55.

Faheem, M., Naseer, M. I., Rasool, M., Chaudhary, A. G., Kumosani, T. A., Ilyas, A. M., ... Jamal, H. S. (2015). Molecular genetics of human primary microcephaly: An overview. *BMC Medical Genomics*, *8*. Available from https://doi.org/10.1186/1755-8794-8-S1-S4.

Flury, P., Vesga, P., Péchy-Tarr, M., Aellen, N., Dennert, F., Hofer, N., ... Siegfried, S. (2017). Antimicrobial and insecticidal: Cyclic lipopeptides and hydrogen cyanide produced by plant-beneficial Pseudomonas strains CHA0, CMR12a, and PCL1391 contribute to insect killing. *Frontiers in Microbiology*, *8*. Available from https://doi.org/10.3389/fmicb.2017.00100.

Flórez, L. V., Scherlach, K., Miller, I. J., Rodrigues, A., Kwan, J. C., Hertweck, C., & Kaltenpoth, M. (2018). An antifungal polyketide associated with horizontally acquired genes supports symbiont-mediated defense in *Lagria villosa* beetles. *Nature Communications*, *9*(1). Available from https://doi.org/10.1038/s41467-018-04955-6.

Garcia-Gonzalez, E., Muller, S., Hertlein, G., Heid, N., Sussmuth, R. D., & Genersch, E. (2014). Biological effects of paenilamicin, a secondary metabolite antibiotic produced by the honey bee pathogenic bacterium *Paenibacillus larvae*. *Microbiology Open*, *3*(5), 642–656. Available from https://doi.org/10.1002/mbo3.195.

Geissler, M., Heravi, K. M., Henkel, M., & Hausmann, R. (2019). *Lipopeptide biosurfactants from* Bacillus *species* (pp. 205–240). Elsevier BV, 10.1016/b978-0-12-812705-6.00006-x.

Gerth, K., Bedorf, N., Irschik, H., H€ofle, G., & Reichenbach, H. (1994). The soraphens: a family of novel antifungal compounds from *Sorangium cellulosum* (Myxobacteria). I. Soraphen A1 alpha: fermentation, isolation, biological properties. *Journal of Antibiotics*, *47*, 23–31.

Giannakou, I. O., Karpouzas, D. G., & Prophetou-Athanasiadou, D. (2004). A novel non-chemical nematicide for the control of root-knot nematodes. *Applied Soil Ecology*, *26*(1), 69–79. Available from https://doi.org/10.1016/j.apsoil.2003.09.002.

Gondil, V. S., Asif, M., & Bhalla, T. C. (2017). Optimization of physicochemical parameters influencing the production of prodigiosin from *Serratia nematodiphila* RL2 and exploring its antibacterial activity. *3 Biotech*, *7*(5). Available from https://doi.org/10.1007/s13205-017-0979-z.

Gong, A. D., Li, H. P., Yuan, Q. S., Song, X. S., Yao, W., He, W. J., ... Liao, Y. C. (2015). Antagonistic mechanism of iturin a and plipastatin a from *Bacillus amyloliquefaciens* S76–3 from wheat spikes against *Fusarium graminearum*. *PLoS One*, *10*(2). Available from https://doi.org/10.1371/journal.pone.0116871.

Gong, L., Tan, H., Chen, F., Li, T., Zhu, J., Jian, Q., ... Duan, X. (2016). Novel synthesized 2, 4-DAPG analogues: Antifungal activity, mechanism and toxicology. *Scientific Reports*, *6*. Available from https://doi.org/10.1038/srep32266.

Gong, Q., Zhang, C., Lu, F., Zhao, H., Bie, X., & Lu, Z. (2014). Identification of bacillomycin D from *Bacillus subtilis* fmbJ and its inhibition effects against *Aspergillus flavus*. *Food Control*, *36*(1), 8–14. Available from https://doi.org/10.1016/j.foodcont.2013.07.034.

Gozari, M., Alborz, M., El-Seedi, H. R., & Jassbi, A. R. (2020). Chemistry, biosynthesis and biological activity of terpenoids and meroterpenoids in bacteria and fungi isolated from different marine habitats. *European Journal of Medicinal Chemistry*, *210*. Available from https://doi.org/10.1016/j.ejmech.2020.112957.

Gross, H., Stockwell, V. O., Henkels, M. D., Nowak-Thompson, B., Loper, J. E., & Gerwick, W. H. (2007). The genomisotopic approach: A systematic method to isolate products of orphan biosynthetic gene clusters. *Chemistry and Biology*, *14*(1), 53–63. Available from https://doi.org/10.1016/j.chembiol.2006.11.007.

Gust, B., Challis, G. L., Fowler, K., Kieser, T., & Chater, K. F. (2003). PCR-targeted Streptomyces gene replacement identifies a protein domain needed for biosynthesis of the sesquiterpene soil odor geosmin. *Proceedings of the National Academy of Sciences of the United States of America*, *100*(4), 1541–1546. Available from https://doi.org/10.1073/pnas.0337542100.

Hamdache, A., Azarken, R., Lamarti, A., Aleu, J., & Collado, I. G. (2013). Comparative genome analysis of *Bacillus* spp. and its relationship with bioactive nonribosomal peptide production. *Phytochemistry Reviews*, *12*(4), 685–716. Available from https://doi.org/10.1007/s11101-013-9278-4.

Hassan, H. M., & Fridovich, I. (1980). Mechanism of the antibiotic action of pyocyanine. *Journal of Bacteriology*, *141*, 156–163.

Hastuti, R. D., Lestari, Y., Suwanto, A., & Saraswati, R. (2012). Endophytic *Streptomyces* spp. as biocontrol agents of rice bacterial leaf blight pathogen (*Xanthomonas oryzae* pv. oryzae). *HAYATI Journal of Biosciences*, *19*(4), 155–162. Available from https://doi.org/10.4308/hjb.19.4.155.

Hazarika, D. J., Gautom, T., Parveen, A., Goswami, G., Barooah, M., Modi, M. K., & Boro, R. C. (2020). Mechanism of interaction of an endofungal bacterium *Serratia marcescens* D1 with its host and non-host fungi. *PLoS One*, *15*(4). Available from https://doi.org/10.1371/journal.pone.0224051.

Hazarika, D. J., Goswami, G., Gautom, T., Parveen, A., Das, P., Barooah, M., & Boro, R. C. (2019). Lipopeptide mediated biocontrol activity of endophytic *Bacillus subtilis* against fungal phytopathogens. *BMC Microbiology*, *19*(1). Available from https://doi.org/10.1186/s12866-019-1440-8.

He, H., Silo-Suh, L. A., Handelsman, J., & Clardy, J. (1994). Zwittermicin A, an antifungal and plant protection agent from *Bacillus cereus*. *Tetrahedron letters*, *35*(16), 2499–2502.

Helfrich, E. J., & Piel, J. (2016). Biosynthesis of polyketides by trans-AT polyketide synthases. *Natural Products Reports*, *33*(2), 231–316.

Henriksen, A., Anthoni, U., Nielsen, T. H., Soerensen, J., Christophersen, C., & Gajhede, M. (2000). Cyclic lipoundecapeptide tensin from *Pseudomonas fluorescens* strain 96.578. *Acta Crystallographica Section C: Crystal Structure Communications*, *56*(1), 113–115.

Hibbing, M. E., Fuqua, C., Parsek, M. R., & Peterson, S. B. (2010). Bacterial competition: Surviving and thriving in the microbial jungle. *Nature Reviews. Microbiology*, *8*(1), 15–25. Available from https://doi.org/10.1038/nrmicro2259.

Hill, D. S., Stein, J. I., Torkewitz, N. R., Morse, A. M., Howell, C. R., Pachlatko, J. P., ... Ligon, J. M. (1994). Cloning of genes involved in the synthesis of pyrrolnitrin from *Pseudomonas fluorescens* and role of pyrrolnitrin synthesis in biological control of plant disease. *Applied Environmental Microbiology*, *60*, 78–85.

Homma, Y., Sato, Z., Hirayama, F., Konno, K., Shirahama, H., & Suzui, T. (1989). Production of antibiotics by *Pseudomonas cepacia* as an agent for biological control of soil borne pathogens. *Soil Biology and Biochemistry*, *21*, 723–728.

Hou, Y., Braun, D. R., Michel, C. R., Klassen, J. L., Adnani, N., Wyche, T. P., & Bugni, T. S. (2012). Microbial strain prioritization using metabolomics tools for the discovery of natural products. *Analytical Chemistry*, *84*(10), 4277–4283. Available from https://doi.org/10.1021/ac202623g.

Houshaymi, B., Awada, R., Kedees, M., & Soayfane, Z. (2019). Pyocyanin, a metabolite of *Pseudomonas aeruginosa*, exhibits antifungal drug activity through inhibition of a pleiotropic drug resistance subfamily FgABC3. *Drug research*, *69*(12), 658–664.

Howell, C. R., & Stipanovic, R. D. (1980). Suppression of *Pythium ultimum*-induced damping-off of cotton seedlings by *Pseudomonas fluorescens* and its antibiotic, pyoluteorin. *Phytopathology*, *70*, 712–715.

Howie, W. J., & Suslow, T. V. (1991). Role of antibiotic biosynthesis in the inhibition of *Pythium ultimum* in the cotton spermosphere and rhizosphere by *Pseudomonas fluorescens*. *Molecular Plant Microbe Interacteractions*, *4*, 393–399.

Huang, H., Hou, L., Li, H., Qiu, Y., Ju, J., & Li, W. (2016). Activation of a plasmid-situated type III PKS gene cluster by deletion of a wbl gene in deepsea-derived *Streptomyces somaliensis* SCSIO ZH66. *Microbial Cell Factories*, *15*(1). Available from https://doi.org/10.1186/s12934-016-0515-6.

Hutchison, M. L., Tester, M. A., & Gross, D. C. (1995). Role of biosurfactant and ion channel-forming activities of syringomycin in transmembrane ion flux: A model for the mechanism of action in the plant–pathogen interaction. *Molecular Plant-Microbe Interactions*, *8*(4), 610–620. Available from https://doi.org/10.1094/MPMI-8-0610.

Indra, R., Clara, W., Jinyu, H., Rebecca, W., GCL, M., & Yit-Heng, C. (2020). CRISPR-mediated activation of biosynthetic gene clusters for bioactive molecule discovery in Filamentous fungi. *ACS Synthetic Biology*, *9*, 1843–1854. Available from https://doi.org/10.1021/acssynbio.0c00197.

Islam, Md. T., Hashidoko, Y., Deora, A., Ito, T., & Tahara, S. (2005). Suppression of damping-off disease in host plants by the rhizoplane bacterium *Lysobacter* sp. strain SB-K88 is linked to plant colonization and antibiosis against soil-borne peronosporomycetes. *Applied Environmental Microbiology*, *71*, 3786–3796.

Jaaffar, A. K. M., Parejko, J. A., Paulitz, T. C., Weller, D. M., & Thomashow, L. S. (2017). Sensitivity of *Rhizoctonia* isolates to phenazine-1-carboxylic acid and biological control by phenazine-producing *Pseudomonas* spp. *Phytopathology*, *107*(6), 692–703.

Jang, J. Y., Yang, S. Y., Kim, Y. C., Lee, C. W., Park, M. S., Kim, J. C., & Kim, I. S. (2013). Identification of orfamide A as an insecticidal metabolite produced by *Pseudomonas protegens* F6. *Journal of Agricultural and Food Chemistry*, *61*(28), 6786–6791. Available from https://doi.org/10.1021/jf401218w.

Jayaprakashvel, M., & Mathivanan, N. (2011). *Management of plant diseases by microbial metabolites. Bacteria in agrobiology: plant nutrient management* (pp. 237–265). Berlin, Heidelberg: Springer.

Jayaprakashvel, M., Muthezhilan, R., Srinivasan, R., Jaffar Hussain, A., Gopalakrishnan, S., Bhagat, J., ... Muthulakshmi, R. (2010). Hydrogen cyanide mediated biocontrol potential of *Pseudomonas* sp. AMET1055 isolated from the rhizosphere of coastal sand dune vegetation. *Advanced Biotech*, *9*, 39–42.

Jespers, A. B. K., & De Waard, M. A. (1995). Effect of fenpiclonil on phosphorylation of glucose in *Fusarium sulphureum*. *Pesticide Science*, *44*(2), 167–175. Available from https://doi.org/10.1002/ps.2780440210.

Jin, P., Wang, H., Tan, Z., Xuan, Z., Dahar, G. Y., Li, Q. X., ... Liu, W. (2020). Antifungal mechanism of bacillomycin D from *Bacillus velezensis* HN-2 against *Colletotrichum gloeosporioides* Penz. *Pesticide Biochemistry and Physiology*, *163*, 102–107. Available from https://doi.org/10.1016/j.pestbp.2019.11.004.

Jin-Su, K., Lee, H. N., Kim, P., Lee, H. S., & Kim, E. S. (2012). Negative role of wblA in response to oxidative stress in *Streptomyces coelicolor*. *Journal of Microbiology and Biotechnology*, *22*(6), 736–741. Available from https://doi.org/10.4014/jmb.1112.12032.

Kaufman, G. (2011). Antibiotics: mode of action and mechanisms of resistance. *Nursing standard (Royal College of Nursing (Great Britain): 1987)*, *25*(42), 49–55. Available from https://doi.org/10.7748/ns.25.42.49.s52.

Kaur, T., & Manhas, R. K. (2014). Antifungal, insecticidal, and plant growth promoting potential of *Streptomyces hydrogenans* DH16. *Journal of Basic Microbiology*, *54*(11), 1175–1185. Available from https://doi.org/10.1002/jobm.201300086.

Keel, C., Weller, D. M., Natsch, A., Défago, G., Cook, R. J., & Thomashow, L. S. (1996). Conservation of the 2, 4-diacetylphloroglucinol biosynthesis locus among fluorescent *Pseudomonas* strains from diverse geographic locations. *Applied and Environmental Microbiology*, *62*(2), 552–563.

Kefi, A., Slimene, I. B., Karkouch, I., Rihouey, C., Azaeiz, S., Bejaoui, M., ... Limam, F. (2015). Characterization of endophytic *Bacillus* strains from tomato plants (*Lycopersicon esculentum*) displaying antifungal activity against *Botrytis cinerea* Pers. *World Journal of Microbiology and Biotechnology*, *31*(12), 1967–1976.

Kerr, J. R., Taylor, G. W., Rutman, A., Høiby, N., Cole, P. J., & Wilson, R. (1999). *Pseudomonas aeruginosa* pyocyanin and 1-hydroxyphenazine inhibit fungal growth. *Journal of Clinical Pathology*, *52*(5), 385–387.

Kevany, B. M., Rasko, D. A., & Thomas, M. G. (2009). Characterization of the complete zwittermicin A biosynthesis gene cluster from *Bacillus cereus*. *Applied and environmental microbiology*, *75*(4), 1144–1155.

Kilani-Feki, O., Khiari, O., Culioli, G., Ortalo-Magné, A., Zouari, N., Blache, Y., & Jaoua, S. (2010). Antifungal activities of an endophytic *Pseudomonas fluorescens* strain Pf1TZ harbouring genes from pyoluteorin and phenazine clusters. *Biotechnology Letters*, *32*(9), 1279–1285. Available from https://doi.org/10.1007/s10529-010-0286-9.

Kim, P. I., Ryu, J., Kim, Y. H., & Chi, Y. T. (2010). Production of biosurfactant lipopeptides iturin A, fengycin, and surfactin A from *Bacillus subtilis* CMB32 for control of *Colletotrichum gloeosporioides*. *Journal of Microbiology and Biotechnology*, *20*(1), 138–145. Available from https://doi.org/10.4014/jmb.0905.05007.

Kimyon, Ö., Das, T., Ibugo, A. I., Kutty, S. K., Ho, K. K., Tebben, J., ... Manefield, M. (2016). Serratia secondary metabolite prodigiosin inhibits *Pseudomonas aeruginosa* biofilm development by producing reactive oxygen species that damage biological molecules. *Frontiers in Microbiology*, *7*. Available from https://doi.org/10.3389/fmicb.2016.00972.

Koumoutsi, A., Chen, X. H., Henne, A., Liesegang, H., Hitzeroth, G., Franke, P., ... Borriss, R. (2004). Structural and functional characterization of gene clusters directing nonribosomal synthesis of bioactive cyclic lipopeptides in *Bacillus amyloliquefaciens* strain FZB42. *Journal of Bacteriology*, *186*(4), 1084–1096.

Laborda, P., Zhao, Y., Ling, J., Hou, R., & Liu, F. (2018). Production of antifungal p-aminobenzoic acid in *Lysobacter antibioticus* OH13. *Journal of Agricultural and Food Chemistry*, *66*(3), 630–636. Available from https://doi.org/10.1021/acs.jafc.7b05084.

Lambowitz, A. M., & Slayman, C. W. (1972). Effect of pyrrolnitrin on electron transport and oxidative phosphorylation in mitochondria isolated from *Neurospora crassa*. *Journal of Bacteriology*, *112*(2), 1020–1022.

Lapenda, J. C., Silva, P. A., Vicalvi, M. C., Sena, K. X. F. R., & Nascimento, S. C. (2015). Antimicrobial activity of prodigiosin isolated from *Serratia marcescens* UFPEDA 398. *World Journal of Microbiology and Biotechnology*, *31*(2), 399–406. Available from https://doi.org/10.1007/s11274-014-1793-y.

Leclere, V., Bechet, M., Adam, A., Guez, J. S., Wathelet, B., Ongena, M., ... Jacques, P. (2005). Mycosubtilin overproduction by *Bacillus subtilis* BBG100 enhances the organism's antagonistic and biocontrol activities. *Applied Environmental Microbiology*, *71* (8), 4577–4584. Available from https://doi.org/10.1128/aem.71.8.4577-4584.

Lee, S. H., Cho, Y. E., Park, S. H., Balaraju, K., Park, J. W., Lee, S. W., & Park, K. (2013). An antibiotic fusaricidin: a cyclic depsipeptide from *Paenibacillus polymyxa E681 induces systemic resistance against Phytophthora* blight of red-pepper. *Phytoparasitica*, *41*(1), 49–58. Available from https://doi.org/10.1007/s12600-012-0263-z.

Lenzi, A., Marvasi, M., & Baldi, A. (2020). Agronomic practices to limit pre-and post-harvest contamination and proliferation of human pathogenic Enterobacteriaceae in vegetable produce. *Food Control*, *119*.

Levenfors, J. J., Hedman, R., Thaning, C., Gerhardson, B., & Welch, C. J. (2004). Broad-spectrum antifungal metabolites produced by the soil bacterium *Serratia plymuthica* A 153. *Soil Biology and Biochemistry*, *36*(4), 677–685. Available from https://doi.org/10.1016/j.soilbio.2003.12.008.

Li, Y., & Chen, S. (2019). Fusaricidin produced by *Paenibacillus polymyxa* WLY78 induces systemic resistance against fusarium wilt of cucumber. *International Journal of Molecular Sciences*, *20*(20). Available from https://doi.org/10.3390/ijms20205240.

Lima, A. O. S., Quecine, M. C., Fungaro, M. H. P., Andreote, F. D., Maccheroni, W., Araújo, W. L., ... Azevedo, J. L. (2005). Molecular characterization of a β − 1,4-endoglucanase from an endophytic *Bacillus pumilus* strain. *Applied Microbiology and Biotechnology*, *68*(1), 57–65. Available from https://doi.org/10.1007/s00253-004-1740-1.

Liu, B., Qiao, H., Huang, L., Buchenauer, H., Han, Q., Kang, Z., & Gong, Y. (2009). Biological control of take-all in wheat by endophytic *Bacillus subtilis* E1R-j and potential mode of action. *Biological Control*, *49*(3), 277–285. Available from https://doi.org/10.1016/j.biocontrol.2009.02.007.

Liu, H., Dong, D., Peng, H., Zhang, X., & Xu, Y. (2006). Genetic diversity of phenazine- and pyoluteorin-producing pseudomonads isolated from green pepper rhizosphere. *Archives of Microbiology*, *185*(2), 91–98. Available from https://doi.org/10.1007/s00203-005-0072-6.

Liu, R. F., Zhang, D. J., Li, Y. G., Tao, L. M., & Tian, L. (2010). A new antifungal cyclic lipopeptide from *Bacillus marinus* B-9987. *Helvetica Chimica Acta*, *93*(12), 2419–2425.

Liu, X., Bimerew, M., Ma, Y., Müller, H., Ovadis, M., Eberl, L., … Chernin, L. (2007). Quorum-sensing signaling is required for production of the antibiotic pyrrolnitrin in a rhizospheric biocontrol strain of *Serratia plymuthica*. *FEMS Microbiology Letters*, *270*(2), 299–305. Available from https://doi.org/10.1111/j.1574-6968.2007.00681.x.

Liu, X., Ruan, L., Peng, D., Li, L., Sun, M., & Yu, Z. (2014). Thuringiensin: A thermostable secondary metabolite from *Bacillus thuringiensis* with insecticidal activity against a wide range of insects. *Toxins*, *6*(8), 2229–2238. Available from https://doi.org/10.3390/toxins6082229.

Lucena, W. A., Pelegrini, P. B., Martins-de-Sa, D., Fonseca, F. C., Gomes, J. E., De Macedo, L. L., … Grossi-de-Sa, M. F. (2014). Molecular approaches to improve the insecticidal activity of *Bacillus thuringiensis* cry toxins. *Toxins*, *6*(8), 2393–2423. Available from https://doi.org/10.3390/toxins6082393.

Ma, Z., Wang, N., Hu, J., & Wang, S. (2012). Isolation and characterization of a new iturinic lipopeptide, mojavensin A produced by a marine-derived bacterium *Bacillus mojavensis* B0621A. *Journal of Antibiotics*, *65*(6), 317–322. Available from http://www.nature.com/ja/journal/v65/n6/suppinfo/ja201219s1.html.

Makarasen, A., Reukngam, N., Khlaychan, P., Chuysinuan, P., Isobe, M., & Techasakul, S. (2018). Mode of action and synergistic effect of valinomycin and cereulide with amphotericin B against *Candida albicans* and *Cryptococcus albidus*. *Journal de Mycologie Medicale*, *28*(1), 112–121. Available from https://doi.org/10.1016/j.mycmed.2017.11.007.

Manhas, R. K., & Kaur, T. (2016). Biocontrol potential of *Streptomyces hydrogenans* strain DH16 toward *Alternaria brassicicola* to control damping off and black leaf spot of *Raphanus sativus*. *Frontiers in Plant Science*, *7*(2016). Available from https://doi.org/10.3389/fpls.2016.01869.

Matsuyama, T., & Tanikawa, T. (2011). *Nakagawa Y. Serrawettins and other surfactants produced by Serratia* (pp. 93–120). .

Mavrodi, D. V., Bonsall, R. F., Delaney, S. M., Soule, M. J., Phillips, G., & Thomashow, L. S. (2001). Functional analysis of genes for biosynthesis of pyocyanin and phenazine-1-carboxamide from *Pseudomonas aeruginosa* PAO1. *Journal of Bacteriology*, *183*(21), 6454–6465.

Mazzola, M., Funnell, D. L., & Raaijmakers, J. M. (2004). Wheat cultivar-specific selection of 2,4-diacetylphloroglucinol-producing fluorescent *Pseudomonas* species from resident soil populations. *Microbial Ecology*, *48*(3), 338–348. Available from https://doi.org/10.1007/s00248-003-1067-y.

McArthur, G. H., IV, & Fong, S. S. (2010). Toward engineering synthetic microbial metabolism. *Journal of Biomedicine and Biotechnology*, *2010*. Available from https://doi.org/10.1155/2010/459760.

Medema, M. H., Blin, K., Cimermancic, P., De Jager, V., Zakrzewski, P., Fischbach, M. A., … Breitling, R. (2011). AntiSMASH: Rapid identification, annotation and analysis of secondary metabolite biosynthesis gene clusters in bacterial and fungal genome sequences. *Nucleic Acids Research*, *39*(2), W339–W344. Available from https://doi.org/10.1093/nar/gkr466.

Melo alda., Soccol, V. T., & Soccol, C. R. (2016). *Bacillus thuringiensis*: Mechanism of action, resistance, and new applications: A review. *Critical Reviews in Biotechnology, 36*(2), 317–326. Available from https://doi.org/10.3109/07388551.2014.960793.

Meyer, S. L. F., Roberts, D. P., Chitwood, D. J., Carta, L. K., Lumsden, R. D., & Mao, W. (2001). Application of *Burkholderia cepacia* and Trichoderma virens, alone and in combinations, against Meloidogyne incognita on bell pepper. *Nematropica., 31*(1), 75–86.

Michelsen, C. F., Watrous, J., Glaring, M. A., Kersten, R., Koyama, N., Dorrestein, P. C., & Stougaard, P. (2015). Nonribosomal peptides, key biocontrol components for *Pseudomonas fluorescens* in 5, isolated from a Greenlandic suppressive soil. *mBio, 6* (2). Available from https://doi.org/10.1128/mBio.00079-15.

Milner, R. J. (1994). History of *Bacillus thuringiensis. Agriculture, Ecosystems and Environment, 49*(1), 9–13. Available from https://doi.org/10.1016/0167-8809(94)90014-0.

Mingma, R., Pathom-aree, W., Trakulnaleamsai, S., Thamchaipenet, A., & Duangmal, K. (2014). Isolation of rhizospheric and roots endophytic actinomycetes from Leguminosae plant and their activities to inhibit soybean pathogen, *Xanthomonas campestris* pv. glycine. *World Journal of Microbiology and Biotechnology, 30*(1), 271–280. Available from https://doi.org/10.1007/s11274-013-1451-9.

Min-Woo, W., Hee-Ju, N., Si-Sun, C., & Eung-Soo, K. (2014). Pikromycin production stimulation through antibiotic down-regulatory gene disruption in *Streptomyces venezuelae. Biotechnology and Bioprocess Engineering, 19*, 973–977. Available from https://doi.org/10.1007/s12257-014-0407-8.

Moffa, M., Pasanisi, D., Scarpa, E., Marra, A. R., Alifano, P., & Pisignano, D. (2017). Secondary metabolite production from industrially relevant bacteria is enhanced by organic nanofibers. *Biotechnology Journal, 12*(11). Available from https://doi.org/10.1002/biot.201700313.

Müller, S., Garcia-Gonzalez, E., Mainz, A., Hertlein, G., Heid, N. C., Mösker, E., ... Süssmuth, R. D. (2014). Paenilamicin: structure and biosynthesis of a hybrid nonribosomal peptide/polyketide antibiotic from the bee pathogen *Paenibacillus larvae. Angewandte Chemie, 53*(40), 10821–10825. Available from https://doi.org/10.1002/anie.201404572.

Nah, J. H., Kim, H. J., Lee, H. N., Lee, M. J., Choi, S. S., & Kim, E. S. (2013). Identification and biotechnological application of novel regulatory genes involved in *Streptomyces polyketide* overproduction through reverse engineering strategy. *BioMed Research International, 2013*. Available from https://doi.org/10.1155/2013/549737.

Nakayama, T., Homma, Y., Hashidoko, Y., Mizutani, J., & Tahara, S. (1999). Possible role of xanthobaccins produced by *Stenotrophomonas* sp. strain SB-K88 in suppression of sugar beet damping-off disease. *Applied Environmental Microbiology, 65*, 4334–4339.

Neipp, P. W., & Becker, J. O. (1999). Evaluation of biocontrol activity of rhizobacteria from Beta vulgaris against Heterodera schachtii. *Journal of Nematology, 31*(1), 54–61.

Netzker, T., Fischer, J., Weber, J., Mattern, D. J., König, C. C., Valiante, V., ... Brakhage, A. A. (2015). Microbial communication leading to the activation of silent fungal secondary metabolite gene clusters. *Frontiers in Microbiology, 6*. Available from https://doi.org/10.3389/fmicb.2015.00299.

Nguyen, X. H., Naing, K. W., Lee, Y. S., Tindwa, H., Lee, G. H., Jeong, B. K., ... Kim, K. Y. (2012). Biocontrol potential of *Streptomyces griseus* H7602 against root rot disease (*Phytophthora capsici*) in pepper. *Plant Pathology Journal, 28*(3), 282–289. Available from https://doi.org/10.5423/PPJ.OA.03.2012.0040.

Niehs, S. P., Kumpfmüller, J., Dose, B., Little, R. F., Ishida, K., Flórez, L. V., ... Hertweck, C. (2020). Insect-associated bacteria assemble the antifungal butenolide gladiofungin by non-canonical polyketide chain termination. *Angewandte Chemie – International Edition*, *59*(51), 23122–23126. Available from https://doi.org/10.1002/anie.202005711.

Nielsen, M. L., Isbrandt, T., Rasmussen, K. B., Thrane, U., Hoof, J. B., Larsen, T. O., & Mortensen, U. H. (2017). Genes Linked to production of secondary metabolites in talaromyces atroroseus revealed using CRISPR-Cas9. *PLoS One*, *12*(1). Available from https://doi.org/10.1371/journal.pone.0169712.

Nielsen, T. H., Sørensen, D., Tobiasen, T., Andersen, J. B., Christophersen, C., Givskov, M., & Sørensen, J. (2002). Antibiotic and biosurfactant properties of cyclic lipopeptides produced by fluorescent *Pseudomonas* spp. from the sugar beet rhizosphere. *Applied Environmental Microbiology*, *68*, 3416–3423.

Nielsen, T. H., Christophersen, C., Anthoni, U., & Sørensen, J. (1999). Viscosinamide, a new cyclic depsipeptide with surfactant and antifungal properties produced by *Pseudomonas fluorescens* DR54. *Journal of Applied Microbiology*, *87*(1), 80–90. Available from https://doi.org/10.1046/j.1365-2672.1999.00798.x.

Nielsen, T. H., Thrane, C., Christophersen, C., Anthoni, U., & Sørensen, J. (2000). Structure, production characteristics and fun gel antagonism of tensin—A new antifungal cyclic lipopeptide from *Pseudomonas fluorescens* strain 96.578. *Journal of Applied Microbiology*, *89*(6), 992–1001. Available from https://doi.org/10.1046/j.1365-2672.2000.01201.x.

Noh, J. H., Kim, S. H., Lee, H. N., Lee, S. Y., & Kim, E. S. (2010). Isolation and genetic manipulation of the antibiotic down-regulatory gene, wblA ortholog for doxorubicin-producing Streptomyces strain improvement. *Applied Microbiology and Biotechnology*, *86*(4), 1145–1153. Available from https://doi.org/10.1007/s00253-009-2391-z.

Nowak-Thompson, B., Chaney, N., Wing, J. S., Gould, S. J., & Loper, J. E. (1999). Characterization of the pyoluteorin biosynthetic gene cluster of *Pseudomonas fluorescens* Pf-5. *Journal of Bacteriology*, *181*(7), 2166–2174. Available from https://doi.org/10.1128/jb.181.7.2166-2174.1999.

Nowak-Thompson, B., Gould, S. J., & Loper, J. E. (1997). Identification and sequence analysis of the genes encoding a polyketide synthase required for pyoluteorin biosynthesis in *Pseudomonas fluorescens* Pf-5. *Gene*, *204*(1–2), 17–24. Available from https://doi.org/10.1016/S0378-1119(97)00501-5.

Olaf, T., de J.V., C. L., Marlies van den, B., Gerards, S., Janssens, T. K., Zaagman, N., ... Besselink, H. (2017). Exploring bacterial interspecific interactions for discovery of novel antimicrobial compounds. *Microbial Biotechnology*, *10*, 910–925. Available from https://doi.org/10.1111/1751-7915.12735.

Ongena, M., & Jacques, P. (2008). Bacillus lipopeptides: versatile weapons for plant disease biocontrol. *Trends in Microbiology*, *16*(3), 115–125. Available from https://doi.org/10.1016/j.tim.2007.12.009.

Paget, M. S. B., Kang, J. G., Roe, J. H., & Buttner, M. J. (1998). σ(R), an RNA polymerase sigma factor that modulates expression of the thioredoxin system in response to oxidative stress in *Streptomyces coelicolor* A3(2). *EMBO Journal*, *17*(19), 5776–5782. Available from https://doi.org/10.1093/emboj/17.190.5776.

Palazzotto, E., & Weber, T. (2018). Omics and multi-omics approaches to study the biosynthesis of secondary metabolites in microorganisms. *Current Opinion in Microbiology*, *45*, 109–116. Available from https://doi.org/10.1016/j.mib.2018.03.004.

Park, C. N., Lee, J. M., Lee, D., & Kim, B. S. (2008). Antifungal activity of valinomycin, a peptide antibiotic produced by *Streptomyces* sp. strain M10 antagonistic to *Botrytis cinerea*. *Journal of Microbiology and Biotechnology*, *18*(5), 880–884, http://www.jmb.or.kr/home/journal/include/downloadPdf.asp?articleuid = {A56ED678-6022-47F2-92CE-D1C450C6AE06}.

Patel, P. S., Huang, S., Fisher, S., Pirnik, D., Aklonis, C., Dean, L., ... Mayerl, F. (1995). Bacillaene, a novel inhibitor of procaryotic protein synthesis produced by *Bacillus subtilis*: Production, taxonomy, isolation, physico-chemical characterization and biological activity. *The Journal of Antibiotics*, *48*(9), 997–1003. Available from https://doi.org/10.7164/antibiotics.48.997.

Peng, H., Zhang, P., Bilal, M., Wang, W., Hu, H., & Zhang, X. (2018). Enhanced biosynthesis of phenazine-1-carboxamide by engineered *Pseudomonas chlororaphis* HT66. *Microbial Cell Factories*, *17*(1), 1–12.

Pettit, R. K. (2011). Small-molecule elicitation of microbial secondary metabolites. *Microbial Biotechnology*, *4*(4), 471–478. Available from https://doi.org/10.1111/j.1751-7915.2010.00196.x.

Pierson, L. S., III, & Pierson, E. A. (1996). Phenazine antibiotic production in *Pseudomonas aureofaciens*: role in rhizosphere ecology and pathogen suppression. *FEMS Microbiology Letters*, *136*, 101–108.

Pimentel, D. (1997). *Techniques for reducing pesticide use: Economic and environmental benefits*.

Prabhu, S., Kumar, S., Subramanian, S., & Sundaram, S. P. (2009). Suppressive effect of *Methylobacterium fujisawaense* against root-knot nematode, *Meloidogyne incognita*. *Indian Journal of Nematology*, *39*(2), 165–169. Available from http://www.indianjournals.com/ijor.aspx?target = ijor:ijn&volume = 39&issue = 2&article = 008.

Qi, G., Zhu, F., Du, P., Yang, X., Qiu, D., Yu, Z., ... Zhao, X. (2010). Lipopeptide induces apoptosis in fungal cells by a mitochondria dependent pathway. *Peptides*, *31*(11), 1978–1986. Available from https://doi.org/10.1016/j.peptides.2010.08.003.

Raaijmakers, J. M., & Weller, D. M. (2001). Exploiting genotypic diversity of 2,4-diacetylphloroglucinol-producing *Pseudomonas* spp.: Characterization of superior root-colonizing *P. fluorescens* strain Q8r1–96. *Applied and Environmental Microbiology*, *67*(6), 2545–2554. Available from https://doi.org/10.1128/AEM.67.6.2545-2554.2001.

Rahul, S., Chandrashekhar, P., Hemant, B., Chandrakant, N., Laxmikant, S., & Satish, P. (2014). Nematicidal activity of microbial pigment from *Serratia marcescens*. *Natural Product Research*, *28*(17), 1399–1404. Available from https://doi.org/10.1080/14786419.2014.904310.

Ramarathnam, R., Bo, S., Chen, Y., Fernando, W. G., Xuewen, G., & de Kievit, T. (2007). Molecular and biochemical detection of fengycin- and bacillomycin D-producing *Bacillus* spp., antagonistic to fungal pathogens of canola and wheat. *Canadian Journal of Microbiology*, *53*(7), 901–911. Available from https://doi.org/10.1139/w07-049.

Ran, P., Ye, W., Wan-wan, F., Yue, X. J., Chen, J. H., Hu, X. Z., ... Li, Y. Z. (2018). CRISPR/dCas9-mediated transcriptional improvement of the biosynthetic gene cluster for the epothilone production in *Myxococcus xanthus*. *Microbial Cell Factories*, *17*. Available from https://doi.org/10.1186/s12934-018-0867-1.

Rangaswamy, V., Jiralerspong, S., Parry, R., & Bender, C. L. (1998). Biosynthesis of the Pseudomonas polyketide coronafacic acid requires monofunctional and multifunctional polyketide synthase proteins. *Proceedings of the National Academy of Sciences*, *95*, 15469–15474. Available from https://doi.org/10.1073/pnas.95.26.15469.

Raza, W., Yang, X. M., Wu, H. S., Wang, Y., Xu, Y. C., & Shen, Q. R. (2009). Isolation and characterisation of fusaricidin-type compound-producing strain of *Paenibacillus polymyxa* SQR-21 active against *Fusarium oxysporum* f. sp *nevium*. *European Journal of Plant Pathology*, *125*(3), 471–483. Available from https://doi.org/10.1007/s10658-009-9496-1.

Reily, M. D., & Lindon, J. C. (2005). *NMR spectroscopy: principles and instrumentation. Metabonomics in toxicity assessment* (pp. 75–104). Boca Raton, FL: CRC Press.

Ren, J. H., Li, H., Wang, Y. F., Ye, J. R., Yan, A. Q., & Wu, X. Q. (2013). Biocontrol potential of an endophytic *Bacillus pumilus* JK-SX001 against poplar canker. *Biological Control*, *67*(3), 421–430. Available from https://doi.org/10.1016/j.biocontrol.2013.09.012.

Ricart, E., Leclère, V., Flissi, A., Mueller, M., Pupin, M., & Lisacek, F. (2019). RBAN: Retro-biosynthetic analysis of nonribosomal peptides. *Journal of Cheminformatics*, *11* (1). Available from https://doi.org/10.1186/s13321-019-0335-x.

Romero, D., de Vicente, A., Rakotoaly, R. H., Dufour, S. E., Veening, J. W., Arrebola, E., ... Pérez-García, A. (2007). The iturin and fengycin families of lipopeptides are key factors in antagonism of *Bacillus subtilis* toward *Podosphaera fusca*. *Molecular Plant-Microbe Interactions*, *20*(4), 430–440.

Ross, C., Opel, V., Scherlach, K., & Hertweck, C. (2014). Biosynthesis of antifungal and antibacterial polyketides by *Burkholderia gladioli* in coculture with *Rhizopus microsporus*. *Mycoses*, *57*(3), 48–55. Available from https://doi.org/10.1111/myc.12246.

Samy, S. J., Xaviar, A., Rahman, A. B., Sharifuddin, H. A. H. (1995). Effect of EM on rice production and methane emission from paddy fields in Malaysia. In *Kyusei nature farming (fourth international conference*.

Sanchis, V. (2012). *Genetic improvement of bt strains and development of novel biopesticides*. Bacillus thuringiensis *biotechnology* (pp. 215–228). the Netherlands: Springer, 9789400730212; doi:10.1007/978-94-007-3021-2_12.

Schroeckh, V., Scherlach, K., Nützmann, H.-W., Shelest, E., Schmidt-Heck, W., Schuemann, J., ... Brakhage, A. A. (2009). Intimate bacterial–fungal interaction triggers biosynthesis of archetypal polyketides in *Aspergillus nidulans*. *Proceedings of the National Academy of Sciences of the United States of America*, *106*, 14558–14563. Available from https://doi.org/10.1073/pnas.0901870106.

Segre, A., Bachmann, R. C., Ballio, A., Bossa, F., Grgurina, I., Iacobellis, N. S., ... Takemoto, J. Y. (1989). The structure of syringomycins A 1, E and G. *FEBS Letters*, *255*, 27–31. Available from https://doi.org/10.1016/0014-5793(89)81054-3.

Seyedsayamdost, M. R., Traxler, M. F., Clardy, J., & Kolter, R. (2012). *Old meets new: Using interspecies interactions to detect secondary metabolite production in actinomycetes. Methods in enzymology* (Vol. 517, pp. 89–109). Academic Press Inc. Available from http://doi.org/10.1016/B978-0-12-404634-4.00005-X.

Sharma, R., Bisaria, V. S., & Sharma, S. (2019). *Rhizosphere: A home for human pathogens. Plant biotic interactions: State of the art* (pp. 113–127). Springer International Publishing. Available from http://doi.org/10.1007/978-3-030-26657-8_8.

Shen, Y., Volrath, S. L., Weatherly, S. C., Elich, T. D., & Tong, L. (2004). A mechanism for the potent inhibition of eukaryotic acetyl-coenzyme A carboxylase by soraphen A, a macrocyclic polyketide natural product. *Molecular Cell*, *16*(6), 881–891.

Siddiqui, I. A., Shaukat, S. S., Khan, G. H., & Ali, N. A. (2003). Suppression of Meloidogyne javanica by *Pseudomonas aeruginosa* IE-6S + in tomato: the influence of NaCl, oxygen and iron levels. *Soil Biology and Biochemistry*, *35*, 1625–1634.

Siddiqui, I. A., & Shaukat, S. S. (2002). Rhizobacteria-mediated induction of systemic resistance (ISR) in tomato against *Meloidogyne javanica*. *Journal of Phytopathology*, *150*(8–9), 469–473. Available from https://doi.org/10.1046/j.1439-0434.2002.00784.x.

Skinnider, M. A., Dejong, C. A., Rees, P. N., Johnston, C. W., Li, H., Webster, A. L., ... Magarvey, N. A. (2015). Genomes to natural products prediction informatics for secondary metabolomes (PRISM). *Nucleic Acids Research*, *43*(20), 9645–9662. Available from https://doi.org/10.1093/nar/gkv1012.

Soerensen, D., Nielsen, T. H., Christophersen, C., Soerensen, J., & Gajhede, M. (2001). Cyclic lipoundecapeptide amphisin from *Pseudomonas* sp. strain DSS73. *Acta Crystallographica Section C: Crystal Structure Communications*, *57*(9), 1123–1124.

Sood, S., Steinmetz, H., Beims, H., Mohr, K. I., Stadler, M., Djukic, M., ... Muller, R. (2014). Paenilarvins: iturin family lipopeptides from the honey bee pathogen *Paenibacillus larvae*. *Chembiochem*, *15*(13), 1947–1955. Available from https://doi.org/10.1002/cbic.201402139.

Srivastava, N., Srivastava, M., Ramteke, P. W., & Mishra, P. K. (2019). *Solid-state fermentation strategy for microbial metabolites production: An overview. New and future developments in microbial biotechnology and bioengineering: Microbial secondary metabolites biochemistry and applications* (pp. 345–354). Elsevier. Available from http://doi.org/10.1016/B978-0-444-63504-4.00023-2.

Su, C., Xiang, Z., Liu, Y., Zhao, X., Sun, Y., Li, Z., ... Zhou, Y. (2016). Analysis of the genomic sequences and metabolites of *Serratia surfactantfaciens* sp. nov. YD25T that simultaneously produces prodigiosin and serrawettin W2. *BMC Genomics*, *17*(1). Available from https://doi.org/10.1186/s12864-016-3171-7.

Sumi, C. D., Yang, B. W., Yeo, I. C., & Hahm, Y. T. (2015). Antimicrobial peptides of the genus *Bacillus*: a new era for antibiotics. *Canadian Journal of Microbiology*, *61*(2), 93–103. Available from https://doi.org/10.1139/cjm-2014-0613.

Sun, L., Lu, Z., Bie, X., Lu, F., & Yang, S. (2006). Isolation and characterization of a coproducer of fengycins and surfactins, endophytic *Bacillus amyloliquefaciens* ES-2, from *Scutellaria baicalensis* Georgi. *World Journal of Microbiology and Biotechnology*, *22* (12), 1259–1266. Available from https://doi.org/10.1007/s11274-006-9170-0.

Suryawanshi, R. K., Patil, C. D., Koli, S. H., Hallsworth, J. E., & Patil, S. V. (2017). Antimicrobial activity of prodigiosin is attributable to plasma-membrane damage. *Natural Product Research*, *31*(5), 572–577. Available from https://doi.org/10.1080/14786419.2016.1195380.

Takemoto, J. Y., Brand, J. G., Kaulin, Y. A., Malev, V. V., Schagina, L. V., & Blasko, K. (2003). 12 The Syringomycins. *Pore-Forming Peptides and Protein Toxins*, *5*.

Takemoto, J. Y. (1992). Bacterial phytotoxin syringomycin and its interaction with host membranes. In D. P. S. Verma (Ed.), *Molecular signals in plant–microbe communications* (pp. 247–260). Boca Raton, FL: CRC Press.

Tamez-Guerra, P., Iracheta, M. M., Pereyra-Alférez, B., Galán-Wong, L. J., Gomez-Flores, R., Tamez-Guerra, R. S., & Rodrıguez-Padilla, C. (2004). Characterization of Mexican *Bacillus thuringiensis* strains toxic for lepidopteran and coleopteran larvae. *Journal of Invertebrate Pathology*, *86*(1–2), 7–18. Available from https://doi.org/10.1016/j.jip.2004.02.009.

Tan, H. M., Cao, L. X., He, Z. F., Su, G. J., Lin, B., & Zhou, S. N. (2006). Isolation of endophytic actinomycetes from different cultivars of tomato and their activities against *Ralstonia solanacearum* in vitro. *World Journal of Microbiology and Biotechnology*, *22*(12), 1275–1280. Available from https://doi.org/10.1007/s11274-006-9172-y.

Tao, Y., Bie, X. m, Lv, F. x, Zhao, H. z, & Lu, Z. X. (2011). Antifungal activity and mechanism of fengycin in the presence and absence of commercial surfactin against *Rhizopus stolonifer*. *Journal of Microbiology*, *49*(1), 146–150. Available from https://doi.org/10.1007/s12275-011-0171-9.

Thomashow, L. S., Weller, D. M., Bonsall, R. F., & Pierson, L. S., III (1990). Production of the antibiotic phenazine-1-carboxylic acid by fluorescent *Pseudomonas* in the rhizosphere of wheat. *Applied Environmental Microbiology*, *56*, 908–912.

Thrane, C., Nielsen, T. H., Nielsen, M. N., Olsson, S., & Sørensen, J. (2000). Viscosinamide-producing *Pseudomonas fluorescens* DR54 exerts biocontrol effect on *Pythium ultimum* in sugar beet rhizosphere. *FEMS Microbiology Ecology*, *33*, 139–146.

Toledo, J., Liedo, P., Williams, T., & Ibarra, J. (1999). Toxicity of *Bacillus thuringiensis* β-exotoxin to three species of fruit flies (Diptera: Tephritidae). *Journal of Economic Entomology*, *92*(5), 1052–1056. Available from https://doi.org/10.1093/jee/92.50.1052.

Toral, L., Rodríguez, M., Béjar, V., & Sampedro, I. (2018). Antifungal activity of lipopeptides from Bacillus XT1 CECT 8661 against *Botrytis cinerea*. *Frontiers in Microbiology*, *9*. Available from https://doi.org/10.3389/fmicb.2018.01315.

Traxler, M. F., Watrous, J. D., Alexandrov, T., Dorrestein, P. C., & Kolter, R. (2013). Interspecies interactions stimulate diversification of the *Streptomyces coelicolor* secreted metabolome. *mBio*, *4*(4). Available from https://doi.org/10.1128/mBio.00459-13.

Troppens, D. M., Dmitriev, R. I., Papkovsky, D. B., O'Gara, F., & Morrissey, J. P. (2013). Genome-wide investigation of cellular targets and mode of action of the antifungal bacterial metabolite 2,4-diacetylphloroglucinol in *Saccharomyces cerevisiae*. *FEMS Yeast Research*, *13*(3), 322–334. Available from https://doi.org/10.1111/1567-1364.12037.

Tsuchiya, S., Kasaishi, Y., Harada, H., Ichimatsu, T., Saitoh, H., Mizuki, E., & Ohba, M. (2002). Assessment of the efficacy of Japanese *Bacillus thuringiensis* isolates against the cigarette beetle, *Lasioderma serricorne* (Coleoptera: Anobiidae). *Journal of Invertebrate Pathology*, *81*(2), 122–126. Available from https://doi.org/10.1016/S0022-2011(02)00148-9.

Van Pée, K. H., & Ligon, J. M. (2000). Biosynthesis of pyrrolnitrin and other phenylpyrrole derivatives by bacteria. *Natural Product Reports*, *17*(2), 157–164. Available from https://doi.org/10.1039/a902138h.

van Rij, E. T., Wesselink, M., Chin-A-Woeng, T. F., Bloemberg, G. V., & Lugtenberg, B. J. (2004). Influence of environmental conditions on the production of phenazine-1-carboxamide by *Pseudomonas chlororaphis* PCL1391. *Molecular Plant-Microbe Interactions*, *17*(5), 557–566.

Vater, J., Niu, B., Dietel, K., & Borriss, R. (2015). Characterization of novel fusaricidins produced by *Paenibacillus polymyxa* M1 using MALDI-TOF mass spectrometry. *Journal of the American Society for Mass Spectrometry*, *26*(9), 1548–1558. Available from https://doi.org/10.1007/s13361-015-1130-1.

Vicente Dos Reis, G., Abraham, W. R., Grigoletto, D. F., de Campos, J. B., Marcon, J., da Silva, J. A., ... Lira, S. P. (2019). Gloeosporiocide, a new antifungal cyclic peptide from *Streptomyces morookaense* AM25 isolated from the Amazon bulk soil. *FEMS Microbiology Letters*, *366*(14). Available from https://doi.org/10.1093/femsle/fnz175.

Voisard, C., Keel, C., Haas, D., & Defago, G. (1989). Cyanide production by *Pseudomonas fluorescens* helps suppress black root of tobacco under gnotobiotic conditions. *The EMBO Journal*, *8*, 351–358.

Wakefield, J., Hassan, H. M., Jaspars, M., Ebel, R., & Rateb, M. E. (2017). Dual induction of new microbial secondary metabolites by fungal bacterial co-cultivation. *Frontiers in Microbiology*, *8*. Available from https://doi.org/10.3389/fmicb.2017.01284.

Wilson, D. J., Patton, S., Florova, G., Hale, V., & Reynolds, K. A. (1998). The shikimic acid pathway and polyketide biosynthesis. *Journal of Industrial Microbiology and Biotechnology*, *20*(5), 299–303. Available from https://doi.org/10.1038/sj.jim.2900527.

Wong, J. W. H., Abuhusain, H. J., McDonald, K. L., & Don, A. S. (2012). MMSAT: Automated quantification of metabolites in selected reaction monitoring experiments. *Analytical Chemistry*, *84*(1), 470–474. Available from https://doi.org/10.1021/ac2026578.

Wu, L., Wu, H., Chen, L., Yu, X., Borriss, R., & Gao, X. (2015). Difficidin and bacilysin from *Bacillus amyloliquefaciens* FZB42 have antibacterial activity against *Xanthomonas oryzae* rice pathogens. *Scientific Reports*, *5*. Available from https://doi.org/10.1038/srep12975.

Xu, Z., Shao, J., Li, B., Yan, X., Shen, Q., & Zhang, R. (2013). Contribution of bacillomycin D in *Bacillus amyloliquefaciens* SQR9 to antifungal activity and biofilm formation. *Applied and Environmental Microbiology*, *79*(3), 808–815. Available from https://doi.org/10.1128/AEM.02645-12.

Xue-Mei, J., Yong-Keun, C., Hag, L. J., & Soon-Kwang, H. (2017). Effects of increased NADPH concentration by metabolic engineering of the pentose phosphate pathway on antibiotic production and sporulation in *Streptomyces lividans* TK24. *Journal of Microbiology and Biotechnology*, *27*, 1867–1876. Available from https://doi.org/10.4014/jmb.1707.07046.

Yang, A., Zeng, S., Yu, L., He, M., Yang, Y., Zhao, X., ... Song, B. (2018). Characterization and antifungal activity against Pestalotiopsis of a fusaricidin-type compound produced by *Paenibacillus polymyxa* Y-1. *Pesticide Biochemistry and Physiology*, *147*, 67–74. Available from https://doi.org/10.1016/j.pestbp.2017.08.012.

Yang, S. Y., Lim, D. J., Noh, M. Y., Kim, J. C., Kim, Y. C., & Kim, I. S. (2017). Characterization of biosurfactants as insecticidal metabolites produced by *Bacillus subtilis* Y9. *Entomological Research*, *47*(1), 55–59. Available from https://doi.org/10.1111/1748-5967.12200.

Yavuzaslanoglu, E., Yamac., & Nicol, J. M. (2011). Influence of actinomycete isolates on cereal cyst nematode *Heterodera filipjevi* juvenile motility. *Nematologia Mediterranea*, *39*, 41–45.

Yu, G. Y., Sinclair, J. B., Hartman, G. L., & Bertagnolli, B. L. (2002). Production of iturin A by *Bacillus amyloliquefaciens* suppressing *Rhizoctonia solani*. *Soil Biology and Biochemistry*, *34*(7), 955–963.

Yu, W. B., & Ye, B. C. (2016). High-level iron mitigates fusaricidin-induced membrane damage and reduces membrane fluidity leading to enhanced drug resistance in *Bacillus subtilis*. *Journal of Basic Microbiology*, *56*(5), 502–509. Available from https://doi.org/10.1002/jobm.201500291.

Yuan, L., Shuangyang, D., Richard, D., Erwin, M., & Kui, Z. (2017). A biosurfactant-inspired heptapeptide with improved specificity to kill MRSA. *Angewandte Chemie*, *56*, 1508–1512. Available from https://doi.org/10.1002/ange.201609277.

Yun, D. C., Yang, S. Y., Kim, Y. C., Kim, I. S., & Kim, Y. H. (2013). Identification of surfactin as an aphicidal metabolite produced by *Bacillus amyloliquefaciens* G1.

Journal of the Korean Society for Applied Biological Chemistry, *56*(6), 751–753. Available from https://doi.org/10.1007/s13765-013-3238-y.

Zhang, B., Dong, C., Shang, Q., Han, Y., & Li, P. (2013). New insights into membrane-active action in plasma membrane of fungal hyphae by the lipopeptide antibiotic bacillomycin L. *Biochimica et Biophysica Acta – Biomembranes*, *1828*(9), 2230–2237. Available from https://doi.org/10.1016/j.bbamem.2013.05.033.

Zhang, L., Yan, K., Zhang, Y., Huang, R., Bian, J., Zheng, C., ... Min, F. (2007). High-throughput synergy screening identifies microbial metabolites as combination agents for the treatment of fungal infections. *Proceedings of the National Academy of Sciences of the United States of America*, *104*(11), 4606–4611. Available from https://doi.org/10.1073/pnas.0609370104.

Zhao, Z., Wang, Q., Wang, K., Brian, K., Liu, C., & Gu, Y. (2010). Study of the antifungal activity of *Bacillus vallismortis* ZZ185 in vitro and identification of its antifungal components. *Bioresource Technology*, *101*, 292–297.

Zouari, I., Jlaiel, L., Tounsi, S., & Trigui, M. (2016). Biocontrol activity of the endophytic *Bacillus amyloliquefaciens* strain CEIZ-11 against *Pythium aphanidermatum* and purification of its bioactive compounds. *Biological Control*, *100*, 54–62.

CHAPTER

Applications of microbial biosurfactants in biocontrol management

10

Pooja Singh[1] and Vinay Rale[2]

[1]*Symbiosis Centre for Waste Resource Management, Symbiosis International (Deemed University), India*

[2]*Symbiosis School of Biological Sciences, Symbiosis International (Deemed University), Pune, India*

10.1 Introduction

Modern agricultural practices being used since last few decades in most of the countries worldwide have made food abundant and affordable for the increasing world population. This includes, among other practices, extensive use of fertilizers and pesticides to deal with the numerous pests attacking the crops and pastures and also for increasing the productivity of the land. Changes and disturbances in the habitat and biodiversity of animals and microorganisms in recent years due to anthropogenic activities have altered multiple host—microbe interactions and have led to an increased susceptibility of various crops to multiple novel pests. Pest infestation instances in crops directly impact the yield and quality of agricultural produce. This, in turn, instigates an increased use of chemical fertilizers and pesticides to combat this threat and maintain the health and productivity of the soil. Indiscriminate use of chemical fertilizers and pesticides has manifested a plethora of environmental challenges in the form of depletion of soil and water quality, loss of soil and water biodiversity, and increased contamination and toxicity to flora and fauna as well as human and environmental health. Also, extensive exposure of insect pests and other microbial pathogens to pesticides leads to the development of resistance in pests and solicits additional use of such agents for pest control thereby worsening the crisis. Apart from extensive negative effect on human and environment health, nonpoint source water pollution caused by pesticide and fertilizer laden agricultural runoffs leads to eutrophication and dead zones in many water bodies (Gold, 2007). In lieu of the multitude of effects on the use of synthetic compounds for pest control, it becomes imperative to revise the existing practices and moves toward a sustainable agriculture. Use of natural compounds and natural pest predators for pest management is one of the primary agricultural practices aiming at an environment-friendly agriculture. Biological surface-active molecules, biosurfactants are such ecologically safe compounds that have been reported to have prominent biocontrol activities and

Biocontrol Mechanisms of Endophytic Microorganisms. DOI: https://doi.org/10.1016/B978-0-323-88478-5.00009-2

immense potential for a sustainable agriculture (Olasanmi & Thring, 2018). Biosurfactants have also been successfully incorporated in synthetic formulations to enhance the effectiveness of chemical or other biological biocontrol agents. This chapter strives to illustrate the various applications of biosurfactants as a component of sustainable agriculture especially their use as a biological pest and pathogen control agent.

Another aspect that is of significance for sustainable pest management is the role played by endophytes. Endophytes are organisms that live in close association with different plant parts and tissues, either in symbiotic association or in a pathogenic role. They have been known to produce a wide range of molecules that play an active and irreplaceable role in plant growth, plant stress and pathogen resistant mechanisms, and protection of plant from other predators like herbivores and pests (Nair and Padmavathy, 2014; Singh, Singh, & Singh, 2019). Another aspect of endophytes that is of significance especially in lieu of the need for sustainable growth and development is the production of bioactive compounds. One such bioactive compound reported to be produced by many endophytes is biosurfactant. The role of biosurfactants and biosurfactant-producing endophytic bacteria in plant defense mechanisms is another aspect explored in this chapter. Various biosurfactant-producing microorganisms have been isolated and studied from various plant surfaces, including rhizosphere and leaves (D'aes, De Maeyer, Pauwelyn, & Hofte, 2010). Many plant-associated endophytes have been found to play an active role in plant immunity, plant resistant to stress factors, plant defense mechanisms, including direct protection from pathogens and pests as well as facilitating biocontrol mechanisms of other plant growth–promoting microbes (Nihorimbere, Marc Ongena, Smargiassi, & Thonart, 2011). Many of these resistant mechanisms can be attributed to the specific secretions of plant-associated microbial population, and hence, the study of plant-associated microbial molecules assumes vital importance. Biosurfactant is one such molecule that has been studied to play a vital role in plant growth and immunity and also in the activity of endophytes for increasing plant resistance and immunity. To fully unravel different aspect of sustainable agriculture, it becomes imperative to study the association of biosurfactant and endophytes in plant biocontrol mechanisms.

10.2 Biosurfactants

Biosurfactants are amphipathic molecules produced by many bacterial and fungal species under varying environmental and biological conditions. They are either produced extracellularly or secreted at the cell surface. The amphipathic nature confers on them the ability to orient themselves at interfaces and bring about lowering of the surface tension. Also of importance is their ability to form micelles (Shekhar, Sundaramanickam, & Balasubramanian, 2015). Both these properties are critical when the organism is growing in a nutrient-limited environment and attachment and detachment from the surface are important for microbial host survival and proliferation. Biosurfactants are amphipathic in nature, that is, they contain both hydrophilic and hydrophobic domains. The hydrophilic part may be ionic (cationic or anionic) or nonionic and consists of molecules like carbohydrate, cyclic peptide, amino acid, phosphate carboxyl acid, or alcohol, while hydrophobic part is mostly a hydrocarbon chain. Depending on the type of the hydrophilic and hydrophobic groups, biosurfactants can be classified as glycolipids, lipopeptides, lipoproteins, phospholipids, and similar such molecules (Cameotra, Makkar, Kaur, & Mehta, 2010). Various different types of biosurfactants have been discovered and reported till date, being produced by a wide range of microbial species in diverse ecological niches (Drakontis & Amin, 2020). Based on their molecular weight, biosurfactants are classified into two main categories: low molecular weight biosurfactants, including glycolipids and lipopeptides and high molecular weight polymers, like some polymeric and particulate biosurfactants. Low molecular weight biosurfactants are primarily implicated in the lowering of the surface and interfacial tensions, while high molecular weight biosurfactants have been reported to be more efficient emulsifying and stabilizing agents (Santos, Rufino, Luna, Santos, & Sarubbo, 2016). Structural diversity of biosurfactants gives them an additional advantage in applicability, and their ability to be produced from diverse substrates confers malleability and cost-effectiveness to the commercial production processes (Singh, Patil, & Rale, 2019). Structure of a biosurfactant is critical for its biological activity. Any alterations in structural diversity also confer on the producing host a diversity in functional attributes. Hence, the study of structure-function relationship of biosurfactants in various niches is important. Various member of many bacterial and yeast species been reported to produce different types of biosurfactants. These include *Pseudomonas, Rhodococcus, Bacillus, Lactobacillus, Acinetobacter, Arthrobacter, Corynebacterium, Enterobacter, Alcanivorax, and Candida* among others (Drakontis & Amin, 2020).

10.2.1 Applications of biosurfactants: a golden molecule for agriculture

Biosurfactants, due to their surface-active properties, increase the availability of hydrophobic and complex compounds to the producing microbe, thus enabling the microbial population to successfully colonize a nutrient-limited environment (Cameotra & Makkar, 2010). Apart from the lowering of surface tension, solubilization and mobilization of compounds are the other primary mechanisms of action of surfactants on hydrophobic substrates. Due to their amphipathic nature and the other abovementioned properties, biosurfactants find extensive use in the bioremediation of various hydrophobic pollutants, oil spills, heavy metals, in microbial enhanced oil recovery (MEOR), food industry, detergent industry, as well as a part of green synthesis for nanoparticle production (Cameotra & Singh, 2008; Fenibo, Ijoma, Selvarajan, & Chikere, 2019; Fracchia et al., 2014;

Marchant & Banat, 2012; Patel, Homaei, Patil, & Daverey, 2019; Singh et al., 2017). Their effectiveness, biodegradable nature, low toxicity, and specificity make them one of the most sought-after molecules in the current century and one of the most promising products for the future decades (Khan & Butt, 2016). Apart from their role in increasing the availability of substrates to the producing host, biosurfactants have also been implicated for their role as an adhesion molecule and as a quorum sensing facilitator and mediator, enabling the host organism to adapt, survive, and proliferate in an oligotrophic surrounding and tolerate other stress conditions and competitions (Dusane et al., 2010). Additionally, biosurfactants have also been implicated in the virulence of the host organism, contributing to their pathogenicity (Cameotra et al., 2010; Ławniczak, Marecik, & Chrzanowski, 2013). Apart from affecting the bioavailability of nutrients to host cells, alterations in the cell surface, interaction with membranes, and alteration in the metabolic properties of the surrounding microbial population by biosurfactants are features of these molecules that facilitate the host to colonize a competitive ecological niche or protect itself from any surrounding stress factor(s) (Cameotra & Singh, 2009; D'aes et al., 2010; Tahseen, Afzal, & Iqbal, 2016). Another sector where biosurfactants find extensive scope of application is agriculture. Owing to their abovementioned properties, various applications of biosurfactants have been proposed and commercialized in this area. Biosurfactants have been found to play an active role in the improvement of soil quality, augmentation of soil health, for plant pathogen elimination, in enhancing plant immunity and also to enhance plant–microbe interactions for better plant health (Drakontis & Amin, 2020; Sachdev & Cameotra, 2013; Singh, Glick, & Rathore, 2018). Biosurfactants have also been found to increase the availability of micronutrients to plants, a property that is exhibited also by many plant-associated microorganisms, thereby reiterating on the involvement of biosurfactants in the activity of endophytes.

Biosurfactants produced by plant-associated microbial population are essential for the microbial host in creating and establishing a symbiotic or pathogenic relation with plant surface. This relationship has also been found to play an active role in microbial swarming processes in rhizosphere, adhesion, and dissociation of microbial cells to plant surfaces, alterations in nutrient availability to plant, and in increasing plant defense mechanisms against pests, pathogens, or other toxic compounds (Thavasi, Marchant, & Banat, 2014). Potential of endophytes as biocontrol agents has been long studied. Along with other molecules such as cytokines, lactones, indole acetic acid, and 1-aminocyclopropane-1-carboxylic acid deaminase, biosurfactants are another class of compounds that have been found to be actively produced by many endophytes. Thus the study of endophytes for biocontrol also encompasses the study of biosurfactants by endophytes and its role in biocontrol. Study of biosurfactants for their applications in agriculture as a biocontrol agent hence will encompass the following aspects:

1. activity of biosurfactants as antimicrobial agents,
2. biosurfactants in plant pest control,

3. role of biosurfactants in plant immunity, and
4. role of biosurfactants in enhancing plant–microbe interactions.

Since many endophytes produce biosurfactants and the function of biosurfactants varies with the structure, it is critical to study the structure-function relationship of different types of biosurfactants produced by different plant-associated endophytes to better understand their role in plant defense mechanisms, including antiphytopathogenic properties. *Pseudomonas* and *Bacillus* are two of the most prominent plant-associated bacterial species and in this chapter, the authors specially discuss the studies on the biosurfactants produced by these two bacterial strains and their involvement and importance in the biocontrol, antimicrobial properties, and plant defense activities.

10.3 Biosurfactants as antimicrobial and biocontrol agents

Although the environmental applications of biosurfactants are the most well studied and exploited aspect, a property of biosurfactant that is gaining unprecedented interest in current times is their antimicrobial properties. Various biosurfactants have been reported to have antimicrobial activities at levels competitive to the traditional antibiotics being used in the health-care sector (Naughton, Marchant, Naughton, & Banat, 2019; Singh & Cameotra, 2004). As mentioned in the previous section, the presence of different types of stresses is one of the reasons for the production of biosurfactant by the microorganisms and hence it is not surprising that many biosurfactants exhibit antimicrobial activities probably as a defense mechanism for the survival of the host in a competitive environment. In plant systems, various theories have been proposed for this activity. Biosurfactants can have a direct antipathogenic effect or stimulate antipathogenic properties of other neighboring microorganisms in the niche. Additionally, biosurfactants have also been reported to show an indirect antiphytopathogenic property by increasing host defense mechanisms enabling them to counter any pathogen attack by antagonistic secretions. A reduction in the rate of deposition and adhesion of pathogens is one of the modes of action for the antimicrobial activity exhibited by many biosurfactants (Fracchia et al., 2014). Also reported is the ability of the biosurfactant to permeate the cell wall, leading to an enhanced entry of the antimicrobial agent in a weak target cell eventually causing cell death. Alterations in cell surface properties are one of the other features of biosurfactant that is actively at play in the natural environmental systems. This is primarily to enhance the adaptability of host to the presence of hydrophobic/toxic compounds and also to facilitate colonization of the microbial species in the environment (Chrzanowski, Ławniczak, & Czaczyk, 2012). Similar activity is also proposed to be involved in the antimicrobial activity of many biosurfactants on different plant pathogens. In one of the reports, antiphytopathogenic activity of biosurfactant was studied for the control of infection on tomato leaves by *Botrytis cinerea*. It was found that a naturally produced compound, ustilagic acid, by a phytopathogenic fungi *Ustilago maydis* exhibited a detergent-like activity and was capable of controlling infection by *B. cinerea* (Teichmann, Linne, Hewald, Marahiel, & Bölker, 2007). Damage to the target cell membrane was the mode of action ascertained. Ustilagic acid is a cellobiose lipid and is also produced by many other phytopathogenic fungi (Kulakovskaya, Golubev, & Tomashevskaya, 2010). Subsequently, various aspects of antimicrobial and antipest nature of biosurfactants have been studied (Thavasi et al., 2014). The hypotheses mostly proposed for the mode of action are as follows:

1. increased membrane permeability by modifications of phospholipids;
2. damage and alterations to the cell membrane by alterations in membrane electrical conductance, disruption of membrane cytoskeletal elements, and insertion of fatty acid components; and
3. adhesion of biosurfactant to cell surface causing deterioration and disruption of cell membrane integrity and nutrition cycle.

Modifications in the profile of phenolic compounds produced by the plants are one such effect observed after pathogen infections. Apart from alterations in phenol composition, a plant enzyme that is most commonly found to be affected upon fungal infections is phenylalanine ammonia-lyase (PAL) which is a key enzyme in phenol biosynthesis and in the biosynthesis of plant phenylpropanoid compounds such as lignin and flavonoids (Schovankova & Opatova, 2011). Activity of PAL has been found to be primarily affected in various stress conditions, including during pathogen attack, and an enhanced enzyme activity has been observed in many resistant cultivars against various plant pathogens (Vanitha, Niranjana, & Umesha, 2009; Zhang et al., 2017). The positive effect of biosurfactant on phenolic compounds in many studies indicates an enhancement of PAL upon biosurfactant application leading to a protective action against pathogen attack. However, more detailed study of the effect of different biosurfactants on PAL is required so as to shed more light on the protective mechanisms of biosurfactants against plant pathogens and also to aid in the development of pathogen resistant plant varieties.

Biosurfactants (primarily glycolipids and lipopeptides) have been implicated in an enhanced immunity in host plants as well as an increase in plant defense mechanisms. One aspect of this is the stimulation of induced systemic resistance (ISR) in plants that is mediated through pattern recognition receptors, recognizing molecular signatures (microbe-associated molecular patterns) that are pathogen-specific (Ongena et al., 2007). Upon induction, specific defense-related genes get activated leading to the accumulation of antimicrobial molecules in plant cells (Boller & Felix, 2009). Mono- and dirhamnolipids secreted by *Pseudomonas aeruginosa* and *Burkholderia plantarii* have been found to be responsible in initiating strong defense responses in the studied host plant, that is grapevine. These responses included Ca^{2+}-mediated cell signaling pathways, reactive oxygen species (ROS), and mitogen-activated protein kinase activation (Varnier, Sanchez,

& Vatsa, 2009). These molecules have also been implicated in inducing multiple defense genes in plants against pathogenesis caused by many microbes, including *B. cinerea*. These included some pathogenesis-related protein genes and genes involved in oxylipins and phytoalexins biosynthesis pathways. However, the mechanisms of interaction between biosurfactant and host plant as well as the involvement of other endophytes in this whole process are an area yet to be fully explored. Also of importance is the study of the structure-specific variations in this activity of biosurfactants and specificity, if any, with respect to other plant metabolites and environmental factors.

Among all the biosurfactants discovered and reported till date, glycolipids and lipopeptides are the two most prominent classes of biosurfactants that have been extensively studied, not only for other economical applications, but also for their antifungal properties against multiple plant pathogens (Bee, Khan, & Sayyed, 2019). In general, the ability to permeate cell wall, create channels, and damage the surface of pathogen has been found to the major modes of action of these antimicrobial agents (Raaijmakers, De Bruijn, & De Kock, 2006). This is especially of importance in the current world scenario when an unprecedented use of fertilizers and pesticides in agricultural practices has irreparably damaged many ecosystems and a sustainable approach to farming is critically required. Also, of grave environmental and health concern is the emergence of many pesticide resistance pest varieties that pose further threat to agricultural output and necessitate the need for eco-friendly vector control tools.

10.3.1 Glycolipids for biocontrol of pathogens

One of the most extensively studied and widely applied biosurfactant, glycolipid, is reported to be produced primarily by many species of bacteria such as *Pseudomonas*, *Burkholderia*, *Mycobacterium*, *Rhodococcus*, *Arthrobacter*, *Nocardia*, *Gordonia*, the yeasts *Starmerella*, *Yarrowia*, and *Pseudozyma*, and fungi such as *Ustilago scitaminea*. Glycolipids contain a glycosyl head group (e.g., sugars like rhamnose and sophorose) as the hydrophilic moiety and different chain saturated or unsaturated fatty acids (like hydroxydecanoic acid) as the hydrophobic fatty acid tail, leading to the formation of biosurfactant, namely, rhamnolipids or sophorolipids, among others. Their structure and function vary according to the variations in the polar head group and nonpolar fatty acid tail. Various different congeners and homologs of rhamnolipids have been discovered over the last many decades. Excellent reviews dealing with the different biosurfactant types and their structures have been published in the last few decades (Al-Ahmad, 2015; Reis, Pacheco, Pereira, & Freire, 2013; Santos et al., 2016). Among all the glycolipids, rhamnolipids produced mainly by members of *Pseudomonas* group followed by sophorolipids produced by members of *Candida* group are the most widely researched and commercially exploited compounds. Conditions of a high carbon to nitrogen ratio in the media or surrounding, low nitrogen source availability, stress conditions, and high cell densities are some of the conditions that have been reported to favor an enhanced production of rhamnolipids (Reis et al., 2013). The ability of rhamnolipids and sophorolipids to reduce surface tension, high biodegradability and their nontoxic and nonmutagenic nature has earned them various applications in different sectors. They can be produced from many inexpensive substrates and hence offer opportunities for a sustainable production (Eslami, Hajfarajollah, & Bazsefidpar, 2020; Singh, Patil, et al., 2019).

Glycolipids have found extensive use in environmental remediation, MEOR, in food industry as well as in cosmetic industry (Fracchia et al., 2014; Liu, Zhong, & Yang, 2018). Naturally, they are reported to be responsible for the motility and biofilm formation activities of the producing host. They also affect the properties of neighboring bacterial population and hence are very effective communication molecules. One property of these compounds that has gained special attention of scientists, agriculturalists, and environmentalists is the antimicrobial property of these compounds. Rhamnolipids have been shown to possess effective antimicrobial activities against a range of human pathogens (Fracchia et al., 2014). They have also shown potential in inhibiting phytopathogens and plant pests thereby suggesting their involvement as a biocontrol agent in agriculture. Rhamnolipids have shown activity against many phytopathogens, including *Botrytis* sp., *Rhizoctonia* sp., *Pythium* sp., *Phytophthora* sp., and *Plasmopara* sp. (Vatsa, Sanchez, Clement, Baillieul, & Dorey, 2010). In one of the earliest reports, crude extracts of *Pseudomonas fluorescens* (6519E01), *P. fluorescens biovar* I (6133D02), and *Serratia plymuthica* (6109D01) were used to treat corn seeds, and a significantly reduced disease impact and increased control of corn seed rot and seed blight disease spread were observed (Haefele, Lamptey, & Marlow, 1991). Subsequently multiple reports have been published regarding the potential of glycolipids for plant disease control as well as control of disease spread. Loss of spore motility, inhibition of mycelial growth, and disruption of zoospore membranes leading to the lysis of zoospores of many fungal strains were the primary modes of antifungal activity ascertained (Ahn et al., 2009; Stanghellini & Miller, 1997; Yoo, Lee, & Kim, 2005). Rhamnolipids and sophorolipids were the primary glycolipids studied for their zoospore-lysis effect on different species of *Pythium* and *Phytophthora* species, including *Pythium aphanidermatum, Phytophthora capsici*, and *Plasmopara lactucae-radicis*. Rhamnolipids have been approved by the United States, Environmental Protection Agency for use as liquid contact biofungicide against zoosporic plant pathogens such as downy mildews and *Phytophthora* blight disease for plant protection, including reduction in disease incidence as well as disease severity (Vatsa et al., 2010). Zonix biofungicide is one such commercially available rhamnolipid preparation by Jeniel Biosurfactant Inc. for use in agriculture and horticulture (Solaiman, 2005). Apart from their direct antiphytopathogenic effect, rhamnolipids have also been implicated in the enhancement of host plant defense against phytopathogens. This includes an induction of the expression of a wide range of plant defense genes and stimulation of plant immunity, including a hypersensitivity-like response (Fracchia et al., 2014; Vatsa et al., 2010). This activity could, in turn, be correlated to an enhancement in the number or activities of the various protective endophyte species also associated with the affected plant. Interestingly, *Pseudomonas* is one of the prominent plant-associated bacterial strains and hence it was proposed that rhamnolipids- or rhamnolipid-like surface-active compounds produced by endophytes play a prominent role in plant defense.

There are multiple reports of many *Pseudomonas* sp. being isolated from rhizosphere as well as other plant surfaces that have exhibited biocontrol of many fungal strains and plant pests. Biosurfactant-producing rhizospheric isolates of *Pseudomonas* and *Bacillus* were reported to show activity against soft rot causing *Pectobacterium* and *Dickeya* spp. (Krzyzanowska et al., 2012). Most of the root isolates belonging to *Pseudomonas* class have been found to produce glycolipid type of biosurfactant with only a few exceptions (Bernat, Nesme, Paraszkiewicz, Schloter, & Plaza, 2019). There have been multiple studies on the biocontrol and disease prevention aspects of various *Pseudomonas* sp. isolated from different plant surfaces, including rhizosphere and leaves. Table 10.1 gives an overview of various *Pseudomonas* biosurfactants and their activities against a wide range of insect pests, bacterial and fungal pathogens. Most of the isolates for biosurfactant production are from plant-associated surfaces or endophytes, thus establishing the role of biosurfactants in the protective effects exhibited by endophytes for biocontrol, disease prevention, and plant defense. The interaction of glycolipids with cell membranes of plants as well as other resident endophytes was proposed to be responsible for the induction of resistance responses in plants as exhibited by glycolipids. The properties are, however, specific to the structure of the glycolipid, and their amphipathic nature was proposed to be responsible for membrane interactions and subsequent elicitor activity (D'aes et al., 2010).

Glycolipids have been proposed to have immense potential as a commercial agent for plant pest control, and they have been shown to be active even against pests resistant to conventional antipest agents. Cuticle membrane damage was one of the modes of action for the antipest activity of glycolipids as exhibited by rhamnolipids from *Pseudomonas* spp. on *Myzus persicae*, green peach aphid (Kim et al., 2011). Similar reports of the use of rhamnolipids against pests establish them as a sustainable agricultural molecule. Apart from rhamnolipids, sophorolipids are another group of glycolipids that offer commercial prospects for plant biocontrol. In a recent report, sophorolipids have shown activity against multiple different fungal strains, including members of *Fusarium, Aspergillus, Alternaria, Aureobasidium, Penicillium, Botrytis*, and *Ustilago* as well as many bacterial plant pathogens like *Acidovorax carotovorum, Erwinia amylovora, Pseudomonas cichorii, Pseudomonas syringae, Pectobacterium carotovorum, Ralstonia solanacearum*, and *Xanthomonas campestris* (Gross & Thavasi, 2013). Rhamnolipids also have shown direct antifungal activities by the inhibition of spore germination and inhibition of mycelial growth process. Their ability to permeabilize biological membranes also was proposed to aid in their pathogen control mechanisms (Sánchez, Aranda, & Teruel, 2010). As mentioned in Section 3.1, rhamnolipids have been found to play an active role in the quorum sensing pathway of the host

Table 10.1 Biocontrol activities of *Bacillus* and *Pseudomonas* strains.

	Producing organism	Biosurfactant	Organism(s) inhibited	Any specific comments	References
1	*Bacillus amyloliquefaciens*	Iturin A2	*Colletotrichum dematium, Rosellinia necatrix, Pyricularia oryzae, Agrobacterium tumefaciens, Xanthomonas campestris pv. campestris*	–	Yoshida, Hiradate, Tsukamoto, Hatakeda, and Shirata (2001)
2	*Bacillus*	Iturin A	*Rhizoctonia solani*	–	Yu, Sinclair, Hartman, and Bertagnolli (2002)
3	*Bacillus subtilis*	Lipopeptide	*Podosphaera fusca*	Inhibition of conidia germination	Romero et al. (2007)
4	*B. amyloliquefaciens*		*Botrytis cinerea* (gray mold) *Phytophthora infestans* (late blight)	Inhibition of mycelial growth and zoospore germination	Ahn et al. (2009)
5	*B. subtilis*	Lipopeptide	*Colletotrichum gloeosporioides* (papaya anthracnose)	–	Kim, Ryu, Kim, and Chi (2010)
6	*Bacillus* spp.	Lipopeptide	*Fusarium, Aspergillus*, and *Bipolaris sorokiniana*	–	Velho, Medina, Segalin, and Brandelli (2011)
7	*B. amyloliquefaciens*		*C. gloeosporioides* (strawberry anthracnose)	ISR	Yamamoto et al. (2015)
8	*B. subtilis*	Crude lipopeptide mixture	*Rhizoctonia bataticola, R. solani*	Loss of sclerotial integrity, granulation, and fragmentation of hyphal mycelia, followed by hyphal shriveling and cell lysis	Mnif et al. (2016)
9	*Bacillus* spp.	Surfactin	*Myzus persicae* (green peach aphid)	Aphid cuticle damage	Yang et al. (2017)
10	*Bacillus mycoides*		*Pythium aphanidermatum*	Zoospore lysis	Peng et al. (2017)
11	*B. amyloliquefaciens* and *B. subtilis*	Lipopeptide extract	*Fusarium moniliforme* (rice bakanae disease), *Fusarium oxysporum* (root rot), *Fusarium*	–	Sarwar et al. (2018)

			solani (root rot), and *Trichoderma atroviride* (ear rot and root rot)		
12	*B. subtilis*		*Tuta absoluta* (tomato leaf minor), *B. cinerea* (gray mold)	Damage to fungal hyphae and tissue damage in larvae midgut	Khedher, Boukedi, Laarif, and Tounsi (2020)
13	*Pseudomonas* spp.	Rhamnolipids	*Phytophthora capsici* and zoospores	Zoospore lysis	Kim, Lee, and Hwang (2000)
14	*Pseudomonas* spp.	Cyclic lipopeptides	*Pythium ultimum, R. solani*	–	Nielsen et al. (2002)
	Pseudomonas fluorescens	Cyclic lipopeptide	Pythium *species*, Albugo candida, *and* P. infestans	Zoospore lysis	De Souza, De Boer, De Waard, Van Beek, and Raaijmakers (2003)
15	*Pseudomonas* spp.	Rhamnolipid	*Phytophthora cryptogea*	Zoospore lysis	De Jonghe, De Dobbelaere, Sarrazyn, and Hoöfte (2005)
16	*Pseudomonas* spp.		*Verticillium microsclerotia* (Verticillium wilt mainly in potatoes)	Pores in membrane leading to loss of germination	Debode et al. (2007)
18	Fluorescent *Pseudomonas*	Unidentified	*P. ultimum* (damping off and root rot of plants), *F. oxysporum* (wilting in crop plants), and *P. cryptogea* (rotting of fruits and flowers)	–	Hultberg, Bergstrand, Khalil, and Alsanius (2008)
19	*Pseudomonas* spp.	Rhamnolipids	*B. cinerea* (Gray mold)	Inhibition of spore germination	Varnier et al. (2009)
20	*Pseudomonas putida*	Cyclic lipopeptides	*P. capsici*	Lysis of zoospores	Kruijt, Tran, and Raaijmakers (2009)
21	*Pseudomonas* spp.	Rhamnolipids (cell-free extract)	2 *Oomycetes*, 3 *Ascomycota*, and 2 *Mucor* sp.	Activity against fungal colony growth and biomass accumulation	Sha, Jiang, Meng, Zhang, and Song (2012)

ISR, *Induced systemic resistance.*

strains. Unraveling the details of this pathway in endophytic *Pseudomonas* and other endophytic populations would aid in the regulatory mechanisms of biosurfactant production for plant defense and immunity. Studying the expression, activation, and involvement of molecules such as quinolones, homoserine lactone, and las R in endophytes would shed more light on the regulation of biosurfactant production in plant-associated microbes and their roles in the biocontrol of plant pathogens and pests as well as in plant disease prevention (Lovaglio, Silva, Ferreira, Hausmann, & Contiero, 2015).

10.3.2 Applications of lipopeptides in agriculture

Gram-positive bacteria, dominated by *Bacillus* spp., have been established to produce a wide variety of antimicrobial metabolites, including antibiotics and various strains have been studied as biocontrol agents of various plant pathogens (Hafeez, Naureen, & Sarwar, 2019). Lipopeptides are a class of biosurfactant molecules being produced primarily by many members of the *Bacillus* sp. along with few reports of other producers such as *Brevibacterium aureum* and *Nocardiopsis alba*. They are controlled by nonribosomal protein syntheses and contain multiple variants depending on the type, sequence, and length of amino acid chain as well as the nature of peptide cyclization and branching. Most of the members are heptapeptides linked with b-hydroxy fatty acids (Ongena & Jacques, 2008). *Bacillus* is one of the prominent members of soil microbial community and is involved dominantly, along with *Pseudomonas*, in plant root colonization and various plant defense mechanisms, including pathogen control, plant disease suppression, and pest management through their larvicidal activities (Edosa, Jo, Keshavarz, & Han, 2018). Lipopeptides from *Bacillus* have been found to play an active role in swarming and colonization features of *Bacillus* and also in biofilm formation for bacterial colony establishment. This colonization and biofilm formation is, in turn, related to the biocontrol activity exhibited by many root-associated *Bacillus* strains (Bais, Fall, & Vivanco, 2004). It remains to be ascertained whether this biocontrol activity is always as a result of direct bactericidal effect of surfactins produced by root-associated *Bacillus* or due to the inhibition of pathogen adhesion or some other plant or microbial factors.

Lipopeptide molecules like lichenysin, fengycin, surfactin, and iturin have been prominently reported to have antifungal and zoosporicidal properties with the ability to permeate membranes being the most prominent mode of action responsible for their toxicity against fungal pathogens (Hultberg, Alsberg, Khalil, & Alsanius, 2010). Lichenysin, fengycin, and iturins are more antibiotic molecules, while surfactin exhibits more surface-active properties along with biocontrol activities. Further, in most of the studies it has been found that more than one of these antimicrobial compounds are coproduced by the bacterial host simultaneously and also act synergistically, for example, surfactin coproduced along with iturin or surfactin along with fengycin (Jacques, 2011). In this chapter, we primarily focus on the biocontrol properties and attributes of surface-active lipopeptide molecules, surfactins. Of all the molecules produced by *Bacillus* sp., Iturin and surfactin are two of the most prominent lipopeptides to be studied for their antifungal properties as well as their prominence in root systems (Ongena & Jacques, 2008). Further, the antifungal activity of these molecules has been found to be structure-specific, as in one report, bacillomycin F (a member of Iturin family) was found to have activity against *A. niger*, while iturin A was found to inhibit the growth of *Aspergillus flavus* and *Fusarium moniliforme* (Gordillo, Navarro, Benitez, Tories de Plaza, & Maldonado, 2009). This indicated a specific interaction between the pathogen and the metabolites of iturin produced by *Bacillus*. A biosurfactant-producing *Bacillus* spp. cell-free extract, with surface-active properties, was reported to exhibit antifungal effect against 10 different fungal strains tested. The greatest activity was reported against *B. cinerea* A 258, *Phomopsis viticola* W 977, *Septoria carvi* K 2082, *Colletotrichum gloeosporioides* A 259, *Phoma complanata* A 233, and *Phoma exigua var. exigua* A 175, and the activity was attributed to the lipoprotein being produced by the *Bacillus* strain (Płaza, Król, Pacwa-Płociniczak, Piotrowska-Seget, & Brigmon, 2012). In a study on the antiphytopathogenic effect of *Bacillus amyloliquefaciens* on Chinese cabbage plants infected with *Rhizoctonia solani*, an increase in phenolic and chlorophyll content, as well as an increase in antioxidant molecules, was proposed to be responsible for effectively mitigating the damaging effects induced by *R. solani* and improving the infected plant growth (Kang, Radhakrishnan, & Lee, 2015). Similar mode of protective action was proposed in another study for the biosurfactant produced by *Bacillus licheniformis* that showed enhanced protection of plant (faba bean) against *R. solani*. A reduction in disease incidence from 62.11% to 20% was observed which was attributed to an enhancement in photosynthetic pigments and endogenous phytohormones in plants upon treatment with the biosurfactant (Akladious, Gomaa, & El-Mahdy, 2019). In another report, lipopeptide surfactin from *B. licheniformis* was found to inhibit the growth of *Magnaporthe grisea* by inducing morphological changes (Tendulkar, Saikumari, & Patel, 2007). Table 10.1 contains few reports of antiphytopathogenic effects of lipopeptides from *Bacillus* strains.

Similar to the antiphytopathogenic properties of externally applied biosurfactants, various reports have been there indicating a similar mode of protective action by many endophytic organisms, including *Bacillus* sp., for increased disease resistance in plants and host plant protection. Many root-, leaves-, and fruit-associated *Bacillus* species have been reported to produce lipoproteins, of which surfactin is reported to be the most prominent. Lipopeptides, mainly surfactin, produced by many Bacilli are responsible for triggering pathogen suppressing effect of the bacilli endophytes (Chowdhury, Hartmann, Gao, & Borriss, 2015). In an earlier report by Snook, Mitchell, Hinton, and Bacon (2009), a surfactin-producing endophytic *Bacillus* strain, *Bacillus mojavensis*, was isolated and found to have activity against *Fusarium verticillioides*. Later, this strain was found to produce a large number of isoforms of surfactin A indicating the role of surfactin A in plant defense (Bacon, Hinton, Mitchell, Snook, & Olubajo, 2012). Similarly, many other plant-associated bacteria have been reported to produce such surface-active molecules that have been subsequently implicated to play a critical role in plant defense mechanisms, including ISR. Lipopeptides have been reported to be perceived by plant cells as signals to initiate defense mechanisms (Ongena et al., 2007). The antimicrobial and ISR activities in *Bacillus* have also been found to be associated with other biocontrol activities in the genus *Bacillus* and are also related to, and affected by, the plant species, plant nutritional constrains, other microbial communities in the ecosystem, and also physiochemical parameters of the soil (Dunlap, Bowman, & Schisler, 2013). The stimulation of ISR activity by *Bacillus* was proposed to be the main mechanism by which *Bacillus* imparts protection to the plants from bacterial and fungal pathogens as well as pests. This was addressed in a review by Chowdhury et al. (2015) on *B. amyloliquefaciens* as well as was confirmed to be applicable for *Bacillus subtilis*. In another report, two molecules of *B. amyloliquefaciens* strain, including surfactin A, were found to trigger ISR in strawberry plants leading to a decreased severity of strawberry anthracnose caused by *C. gloeosporioides*, thus reiterating the role of lipopeptides in plant defense response (Yamamoto, Shiraishi, & Suzuki, 2015). In another report, cyclic lipopeptide from *Bacillus* sp. was found to lead to a suppression of, and enhanced protection against, gray leaf spot caused by *Magnaporthe oryzae* in perennial ryegrass. Upon the application of cell extract containing surfactin, a multilayered, H_2O_2-mediated, ISR defense response was triggered. An accumulation of H_2O_2, elevated peroxidase activity, and deposition of callose and phenolic/polyphenolic compounds underneath the fungal appressoria in naive leaves was observed. H_2O_2 is a signaling molecule as well as a direct antipathogenic agent. Also reported was the activation and enhanced expression of several enzymes like peroxidase, oxalate oxidase, phenylalanine ammonia lyase, and lipoxygenase (Rahman, Uddin, & Wenner, 2015). Induced cell wall fortification, and the accumulation of antimicrobial compounds followed by ROS like H_2O_2 were the observed hypersensitive response of plants. Since the role and expression of endophytes are linked to microbial profile of the ecological niche and other host plant-associated factors, the exact sequence of events and molecular determinants involved in lipopeptide-induced ISR is yet to be completely deciphered in different plant species with diverse biosurfactant endophytes. However, few reports, as earlier, suggest that the effect of lipopeptides on phenolic content of plant as well as two prominent enzymes is involved in plant defense, PAL, peroxidase and lipoxygenase (Ongena & Jacques, 2008). Also, similar to antiphytopathogenic effect, the stimulation of ISR activity is related to the structure of lipopeptide since fengycin and not surfactin-induced ISR on potato tuber cells, while both surfactants were able to stimulate defensive changes in tobacco cells.

Rhizosphere colonization is affected by root exudates and nutritional status and it was proposed that as observed in the in vitro conditions, rhizosphere conditions can modulate associated bacterial cell physiology toward the secretion of specific compounds over other molecules (Bais, Fall, & Vivanco, 2004). Further, surfactin production is related to quorum sensing processes and along with several transcription factors, also involves two pheromones, ComX and PhrC, that are also involved in the development of competence in the *Bacillus* host (Hamoen, Venema, & Kuipers, 2003). The type and amount of lipoprotein produced by the *Bacillus* strains in rhizosphere can be diverse and vary on abovementioned plant-associated factors (Ongena & Jacques, 2008). Effect of lipoprotein on the host defense mechanisms can, in turn, be varied and can include direct antimicrobial activity, alterations in rhizosphere competence, or affecting host plant immune responses. Hence, like rhamnolipids, lipopeptides can also be perceived by the plant cells as trigger and signal to initiate plant defense mechanisms and pathways. However, further studies on the specific plant receptors involved, effect, and the role played by neighboring endophytes and other physiochemical factors of the rhizosphere are required to actively employ lipopeptides as biocontrol agents and in understanding the protective role played by some *Bacillus* species in plant disease control. A lot of work has been done and a lot remains to be explored to understand the genetics, synthesis, structure, and regulation of biosurfactant production by many plant-associated bacteria, including *Bacillus* to optimize biocontrol strategies.

10.4 Conclusion: challenges and opportunities

Unsolicited, indiscriminate, and extensive use of chemical fertilizers and pesticides have invariably introduced large amounts of toxic compounds into various ecological niches thereby posing a threat to human and animal life. Biosurfactants have been established to have competitive antimicrobial activity and also enhance soil properties, thereby promoting enhanced crop yield and improved crop and pasture health. The prominent challenge that limits their extensive application in agriculture is their cost of production. Economizing the application of biosurfactant in routine agricultural practices in various countries across the globe, for pest and pathogen control, would open a new era of sustainable agriculture. The role of biosurfactants as elicitors of systemic resistance opens a promising research area in making of genetically engineered plants that are pest and pathogen resistance. Detailed studies on the presence and distribution of endophytes, their ability to produce biosurfactant, and the nature and function of biosurfactants isolated from endophytes would go a long way in the establishment of biosurfactants as sustainable agents of biocontrol. More information on details of biocontrol mechanisms of biosurfactants will aid in the making of bioformulations containing biosurfactant with more specificity and wider applicability. Extensive studies on biosurfactants from plant-associated endophytes might also lead to discovery and isolation of novel biosurfactant molecules of wider metabolic functions. Also of importance is a detailed study of metabolism of endophytes, controlling pathways and metabolites as well as the effect of environment and other physiological factors on the biosurfactant production ability of endophytic microorganisms and the variation, if any, among different plant varieties and cultivars. This would further help in monitoring and strategizing the role and activity of external biosurfactant applications in agricultural practices, thereby strengthening the productivity and economic profitability of agricultural processes. Nevertheless, biosurfactants, as externally applied agents or as molecules produced by plant-associated endophytes, are safe, cost-effective, efficient, and sustainable biocontrol agents.

References

Ahn, J. Y., Park, M. S., Kim, S. K., Choi, G. J., Jang, K. S., Choi, Y. H., ... Kim, J. C. (2009). Suppression effect of gray mold and late blight on tomato plants by rhamnolipid B. *Research in Plant Disease*, *15*(3), 222–229.

Akladious, S. A., Gomaa, E. Z., & El-Mahdy, O. M. (2019). Efficiency of bacterial biosurfactant for biocontrol of *Rhizoctonia solani* (AG-4) causing root rot in faba bean (*Vicia faba*) plants. *European Journal of Plant Pathology*, *153*, 1237–1257.

Al-Ahmad, K. (2015). The definition, preparation and application of rhamnolipids as biosurfactants. *International Journal of Food Sciences and Nutrition*, *4*(6), 613–623.

Bacon, C. W., Hinton, D. M., Mitchell, T. R., Snook, M. E., & Olubajo, B. (2012). Characterization of endophytic strains of *Bacillus mojavensis* and their production of surfactin isomers. *Biological Control*, *62*(1), 1–9.

Bais, H. P., Fall, R., & Vivanco, J. M. (2004). Biocontrol of *Bacillus subtilis* against infection of *Arabidopsis* roots by *Pseudomonas syringae* is facilitated by biofilm formation and surfactin production. *Plant Physiology*, *134*, 307–319.

Bee, H., Khan, M. Y., & Sayyed, R. Z. (2019). *Microbial surfactants and their significance in agriculture. Plant growth promoting rhizobacteria (PGPR): Prospects for sustainable agriculture* (pp. 205–215). Singapore: Springer.

Bernat, P., Nesme, J., Paraszkiewicz, K., Schloter, M., & Plaza, G. (2019). Characterization of extracellular biosurfactants expressed by a *Pseudomonas putida* strain isolated from the interior of healthy roots from *Sida hermaphrodita* grown in a heavy metal contaminated soil. *Current Microbiology*, *76*, 1320–1329.

Boller, T., & Felix, G. (2009). A renaissance of elicitors: Perception of microbe-associated molecular patterns and danger signals by pattern-recognition receptors. *Annual Review of Plant Physiology*, *60*, 379–406.

Cameotra, S. S., & Makkar, R. S. (2010). Biosurfactant-enhanced bioremediation of hydrophobic pollutants. *Pure and Applied Chemistry. Chimie Pure et Appliquee*, *82*(1), 97–116.

Cameotra, S. S., & Singh, P. (2008). Bioremediation of oil sludge using crude biosurfactants. *International Biodeterioration & Biodegradation*, *62*(3), 274–280.

Cameotra, S. S., & Singh, P. (2009). Synthesis of rhamnolipid biosurfactant and mode of hexadecane uptake by *Pseudomonas* species. *Microbial Cell Factories*, *8*(1), 16.

Cameotra, S. S., Makkar, R. S., Kaur, J., & Mehta, S. K. (2010). *Synthesis of biosurfactants and their advantages to microorganisms and mankind. Biosurfactants* (pp. 261–280). New York: Springer.

Chowdhury, S. P., Hartmann, A., Gao, X., & Borriss, R. (2015). Biocontrol mechanism by root-associated *Bacillus amyloliquefaciens* FZB42—a review. *Frontiers in Microbiology*, *6*, 780.

Chrzanowski, Ł., Ławniczak, Ł., & Czaczyk, K. (2012). Why do microorganisms produce rhamnolipids? *World Journal of Microbial Biotechnology*, *28*, 401—419.

D'aes, J., De Maeyer, K., Pauwelyn, E., & Hofte, M. (2010). Biosurfactants in plant—Pseudomonas interactions and their importance to biocontrol. *Environmental Microbiology*, *2*, 359—372.

De Jonghe, K., De Dobbelaere, I., Sarrazyn, R., & Hoöfte, M. (2005). Control of *Phytophthora cryptogea* in the hydroponic forcing of Witloof chicory with the rhamnolipid-based biosurfactant formulation PRO1. *Plant Pathology*, *54*, 219—226.

De Souza, J. T., De Boer, M., De Waard, P., Van Beek, T. A., & Raaijmakers, J. M. (2003). Biochemical, genetic and zoosporicidal properties of cyclic lipopeptide surfactants produced by *Pseudomonas fluorescens*. *Applied Environmental Microbiology*, *69*, 7161—7172.

Debode, J., De Maeyer, K., Perneel, M., Pannecoucque, J., De Backer, G., & Hoöfte, M. (2007). Biosurfactants are involved in the biological control of *Verticillium microsclerotia* by *Pseudomonas* spp. *Journal of Applied Microbiology*, *103*, 1184—1196.

Drakontis, C. E., & Amin, S. (2020). Biosurfactants: Formulations, properties, and applications. *Current Opinion in Colloid & Interface Science*, *48*, 77—90.

Dunlap, C. A., Bowman, M. J., & Schisler, D. A. (2013). Genomic analysis and secondary metabolite production in *Bacillus amyloliquefaciens* AS 43.3: A biocontrol antagonist of Fusarium head blight. *Biological Control*, *64*, 166—175.

Dusane, D., Rahman, P., Zinjarde, S., Venugopalan, V., McLean, R., & Weber, M. (2010). Quorum sensing: Implication on rhamnolipid biosurfactant production. *Biotechnology Genetic Engineering Reviews*, *27*, 159—184.

Edosa, T. T., Jo, Y. H., Keshavarz, M., & Han, Y. S. (2018). Biosurfactants: Production and potential application in insect pest management. *Trends in Entomology*, *14*, 79—87.

Eslami, P., Hajfarajollah, H., & Bazsefidpar, S. (2020). Recent advancements in the production of rhamnolipid biosurfactants by *Pseudomonas aeruginosa*. *RSC Advances*, *10* (56), 34014—34032.

Fenibo, E. O., Ijoma, G. N., Selvarajan, R., & Chikere, C. B. (2019). Microbial surfactants: The next generation multifunctional biomolecules for applications in the petroleum industry and its associated environmental remediation. *Microorganisms*, *7*, 581.

Fracchia, L., Ceresa, C., Franzetti, A., Cavallo, M., Gandolfi, I., Van Hamme, J., ... Banat, I. M. (2014). Industrial applications of biosurfactants. In N. Kosaric, & S. Sukan (Eds.), *Biosurfactants: Production and utilization—Processes, technologies, and economics* (pp. 245—260). Florida: CRC Press, Taylor and Francis.

Gold, M. V. (2007). *Sustainable agriculture: Definitions and terms*. https://www.nal.usda.gov/afsic/sustainable-agriculture-definitions-and-terms#toc1. Assessed 12.11.20.

Gordillo, M. A., Navarro, A. R., Benitez, L. M., Tories de Plaza, M. I., & Maldonado, M. C. (2009). Preliminary study and improve the production of metabolites with antifungal activity by a *Bacillus* sp. strain IBA 33. *Microbiology. Insights*, *2*(1), 15—24.

Gross, R. A. & Thavasi, R. (2013). Modified sophorolipids combinations as antimicrobial agents. *United States patent no. 2013/0142855 Al.*

Haefele, D. M., Lamptey, J. C., & Marlow, J. L. (1991). *United States patent no. 4,996,049*. Washington, DC: United States Patent and Trademark Office.

Hafeez, F. Y., Naureen, Z., & Sarwar, A. (2019). Surfactin: An emerging biocontrol tool for agriculture sustainability. In A. Kumar, & V. Meena (Eds.), *Plant growth promoting rhizobacteria for agricultural sustainability* (pp. 203–213). Singapore: Springer.

Hamoen, L. W., Venema, G., & Kuipers, O. P. (2003). Controlling competence in *Bacillus subtilis*: Shared use of regulators. *Microbiology (Reading, England)*, *149*(1), 9–17.

Hultberg, M., Alsberg, T., Khalil, S., & Alsanius, B. (2010). Suppression of disease in tomato infected by *Pythium ultimum* with a biosurfactant produced by *Pseudomonas koreensis*. *Bio Control*, *55*(3), 435–444.

Hultberg, M., Bergstrand, K. J., Khalil, S., & Alsanius, B. (2008). Characterization of biosurfactant-producing strains of fluorescent pseudomonads in a soilless cultivation system. *Antonie Van Leeuwenhoek*, *94*(2), 329–334.

Jacques, P. (2011). Surfactin and other lipopeptides from *Bacillus spp*. In G. Soberon-Chavez (Ed.), *Biosurfactants. From genes to applications* (pp. 57–91). Berlin Heidelberg: Springer-Verlag.

Kang, S. M., Radhakrishnan, R., & Lee, I. J. (2015). *Bacillus amyloliquefaciens* subsp. *plantarum GR53*, a potent biocontrol agent resists *Rhizoctonia* disease on Chinese cabbage through hormonal and antioxidants regulation. *World Journal of Microbiology and Biotechnology*, *31*(10), 1517–1527.

Khan, A., & Butt, A. (2016). Biosurfactants and their potential applications for microbes and mankind: An overview. *Middle East Journal of Business*, *55*(3034), 1–10.

Khedher, S. B., Boukedi, H., Laarif, A., & Tounsi, S. (2020). Biosurfactant produced by *Bacillus subtilis* V26: A potential biological control approach for sustainable agriculture development. *Organic Agriculture*, *10*, 1–8.

Kim, B. S., Lee, J. Y., & Hwang, B. K. (2000). *In vivo* control and *in vitro* antifungal activity of rhamnolipid B, a glycolipid antibiotic, against *Phytophthora capsici* and *Colletotrichum orbiculare*. *Pest Management Science*, *56*, 1029–1035.

Kim, P. I., Ryu, J., Kim, Y. H., & Chi, Y. T. (2010). Production of biosurfactant lipopeptides Iturin A, fengycin and surfactin A from *Bacillus subtilis* CMB32 for control of *Colletotrichum gloeosporioides*. *Journal of Microbiology and l Biotechnology*, *20*, 138–145.

Kim, S. K., Kim, Y. C., Lee, S., Kim, J. C., Yun, M. Y., & Kim, I. S. (2011). Insecticidal activity of rhamnolipid isolated from *Pseudomonas* sp. EP-3 against green peach aphid (*Myzus persicae*). *Journal of Agriculture and Food Chemistry*, *59*, 934–938.

Kruijt, M., Tran, H., & Raaijmakers, J. M. (2009). Functional, genetic and chemical characterization of biosurfactants produced by plant growth-promoting *Pseudomonas putida* 267. *Journal of Applied Microbiology*, *107*, 546–556.

Krzyzanowska, D. M., Potrykus, M., Golanowska, M., Polonis, K., Gwizdek-Wisniewska, A., Lojkowska, E., & Jafra, S. (2012). Rhizosphere bacteria as potential biocontrol agents against soft rot caused by various *Pectobacterium* and *Dickeya* spp. strains. *Journal of Plant Pathology*, *94*, 353–365.

Kulakovskaya, T. V., Golubev, W. I., Tomashevskaya, M. A., Kulakovskaya, E. V., Shashkov, A. S., Grachev, A. A., … Nifantiev, N. E. (2010). Production of antifungal cellobiose lipids by *Trichosporon porosum*. *Mycopathologia*, *169*, 117–123.

Ławniczak, Ł., Marecik, R., & Chrzanowski, Ł. (2013). Contributions of biosurfactants to natural or induced bioremediation. *Applied Microbiology and Biotechnology*, *97*(6), 2327–2339.

Liu, G., Zhong, H., Yang, X., Liu, Y., Shao, B., & Liu, Z. (2018). Advances in applications of rhamnolipids biosurfactant in environmental remediation: A review. *Biotechnology and Bioengineering*, *115*(4), 796–814.

Lovaglio, R. B., Silva, V. L., Ferreira, H., Hausmann, R., & Contiero, J. (2015). Rhamnolipids know-how: Looking for strategies for its industrial dissemination. *Biotechnology Advances*, *33*(8), 1715–1726.

Marchant, R., & Banat, I. M. (2012). Biosurfactants: A sustainable replacement for chemical surfactants? *Biotechnology Letters*, *34*(9), 1597–1605.

Mnif, I., Grau-Campistany, A., Coronel-León, J., Hammami, I., Triki, M. A., Manresa, A., & Ghribi, D. (2016). Purification and identification of *Bacillus subtilis* SPB1 lipopeptide biosurfactant exhibiting antifungal activity against *Rhizoctonia bataticola* and *Rhizoctonia solani*. *Environmental Science and Pollution Research*, *23*(7), 6690–6699.

Nair, D. N., & Padmavathy, S. (2014). Impact of endophytic microorganisms on plants, environment and humans. *The Scientific World Journal*, *2014*, 1–11.

Naughton, P. J., Marchant, R., Naughton, V., & Banat, I. M. (2019). Microbial biosurfactants: Current trends and applications in agricultural and biomedical industries. *Journal of Applied Microbiology*, *127*(1), 12–28.

Nielsen, T. H., Sørensen, D., Tobiasen, C., Andersen, J. B., Christophersen, C., Givskov, M., & Sørensen, J. (2002). Antibiotic and biosurfactant properties of cyclic lipopeptides produced by fluorescent *Pseudomonas* spp. from the sugar beet rhizosphere. *Applied and Environmental Microbiology*, *68*(7), 3416–3423.

Nihorimbere, V., Marc Ongena, M., Smargiassi, M., & Thonart, P. (2011). Beneficial effect of the rhizosphere microbial community for plant growth and health. *Biotechnologie, Agronomie, Société et Environnement*, *15*, 327–337.

Olasanmi, I. O., & Thring, R. W. (2018). The role of biosurfactants in the continued drive for environmental sustainability. *Sustainability*, *10*(12), 4817.

Ongena, M., & Jacques, P. (2008). *Bacillus* lipopeptides: Versatile weapons for plant disease biocontrol. *Trends in Microbiology*, *16*(3), 115–125.

Ongena, M., Jourdan, E., Adam, A., Paquot, M., Brans, A., Joris, B., Arpigny, J. L., & Thonart, P. (2007). Surfactin and fengycin lipopeptides of *Bacillus subtilis* as elicitors of induced systemic resistance in plants. *Environmental Microbiology*, *9*, 1084–1090.

Patel, S., Homaei, A., Patil, S., & Daverey, A. (2019). Microbial biosurfactants for oil spill remediation: Pitfalls and potentials. *Applied Microbiology and Biotechnology*, *103*(1), 27–37.

Peng, Y. H., Chou, Y. J., Liu, Y. C., Jen, J. F., Chung, K. R., & Huang, J. W. (2017). Inhibition of cucumber *Pythium* damping-off pathogen with zoosporicidal biosurfactants produced by *Bacillus mycoides*. *Journal of Plant Diseases and Protection*, *124*(5), 481–491.

Płaza, G. A., Król, E., Pacwa-Płociniczak, M., Piotrowska-Seget, Z., & Brigmon, L. R. (2012). Study of antifungal activity of *Bacillus* species cultured on agro-industrial wastes. *Acta Scientiarum Polonorum Hortorum Cultus*, *11*(5), 169–182.

Raaijmakers, J. M., De Bruijn, I., & De Kock, M. J. D. (2006). Cyclic lipopeptide production by plant-associated *Pseudomonas* spp.: Diversity, activity, biosynthesis, and regulation. *Molecular Plant-Microbe Interactions*, *19*, 699–710.

Rahman, A., Uddin, W., & Wenner, N. G. (2015). Induced systemic resistance responses in perennial ryegrass against *Magnaporthe oryzae* elicited by semi-purified surfactin lipopeptides and live cells of *Bacillus amyloliquefaciens*. *Molecular Plant Pathology*, *16*(6), 546–558.

Reis, R. S., Pacheco, G. J., Pereira, A. G., & Freire, D. M. G. (2013). Biosurfactants: Production and applications. *Biodegradation-life of Science*, *3*, 31–61.

Romero, D., de Vicente, A., Rakotoaly, R. H., Dufour, S. E., Veening, J. W., Arrebola, E., & Pérez-García, A. (2007). The iturin and fengycin families of lipopeptides are key factors in antagonism of *Bacillus subtilis* toward *Podosphaera fusca*. *Molecular Plant-Microbe Interactions*, *20*(4), 430–440.

Sachdev, D. P., & Cameotra, S. S. (2013). Biosurfactants in agriculture. *Applied Microbiology and Biotechnology*, *97*(3), 1005–1016.

Saánchez, M., Aranda, F. J., Teruel, J. A., Espuny, M. J., Marques, A., Manresa, A., & Ortiz, A. (2010). Permeabilization of biological and artificial membranes by a bacterial dirhamnolipid produced by *Pseudomonas aeruginosa*. *Journal of Colloid and Interface Science*, *341*, 240–247.

Santos, D. K. F., Rufino, R. D., Luna, J. M., Santos, V. A., & Sarubbo, L. A. (2016). Biosurfactants: Multifunctional biomolecules of the 21st century. *International Journal of Molecular Sciences*, *17*(3), 401.

Sarwar, A., Brader, G., Corretto, E., Aleti, G., Abaidullah, M., Sessitsch, A., & Hafeez, F. Y. (2018). *PLoS One*, *13*(6), e0198107.

Schovankova, J., & Opatova, H. (2011). Changes in phenol composition and activity of phenylalanine-ammonia lyase in apples after fungal infections. *Horticultural Science*, *38*(1), 1–10.

Sha, R. Y., Jiang, L. F., Meng, Q., Zhang, G. L., & Song, Z. R. (2012). Producing cell-free culture broth of rhamnolipids as a cost-effective fungicide against plant pathogens. *Journal of Basic Microbiology*, *52*, 458–466.

Shekhar, S., Sundaramanickam, A., & Balasubramanian, T. (2015). Biosurfactant producing microbes and their potential applications: A review. *Critical Reviews in Environmental Science and Technology*, *45*(14), 1522–1554.

Singh, D., Singh, V. K., & Singh, A. K. (2019). Endophytic Microbes: Prospects and Their Application in Abiotic Stress Management and Phytoremediation. In S. Verma, & J. White Jr (Eds.), *Seed Endophytes*. Springer.

Singh, P., & Cameotra, S. S. (2004). Potential applications of microbial surfactants in biomedical sciences. *Trends in Biotechnology*, *22*(3), 142–146.

Singh, P., Patil, Y., & Rale, V. (2019). Biosurfactant production: Emerging trends and promising strategies. *Journal of Applied Microbiology*, *126*(1), 2–13.

Singh, P., Ravindran, S., Suthar, J., Deshpande, P., Rokhade, R., & Rale, V. (2017). Production of biosurfactant stabilized nanoparticles. *International Journal of Pharma and Biosciences*, *8*(2), 701–707, B.

Singh, R., Glick, B. R., & Rathore, D. (2018). Biosurfactants as a biological tool to increase micronutrient availability in soil: A review. *Pedosphere*, *28*(2), 170–189.

Snook, M. E., Mitchell, T., Hinton, D. M., & Bacon, C. W. (2009). Isolation and characterization of Leu7-surfactin from the endophytic bacterium *Bacillus mojavensis* RRC 101, a biocontrol agent for *Fusarium verticillioides*. *Journal of Agricultural and Food Chemistry*, *57*, 4287–4292.

Solaiman, D. K. Y. (2005). Applications of microbial biosurfactants. *Inform (Champaign)*, *16*(7), 408. Available from https://www.ars.usda.gov/research/publications/publication/?seqNo115 = 180277, Assessed, December 10, 2020.

Stanghellini, M. E., & Miller, R. M. (1997). Biosurfactants: Their identity and potential efficacy in the biological control of zoosporic plant pathogens. *Plant Disease*, *81*, 4–12.

Tahseen, R., Afzal, M., Iqbal, S., Shabir, G., Khan, Q. M., Khalid, Z. M., & Banat, I. M. (2016). Rhamnolipids and nutrients boost remediation of crude oil-contaminated soil by enhancing bacterial colonization and metabolic activities. *International Biodeterioration & Biodegradation*, *115*, 192–198.

Teichmann, B., Linne, U., Hewald, S., Marahiel, M. A., & Boölker, M. (2007). A biosynthetic gene cluster or a secreted cellobiose lipid with antifungal activity from *Ustilago maydis*. *Molecular Microbiology*, *66*, 525–533.

Tendulkar, S. R., Saikumari, Y. K., Patel, V., Raghotama, S., Munshi, T. K., Balaram, P., & Chattoo, B. B. (2007). Isolation, purification and characterization of an antifungal molecule produced by *Bacillus licheniformis* BC98, and its effect on phytopathogen *Magnaporthe grisea*. *Journal of Applied Microbiology*, *103*, 2331–2339.

Thavasi, R., Marchant, R., & Banat, I. M. (2014). 15 Biosurfactant applications in agriculture. *Biosurfactants: Production and Utilization—Processes, Technologies, and Economics*, *159*, 313.

Vanitha, S. C., Niranjana, S. R., & Umesha, S. (2009). Role of phenylalanine ammonia lyase and polyphenol oxidase in host resistance to bacterial wilt of tomato. *Journal of Phytopathology*, *157*(9), 552–557.

Varnier, A. L., Sanchez, L., Vatsa, P., Boudesocque, L., Garcia-Brugger, A., Rabenoelina, F., ... Dorey, S. (2009). Bacterial rhamnolipids are novel MAMPs conferring resistance to *Botrytis cinerea* in grapevine. *Plant, Cell and Environment*, *32*, 178–193.

Vatsa, P., Sanchez, L., Clement, C., Baillieul, F., & Dorey, S. (2010). Rhamnolipid biosurfactants as new players in animal and plant defense against microbes. *International Journal of Molecular Sciences*, *11*, 5095–5108.

Velho, R. V., Medina, L. F., Segalin, J., & Brandelli, A. (2011). Production of lipopeptides among *Bacillus* strains showing growth inhibition of phytopathogenic fungi. *Folia Microbiologica*, *56*, 297–303.

Yamamoto, S., Shiraishi, S., & Suzuki, S. (2015). Are cyclic lipopeptides produced by *Bacillus amyloliquefaciens* S13-3 responsible for the plant defence response in strawberry against *Colletotrichum gloeosporioides*? *Letters in Applied Microbiology*, *60*(4), 379–386.

Yang, S. Y., Lim, D. J., Noh, M. Y., Kim, J. C., Kim, Y. C., & Kim, I. S. (2017). Characterization of biosurfactants as insecticidal metabolites produced by *Bacillus subtilis* Y9. *Entomological Research*, *47*(1), 55–Y59.

Yoo, D. S., Lee, B. S., & Kim, E. K. (2005). Characteristics of microbial biosurfactant as an antifungal agent against plant pathogenic fungus. *Journal of Microbiology and Biotechnology*, *15*, 1164–1169.

Yoshida, S., Hiradate, S., Tsukamoto, T., Hatakeda, K., & Shirata, A. (2001). Antimicrobial activity of culture filtrate of *Bacillus amyloliquefaciens* RC-2 isolated from mulberry leaves. *Phytopathology*, *91*, 181–187.

Yu, G., Sinclair, J. B., Hartman, G. L., & Bertagnolli, B. L. (2002). Production of iturin A by *Bacillus amyloliquefaciens* suppressing *Rhizoctonia solani*. *Soil Biology & Biochemistry*, *34*(2), 955–963.

Zhang, C., Wang, X., Zhang, F., Dong, L., Wu, J., Cheng, Q., ... Li, N. (2017). Phenylalanine ammonia-lyase2. 1 contributes to the soybean response towards Phytophthora sojae infection. *Scientific Reports*, *7*(1), 1–13.

CHAPTER

Microbial biofilms in plant disease management 11

Amrita Patil, Rashmi Gondi, Vinay Rale and Sunil D. Saroj
Symbiosis School of Biological Sciences, Symbiosis International (Deemed University), Pune, India

11.1 Introduction

The plant diseases such as wilting, browning, molding, and rotting initiate within the plant much before manifesting phenotypic effects. In nature, plants may be affected by more than one disease-causing agent at a time (Zhang et al., 2016). Many plant diseases annually cause dramatic losses worldwide and collectively constitute significant losses to farmers and can reduce the esthetic values of landscape plants and home gardens. Many valuable crops and ornamental plants are vulnerable to diseases and would have difficulty surviving in nature without human intervention (Gepts, 2002). Such losses from plant diseases can have a significant economic effect, contributing to a reduction in income for crop producers and distributors and higher prices for consumers. To ensure food safety, alternative sustainable ways of managing plant diseases are needed (Gupta, Debnath, Sharma, Sharma, & Purohit, 2017). The central goal of revisiting strategies in controlling plant diseases is to achieve significant reduction in the associated economic and esthetic damages.

Plant diseases are managed using physical, chemical, and biological methods (Stone et al., 2010). Several agrochemicals have been developed so far and are used in either powder or liquid forms to control plant diseases. Some of them can inhibit the growth of pathogens, and some can kill the pathogens responsible for plant diseases (Raaijmakers, Paulitz, Steinberg, Alabouvette, & Moënne-Loccoz, 2009). However, biological methods to regulate plant diseases involve the use of microorganisms to reduce or prevent diseases (O'Brien, 2017). These microorganisms can be present inside the host or can be applied to the plant parts where it can directly or indirectly interfere with the pathogen causing infection. The use of plant growth–promoting rhizobacteria (PGPR) is significant and is increasingly being applied in the field as they are devoid of the harmful effects (Walsh, Ikeda, & Boland, 1999). Several PGPRs have been documented to have well-developed quorum sensing (QS) system and naturally form biofilms in the rhizosphere. The ability to form biofilms imparts PGPRs selective advantages over the planktonic mode of bacterial existence. Hence, in this chapter, we provide evidence that both

Biocontrol Mechanisms of Endophytic Microorganisms. DOI: https://doi.org/10.1016/B978-0-323-88478-5.00005-5

natural and in vitro developed biofilm lifestyle is advantageous to manage plant diseases, with a focus on plant PGPR, the role of PGPRs in plant health with an emphasis on QS as well as biofilm formation phenomenon in PGPR and its advantages over other strategies. We also discuss the utilization of PGPRs in bacterial and fungal disease management. This might provide useful solutions to reduce the incidence of disease and protect the ecosystem.

11.2 Plant growth–promoting bacteria and plant health

Microorganisms are often associated with plant parts or tissues (Partida-Martínez & Heil, 2011). These associated bacterial populations can closely control the structure, behavior, and growth of the plant. Rhizomicrobiome is of immense importance to agriculture (Shi et al., 2018). PGPR are naturally occurring soil bacteria that actively colonize plant roots, minimize disease incidence, and benefit the overall plant health by producing various plant growth–promoting substances. As the adverse effects of chemical fertilizers and their costs are increasing, the usage of beneficial soil microorganisms such as PGPR for healthy cultivation has increased globally over the last few decades (Gupta, Parihar, Ahirwar, Snehi, & Singh, 2015). Biological control methods, utilizing beneficial microbes, are an excellent approach to limit disease-causing microbes' adverse effects on plant health and productivity (Syed Ab Rahman, Singh, Pieterse, & Schenk, 2018). PGPRs are well-recognized biofertilizers for sustainable agriculture and hold great promise in the improvement of agriculture yields. PGPRs regulate plant diseases and improve plant health through various modes such as antagonism, induction of systemic resistance (ISR), space, nutrient competition, improving soil texture, acquisition, and assimilation of nutrients, secretion, and modulation of extracellular components such as secondary metabolites, hormones, signaling compounds, and antibiotics (Siddiqui & Shaukat, 2002; Singh, Sarma, & Keswani, 2017).

Associative nitrogen fixers or free-living nitrogen-fixing bacteria, such as *Pseudomonas* spp., and symbiotic nitrogen-fixing bacteria, such as *Bacilli* spp., have been shown to attach to the root and efficiently colonize the root surfaces. *Pseudomonas* spp. and *Bacilli* spp. function as predominant PGPRs (Gupta et al., 2015). These species can form spores that can persist in the soil for a long time, even under harsh environmental conditions. Hence, they possess great potential in plant disease management on a commercial scale (Yasmin et al., 2017).

11.2.1 Significance of PGPR

The agrochemicals and their degradation products might have harmful effects on the environment and human beings. This has driven researchers to seek an environmentally sustainable approach to control plant diseases (Gupta et al., 2017). Intensive use of herbicides, insecticides, and agrochemicals might directly or indirectly affect human health and the environment. The misuse of agrochemicals has led to problems associated with the contamination of such chemicals in the soil, freshwater reservoirs, and in the food chain (Capita & Alonso-Calleja, 2013). The intensive use of these substances has also led to decreased soil fertility, low germination of crops, and increased disease-resistant varieties, which impose a direct or indirect risk to consumers.

Most of the problems raised due to chemical methods can be addressed through the use of PGPRs (Zheng et al., 2018). PGPRs are considered very important in agriculture because of their role in regulating plant growth, sustainable agriculture, increased crop production, increased germination of seeds, and nutritional and hormonal balance (Nadeem, Zahir, Naveed, & Nawaz, 2013). PGPR also plays an important role in the induction of resistance against phytopathogens. Hence, considering the ecological and economic significance of PGPRs, they should be widely used in agriculture (Bhattacharyya & Jha, 2012; Bresson, Varoquaux, Bontpart, Touraine, & Vile, 2013; Liu, Garrett, Fadamiro, & Kloepper, 2016).

11.2.2 PGPR and quorum sensing

QS is a signaling mechanism in bacteria regulated by the extracellular concentration of signal molecules (Cornforth et al., 2014). It has been found that QS regulates various phenotypes, including the development of biofilm in several PGPRs, for example, *Bacillus* spp., *Pseudomonas segetis*, and *Pseudomonas syringae* (Quiñones, Dulla, & Lindow, 2005). *N*-Acyl-homoserine lactone (AHL), autoinducer-2, diffusible signal factor, cyclic depsipeptide, pyrroloquinoline, cyclodipeptides, surfactin lipopeptide, and ComX are a few major QS signal molecules produced by PGPRs (*Pseudomonas* spp. and *Bacillus* spp.) (Table 11.1). AHL-mediated QS is commonly detected in *Pseudomonas* spp. than other root-colonizing bacteria (Rodríguez et al., 2020) (Table 11.1).

In contrast, several studies have explained the role of PGPRs in silencing QS-mediated virulence in phytopathogenic bacteria. *P. segetis* has also been successfully used to minimize the development of QS-controlled virulence factors in plant bacterial pathogens (Vega, Rodríguez, Llamas, Béjar, & Sampedro, 2020).

11.2.3 PGPR biofilms and plant health

PGPR plays an essential role in improving plant growth through a variety of beneficial effects (Gupta et al., 2015). This is often achieved by the formation of biofilms in the rhizosphere, which has advantages over the planktonic mode of bacterial existence. PGPR biofilms have a positive impact on plants' overall growth compared to the planktonic counterpart (Alavi, Starcher, Zachow, Müller, & Berg, 2013). PGPRs efficiently do QS within the biofilm and form stable biofilm on the plant roots (Bhattacharyya & Jha, 2012). PGPR biofilms enhance plant growth by phosphate solubilization, siderophore production,

Table 11.1 Quorum sensing molecules involved in biofilm formation in *Bacillus* and *Pseudomonas* spp. plant growth–promoting rhizobacteria (PGPR).

PGPR	Quorum sensing molecule	References
Bacillus cereus	*N*-acyl-L-homoserine lactone (AHL)	Medina-Martínez, Uyttendaele, Rajkovic, Nadal, and Debevere (2007)
Pseudomonas aeruginosa MMA83	*N*-octadecanoylhomoserine lactone (C_{18}-HSL)	Malešević et al. (2019)
Pseudomonas segetis strain P6	*N*-acyl-homoserine lactone (AHL)	Rodríguez et al. (2020)
Pseudomonas syringae pv. *tabaci*	*N*-acyl-homoserine lactone (AHL)	Ryu et al. (2013)
Pseudomonas corrugate	*N*-hexanoyl L-homoserine lactone	Palmer, Senechal, Mukherjee, Ané, and Blackwell (2014)
Biocontrol strain, Pseudomonas chlororaphis	*N*-hexanoyl homoserine lactone	Bauer et al. (2016)
Bacillus subtilis	*N*-hexanoyl homoserine lactone	Palmer et al. (2014)
Pseudomonas aureofaciens	AHL mimic compounds	Teplitski, Robinson, and Bauer (2000)
Paraburkholderia phytofirmans	3-OH-C8-AHL	Teplitski et al. (2000)
Bacillus thuringiensis	AHL mimic compounds	Dong et al. (2001)
P. aeruginosa PUPa3	3-oxo-C12-HSL	Steindler et al. (2009)
Pseudomonas fluorescens (PF-04)	AHL molecules	Li et al. (2018)
PGPR *P. fluorescens* F113	Pyrroloquinoline	Redondo-nieto et al. (2013)
P. aeruginosa	Cyclodipeptides	Holden et al. (1999); Ortiz-Castro et al. (2011)
P. aeruginosa	Diffusible signal factor	Ryan et al. (2008)
Bacillus velezensis SQR9	Autoinducer-2	Xiong et al. (2020)
B. cereus	Autoinducer-2	Auger, Krin, Aymerich, and Gohar (2006)
Paenibacillus polymyxa	Autoinducer-2	Luo et al. (2018)
P. polymyxa E681	Cyclic depsipeptide	Li, Song, Yi, and Kuipers (2020)
Bacillus amyloliquefaciens	Surfactin lipopeptide	Li et al. (2020)
B. subtilis	ComX	Dogsa et al. (2014)

biological nitrogen fixation, rhizosphere engineering, production of 1-aminocyclopropane-1-carboxylate deaminase (ACC), QS signal interference, and inhibition of biofilm formation, phytohormone production, exhibiting antifungal activity, production of volatile organic compounds, ISR, promoting beneficial plant—microbe symbiosis, interference with pathogen toxin production, etc. The growth-promoting substances are likely to be produced in large quantities by PGPR biofilms that indirectly influence the plants' overall morphology. Recent progress in our understanding of PGPR diversity in the rhizosphere, along with their colonization ability and mechanism of action, should facilitate their application as a reliable plant disease management strategy (Bhattacharyya & Jha, 2012). *Pseudomonas* and *Bacillus* spp. form single- and dual-species biofilms. *Pseudomonas putida* strains also play a role in protecting the plants against stress (Giaouris, Chorianopoulos, Doulgeraki, & Nychas, 2013). The in vitro production of PGPR biofilm can provide improved crop yields through a range of plant growth mechanisms. The microbial inoculant industry would also be immensely benefitted from the production of PGPR biofilm with N2 fixing microbes (Gupta et al., 2015). The efficacy of PGPR biofilm in plant disease management needs further investigation.

Bacillus species can form endospores that are highly resilient to harsh environmental conditions and may also secrete metabolites that promote plant growth and health. *Bacillus subtilis* also plays a significant role in improving resistance to biotic stresses. This induction of disease resistance involves the expression of specific genes and hormones, such as ACC. Similar results have been noted in *Pseudomonas* spp. strain ACP (Ma, Guinel, & Glick, 2003). Biofilm formation under different stress conditions is a strategy followed by bacterial strains for their efficient survival in the plant rhizosphere (Bais, Fall, & Vivanco, 2004). The timing of biofilm formation (mostly 24 hours) on host roots also depends on the promoter of the genes responsible for producing the matrix when the bacterium is initially in contact with the root. *Bacillus* spp. produce antimicrobial metabolites that can be used as a substitute for synthetic chemicals or as a supplement to the use of biopesticides, and biofertilizers, to combat plant diseases (Roy et al., 2018). *B. subtilis* 6051 exhibits strong, very stable biofilm formation and also produces surfactin, indicating that it would be a good biocontrol agent against pathogenic bacteria. *B. subtilis* also forms biofilms on plant roots that help in the production of lipopeptides and increase their antimicrobial activity in the soil (Kinsella, Schulthess, Morris, & Stuart, 2009). It was speculated that *Bacillus amyloliquefaciens* and *Pseudomonas fluorescens* biofilm are beneficial in enhancing the salt stress tolerance in barley. The findings revealed that the biofilm formation in these stains increased with an increase in salt concentration. Moreover, *B. amyloliquefaciens*, a beneficial bacterium, has been used as an exogenous strain in a commercial bioorganic fertilizer to stimulate plant growth and to eradicate soil-borne diseases. Furthermore, the phosphorus solubilizing microorganisms *Bacillus megaterium* have been commercialized as BioPhos (BioPower Lanka, Sri Lanka) and can reduce phosphate fertilizer requirements of plantation crops up to 75% (Bais et al., 2004).

11.2.3.1 PGPRs and bacterial disease management

PGPR biofilms inhibit several bacterial diseases. In a study a total of 60 *Bacillus* isolates were collected from different locations around China; 6 strains out of them were found to have more than 50% biocontrol efficacy on tomato plants against the *Ralstonia solanacearum* plant pathogen under greenhouse conditions (Chen et al., 2013). With respect to biofilm formation, wild strains of *B. subtilis* are more robust than that of laboratory strains both in defined medium and tomato plant roots. It was revealed in plate assays that these strains also exhibited strong antagonistic behavior against plant pathogens. This study also states that matrix development is essential for bacterial colonization on plant root surfaces (Chen et al., 2013). It is speculated that the biofilm-forming ability of PGPR on plant root will be an added advantage to plant health. It has been documented that *Bacillus* spp. and *Pseudomonas* spp. control the diseases caused by rice pathogens, that is, *Xanthomonas oryzae* pv. *oryzae, Rhizoctonia solani*, and *Magnaporthe oryzae* up to 90% depending on the bacteria used, pathogen, and the rice variety (Yasmin et al., 2017). The culture media characterization in *B. subtilis* strain QST 713 revealed that this strain can produce antibiotics that can suppress pathogens in vitro. Two important *B. subtilis* antibiotics, surfactin and iturin A, were extracted from root and rhizosphere soil. Rhizosphere concentrations of both antibiotics increased with plant age (Kinsella et al., 2009). Another study showed that *B. subtilis 6051* forms a stable biofilm on roots and produces surfactin to protect plants from pathogenic bacteria (Bais et al., 2004). Common plant-associated bacteria found on leaves, roots, and the soil such as *P. putida, P. fluorescens*, and other related pseudomonads along with the majority of other natural isolates have been reported to form effective biofilms. The transformant *P. fluorescens P3/pME6863* can degrade the AHLs. *P. fluorescens P3/pME6863* greatly reduced crown gall of tomato caused by *Agrobacterium tumefaciens* and potato soft rot caused by *Erwinia carotovora* to a similar level as *Bacillus* spp. *A24*. Suppression of potato soft rot was observed even when the AHL-degrading *P. fluorescens P3/pME6863* was added to tubers 2 days after the pathogen infection, suggesting that biocontrol was not only preventive but also curative. When the antagonists were added individually with the bacterial plant pathogens, the biocontrol efficacy of the AHL degraders was higher than that observed with many other *Pseudomonas* spp. 2,4-diacetylphloroglucinol-producing strains with *Pseudomonas chlororaphis PCL1391*, which depends on phenazine antibiotic production for disease suppression. The production of phenazine by this well-characterized biological strain *P. chlororaphis PCL1391* is regulated by AHL-mediated QS (Molina et al., 2003). The root-associated biocontrol agent *P. fluorescens* 2P24 requires AHLs for the formation of biofilm and also controls all wheat-borne diseases (Shen, Hu, Peng, Wang, & Zhang, 2013; Wang, Knill, Glick, & Defago, 2000). *P. fluorescens* strain CHA0 has a broad spectrum of biocontrol activity in plant diseases. The introduction of ACC deaminase genes into strain CHA0 enhanced its ability to protect potato tubers from and Erwinia soft rot cucumber plants from Pythium-damping off (Wang et al., 2000).

Several successful studies have focused primarily on the isolation and identification of members of the fluorescent pseudomonad population present in the rhizosphere due to their ability to provide plant protection against diseases through the production of a variety of secondary metabolites (Weller et al., 2007). The most effective bioagent studied till date appears to antagonize pathogen using multiple mechanisms as in *Pseudomonas* spp., utilizing both antibiosis and induction of host resistance to suppress the disease-causing microorganisms. A study evaluated two characterized PGPR (*P. fluorescens* FAP2 and *Bacillus licheniformis* B642) for their biofilm-related functions using standard protocols. Positive interaction between *P. fluorescens* FAP2 and *B. licheniformis* B642 was observed when they were cocultured. Both strains exhibited several plant growth–promoting (PGP) characteristics (production of IAA (indole-3-acetic acid), siderophore, and ammonia; phosphate solubilization) and biofilm-related functions such as the production of EPS (extrapolysacchardides), alginate, cell surface hydrophobicity, and swarming motility. Both strains formed strong biofilms on a glass coverslip in vitro. Indigenous *P. fluorescens* FAP2 strain and *B. licheniformis* B642 are compatible PGPR in both planktonic and biofilm modes of growth and an efficient PGPR consortium may be formed. However, compatible interaction between widely used PGPR in biofilm mode in vitro and in the rhizosphere will provide a better understanding of the development of an effective consortium. A further in-depth investigation is required to clarify the molecular mechanism of the interaction in the biofilm mode of growth under natural conditions (Ansari & Ahmad, 2019; Weller et al., 2007) (Table 11.2). *P. fluorescens* and *Bacillus* spp. are widely used as bioinoculant for crop production and as biocontrol agents. The positive influence of inoculation of PGPR (*Pseudomonas* spp. and *Bacillus* spp.) to wheat has been reported on plant growth, yield, and physiological attributes by several researchers (Ansari & Ahmad, 2019). These mechanisms directly influence the stimulation of systemic disease-resistance mechanisms. The use of molecular approaches can be beneficial in studying such interactions in situ as the nature of the interaction is multispecies. A comparative genomic study was performed with *P. chlororaphis* GP72 with three other pseudomonad PGPRs: *P. fluorescens Pf-5*, Pseudomonas *aeruginosa M18*, and the nitrogen-fixing strain *Pseudomonas stutzeri A1501*. Comparisons between the four species of *Pseudomonas* spp. showed 603 conserved genes in GP72, demonstrating similar plant growth–promoting traits shared between these PGPR. PGPR isolates related to genera *Pseudomonas* were found to be effective in phytoremediation of Fe (3 +)-contaminated soil where *Pennisetum glaucum* and *Sorghum bicolor* were grown as host plants. Phylogenetic analysis of 16S rDNA and *phlD*, *phzF*, and *acdS* genes demonstrated that some strains of *Pseudomonas* spp. were identical to the model PGPR strains *Pseudomonas protegens Pf-5, P. chlororaphis* subsp. *aureofaciens 30a "84*, and *Pseudomonas brassicacearuma Q8r1–96*. Moreover, *P. protegens*– and *P. chlororaphis*–like strains had the highest biocontrol activity against Rhizoctonia root rot in wheat. This PGPR further accelerated the symbiotic development of the AMF (Arbuscular Mycorrhizal Fungi) in potato plants.

Table 11.2 Effect of biofilm-forming *Bacillus* and *Pseudomonas* spp. plant growth–promoting rhizobacteria (PGPR) on bacterial plant diseases.

PGPR	Inhibit bacterial disease	References
Bacillus amyloliquefaciens SQR9	Soil-borne diseases	Domenech et al. (2006); Kasim et al. (2016)
Pseudomonas fluorescence	Wheat leaf rust disease	Flaishman et al. (1996)
Pseudomonas entemophila	Foliar diseases	Ansari and Ahmad (2019)
Bacillus	Foliar diseases	Ansari and Ahmad (2019)
Bacillus subtilis FB17	Foliar diseases	Bais and Rudrapp (2013)
B. subtilis	Tomato wilt disease	Chen et al. (2013); Roberts et al. (2008)
B. subtilis M1	Arabidopsis roots by *Pseudomonas syringae*	Bais et al. (2004)
Trichoderma viride–B. subtilis	*Macrophomina phaseolina*–infected cotton crop	Triveni et al. (2015)
Anabaena–B. subtilis	*M. phaseolina*-infected cotton crop	Triveni et al. (2015)
Pseudomonas fluorescens strain Pf7–14	Rice blast	Krishnamurthy and Gnanamanickam (1998)
P. fluorescens strain Pf1	Leaf spot disease of groundnut	Meena (2011)
B. amyloliquefaciens strain SQR9	Cucumber wilt disease	Xu et al. (2014)
B. subtilis	Bacterial spot disease in tomato	Roberts et al. (2008)
Bacillus pumilus HR10	Pine seedling damping-off disease caused by *Rhizoctonia solani*	Zhu, Wu, Wang, and Dai (2020)
B. amyloliquefaciens	*Myzus persicae*	Choudhary et al. (2016)
P. fluorescens	Stem rot in groundnut	Manjula et al. (2004)
Bacillus licheniformis	Control *M. phaseolina* in chickpea	
B. subtilis	Control *M. phaseolina* in chickpea	
B. subtilis	Control *Heterodera cajani* and *Fusarium udum* in pigeon pea	
B. pumilus	Control colletotrichum orbiculare in cucumber	
Paenibacillus polymyxa E681	Wilt caused by a complex of soil-borne pathogens	Ryu et al. (2006)
B. amyloliquefaciens NJN-6	Soil-borne pathogens from root exudates of banana	Yuan et al. (2015)

Pyoluteorin, pyrrolnitrin, tropolone, and cyclic lipopeptides secreted by *Pseudomonads* have antibiotic properties, and strains of these beneficial microorganisms have been applied in the control of plant diseases. A research performed with denaturing gradient gel electrophoresis of DNA isolated from rhizosphere soils of *Rauwolfia* spp. collected from Western Ghat regions of Karnataka revealed the presence of *Pseudomonas* spp. and *Bacillus* spp. All these rhizobacteria showed disease suppression traits when screened. The key role that this antibiotic plays in disease suppression has been demonstrated both by studies using genetic mutational analysis with *P. fluorescens* strains CHA0, F113, Q8r1–96, Q2–87, SSB17, and by direct isolation of the antibiotic from rhizospheres colonized by these bacteria (Weller et al., 2007, 2012). *P. putida WCS358r* was genetically engineered for the production of phenazine and 2,4-diacetyl-phloroglucinol (2,4-DAPG) shows increased ability to suppress diseases in wheat. The two rhizobacteria *P. aeruginosa* strain IE-6S$^+$ and *P. fluorescens* strain CHA0 are used to significantly reduce *Meloidogyne javanica* juvenile penetration into tomato roots under glasshouse conditions. Application of IE-6S$^+$ and CHA0 to each half of the split-root system caused a significant 42% and 29% systemic reduction in nematode penetration in the other half of the split-root system. IE-6S$^+$ bacteria induce systemic resistance in the plant to a relatively higher degree against root-knot nematode compared to the rhizosphere colonizer (CHA0) (Siddiqui & Shaukat, 2002). The research emphasized the possible association between strain-specific molecular diversity of fluorescent *P. aeruginosa* (FP) and presented an account of their potential biological control activity against *Fusarium oxysporum.* All *Pseudomonas* isolates were used to prime tomato seeds and plants were inoculated with *F. oxysporum* and the disease incidence was assessed under greenhouse condition. Furthermore, it was shown that the three FPs, M80 (SUB1688209 Seq1 KX570929), T109 (SUB1688209 Seq1 KX570931M96), and (SUB1688209 Seq1 KX570930) were able to induce systemic resistance. The lowest prevalence of disease was seen by priming with M80 (mean $\pm$ SE, 0.75 $\pm$ 0.02) followed by M96 (0.93 $\pm$ 0.00) and T109 (0.93 $\pm$ 0.07), almost no visible symptoms observed on tomato leaves, suggesting a systemic resistance induction (Jayamohan, Patil, & Kumudini, 2018). The PGPRs were also screened in vitro, for the inhibition of *Pythium myriotylum* inducing soft rot in ginger. Results showed that only five PGPRs had >70% suppression of *P. myriotylum*, namely, *B. amyloliquefaciens and P. aeruginosa*. The greenhouse analysis showed that GRB35 (*B. amyloliquefaciens*) had lower disease incidence (48.1%). Overall, the results suggested that for growth promotion and management of soft rot disease in ginger, GRB35 *B. amyloliquefaciens* could be a good alternative to chemical measures. The use of *B. amyloliquefaciens* for integration into nutritional and disease management for ginger cultivation is recommended (Dinesh, Anandaraj, Kumar, Bini, & Subila, 2015). *P. fluorescens* strains that produce the polyketide antibiotic 2,4-DAPG are among the most effective rhizobacteria that suppress root and crown rots, wilts, and damping-off diseases of a variety of crops, and they suppress soil-borne pathogens. *P. syringae* pv. tomato. Strain Q2–87 induced resistance on transgenic NahG plants but not on jar1, npr1−1, and etr1 *Arabidopsis* mutants. These findings suggest that the antibiotic 2,4-DAPG is a major determinant of ISR in 2,4-DAPG producing *P. fluorescens*, and that the activity induced by these bacteria operates through the ethylene- and jasmonic acid−dependent signal transduction pathway (Weller et al., 2007, 2012).

Biofilm formation across the root surface by *Bacillus* spp. and their release of toxins (bacillomycin, iturin, surfactin, macrolactin, and fengycin) kills the pathogenic bacterial species and decreases the occurrence of diseases in plants (Galelli, Sarti, & Miyazaki, 2015). The formation of biofilm and the production of surfactin from *B. amyloliquefaciens plantarum* defend the viral disease in plants by stimulating ISR machinery. However, the efficiency and inconsistent vitality of PGPR are technical limitations in the use of PGPR as a biofertilizer (Ji et al., 2019). To overcome these problems, the micro-dielectric barrier discharge plasma to increase the vitality and functionality of a PGPR, *B. subtilis* CB-R05, was used. Rice seedlings infected with plasma-treated bacteria showed improved immunity to fungal infection. Scanning electron microscope (SEM) analysis showed that plasma-treated bacteria colonized the rice plant roots more densely than untreated bacteria (Table 11.2).

Bacillus spp. and *Pseudomonas* spp. are prevalent in most rhizosphere soils and help in the protection of plant species against various pathogens (Ansari & Ahmad, 2019; Bais & Rudrapp, 2013; Bais et al., 2004; Flaishman, Zahir, Zilberstein, Voisard, & Haas, 1996; Nissipaul, Triveni, Subhashreddy, & Suman, 2017; Triveni et al., 2015; Zheng et al., 2018) (Table 11.2). Studies on the pathways of plant growth promotion by PGPR have provided a deeper understanding of the various aspects of disease suppression by these biocontrol agents. Still, much of the emphasis was on biofilm-forming rhizobacterial strains, particularly *Pseudomonas* spp. and *Bacillus* spp. (Domenech, Reddy, Kloepper, Ramos, & Gutierrez-Mañero, 2006; Molina et al., 2003). *Bacillus* spp. are among the most commonly isolated PGPRs and some of them can increase crop yield and eradicate plant disease. In addition, *B. subtilis* forms biofilm-like structures that guard against tomato wilt disease (Chen et al., 2013; Luo et al., 2018) (Table 11.2). It was stated that a beneficial bacterium, *B. amyloliquefaciens* SQR9, has been used as an exogenous strain in commercial bioorganic fertilizer to enhance plant growth and suppress soil-borne diseases in the field (Kasim, Gaafar, Abou-Ali, Omar, & Hewait, 2016). It could be inferred that the *Pseudomonas fluorescence* and *Bacillus polymyxa* significantly reduced leaf rust pathogen and improved growth and grain yield. PGPR that colonizes root systems through seed applications and prevents plants from foliar diseases includes *Pseudomonas* spp., *Paenibacillus* spp., and *Bacillus* spp. (Ryu, Kim, Choi, Kim, & Park, 2006). *P. fluorescence* inoculation is known to be the most effective bacterial inoculation in reducing wheat leaf rust (Manjula, Kishore, Girish, & Singh, 2004). Multiple gene clusters encoding for secondary metabolites, for example, bacilysin, bacillibactin, and microcin, have been found in each of these strains (Dunlap, Bowman, & Schisler, 2013) (Table 11.2). Biocontrol activity by PGPR strains is often mediated by bacterial secondary metabolites, such as zwittermicin A and kanosamine for biological control of fengycin by *B. subtilis* strain F-29–3 for Rhizoctonia disease. Related findings have been documented in previous studies with *P. fluorescens strain Pf7–14* used as seed and foliar therapy spray for the biological control of rice blast and *P. fluorescens strain Pf1* added to the rhizosphere and phylloplane for biological control of late leaf spot disease in groundnut (Krishnamurthy & Gnanamanickam, 1998; Meena, 2011). Previously, it has been shown that overexpression of genes involved in phosphorylation of DegU, a two-component response regulator of *B. amyloliquefaciens* strain SQR9, has a beneficial impact on root colonization as well as other growth-promoting activities of PGPR strains to control wilt disease in cucumber (Xu et al., 2014). *B. subtilis* can suppress plant pathogens in vitro and control bacterial spot disease in tomato (Roberts et al., 2008) (Table 11.2).

11.2.3.2 PGPRs and fungal disease management

PGPR biofilms inhibit several fungal diseases. *F. oxysporum* subspp. *Cucumerinum* was successfully controlled by a newly isolated strain, *B. subtilis SQR 9*, both in vivo and in vitro. Greenhouse studies have been performed to determine the effect of *B. subtilis SQR 9*, known as bioorganic fertilizer (BIO), to control *Fusarium* wilt. The occurrence of wilt decreased by up to 49%–61% by BIO. These findings reveal that the strain was able to survive in the rhizosphere of cucumber, suppressing the growth of *F. oxysporum* subssp. *Cucumerinum* and protected the host from a pathogen (Cao et al., 2011). Two strains of *Bacillus* spp., *Bacillus* cereus and *Bacillus safensis*, show antagonism against *Alternaria* spp. Both of these strains secrete IAA, chitinase, siderophores and exhibit the property of phosphate solubilization. These strains are shown to minimize lentil leaf spot and blight disease with significant increase in plant growth (Roy et al., 2018). *Pseudomonas* spp. colonizing plant roots inhibited the growth of several fungal pathogens such as *Fusarium* spp., *Penicillium verrucosum*, *Aspergillus niger*, *Aspergillus fumigates*, and *Trichoderma viride*. However, *Bacillus* spp. associated with plant roots reduced the growth of *Fusarium verticillioides*, *Rhizoctonia*, and *A. niger* (Table 11.3). Many antibiotics have been derived from bacteria of the genera *Bacillus* and *Pseudomonas*. They contain a variety of metabolites that serve as antifungal agents. For *Bacillus*, they are either derived from the ribosome or the nonribosomal peptide and/or polyketide synthetases. One such antifungal agent is caspofungin. It has been reported that plant-associated *B. amyloliquefaciens* strains of subsp. *plantarum* are distinct from other representatives of endospore-forming *B. amyloliquefaciens* due to their ability to colonize plant rhizosphere, to promote plant growth, and to suppress competing phytopathogenic bacteria and fungi (Brannen & Kenney, 1997). Some strains of fluorescent pseudomonads are essential biological components of agricultural soils that are suppressive to diseases caused by pathogenic fungi on crop plants (Naik & Sakthivel, 2006). *Pseudomonas aureofaciens* is antagonistic to many species, including pathogenic fungi (Ranjbariyan, Shams-Ghahfarokhi, Kalantari, & Razzaghi-Abyaneh, 2011).

Table 11.3 Effect of biofilm-forming plant growth–promoting rhizobacterias (PGPRs) (*Bacillus* spp. and *Pseudomonas* spp.) on fungal plant diseases.

PGPR	Inhibit fungi/fungal disease	References
Pseudomonas aureofaciens	Plant pathogenic fungi	Ranjbariyan et al. (2011)
Pseudomonas strain PUP6	Plant pathogenic fungi	Naik and Sakthivel (2006)
Pseudomonas chlororaphis S105	Plant pathogenic fungi	Ranjbariyan et al. (2011)
Pseudomonas fluorescens	Fusarium wilt control in chickpea	Nissipaul et al. (2017)
Pseudomonas Q16	*Trichoderma viride*	Jošić et al. (2015)
Pseudomonas B25	*T. viride*	Jošić et al. (2015)
Pseudomonas Q16	*Aspergillus fumigatus*	Jošić et al. (2015)
Pseudomonas B25	*A. fumigatus*	Jošić et al. (2015)
Pseudomonas Q16	*Aspergillus niger*	Jošić et al. (2015)
Pseudomonas B25	*A. niger*	Jošić et al. (2015)
Pseudomonas Q16	*Penicillium verrucosum*	Jošić et al. (2015)
Pseudomonas B25	*P. verrucosum*	Jošić et al. (2015)
Pseudomonas spp.	Seed borne fungus, *A. niger*	Yuttavanichakul et al. (2012)
Bacillus amyloliquefaciens SQR9	Plant pathogenic fungi	Zhang et al. (2015)
Bacillus toyonensis COPE52	Plant pathogenic fungi	Rojas-Solis et al. (2020)
Bacillus *mojavensis*	*Fusarium verticillioides (Sacc.)*	Camele, Elshafie, Caputo, Sakr, and De Feo (2019)
Bacillus subtilis	Fusarium wilt control in chickpea	Nissipaul et al. (2017)
B. amyloliquefaciens AS 43.3	Fusarium head blight	Dunlap et al. (2013)
B. subtilis strain GB03	*Rhizoctonia*	Brannen and Kenney (1997)
B. subtilis strain GB03	*Fusarium*	Brannen and Kenney (1997)
Bacillus megaterium	Seed borne fungus, *A. niger*	Yuttavanichakul et al. (2012)
B. subtilis	Seed borne fungus, *A. niger*	Yuttavanichakul et al. (2012)

Bacillus spp. secrete compounds such as oligomycin A, kanosamine, zwittermicin, and xanthobaccin that inhibit the growth of pathogens, particularly fungi (Milner, Raffel, Lethbridge, & Handelsman, 1995). ISR has been studied in many rhizobacteria-inoculated plants and, as initially shown that *P. fluorescens* strain

WCS417r protected plants against growth inhibition, and against the fungal pathogen *F. oxysporum* f. sp. *dianthi*. In this study, *Pseudomonas* isolates from the rhizosphere of alfalfa and clover plants showing significantly were selected and tested for their PGP traits (Jošić et al., 2015). Antifungal metabolites, as natural products produced by selected *Pseudomonas* strains, may be useful fungitoxicants without harmful side effects. *Pseudomonas* can produce several extracellular metabolites, including antibiotics and lytic enzymes, which inhibit the growth of different fungi. The most promising PGPR isolates, *Pseudomonas Q16* and B25, demonstrated the strongest antifungal activity against *T. viride*, and good antifungal effect against *Aspergillus fumigatus* and *A. niger*, while *P. verrucosum* was the most resistant fungus. *Pseudomonas* spp. B25 had higher antifungal potential than Q16. *Pseudomonas pseudoalcaligenes*, *P. fluorescens*, *P. putida*, and *Stenotrophomonas maltophilia* produce cell-wall degrading enzymes such as protease, pectinase, cellulose, and chitinase, in addition to other antifungal metabolites [hydrogen cyanide (HCN), antibiotics]. It was reported that the application of liquid cultures of *Pseudomonas* sp. Ps255 and *Bacillus* sp. B73 is not only improving soybean health and yield but also protecting soybean from soil-borne fungi. This report shows that *Bacillus* and *Pseudomonas* spp. can be used as an alternative to chemical protection against *Fusarium culmorum*, *F. oxysporum*, *R. solani*, *Sclerotinia sclerotiorum*, and *Phomopsis sojae* (Jošić et al., 2015). PGPR-elicited ISR was first observed on carnation (*Dianthus caryophyllus*) with decreased susceptibility to wilt caused by *Fusarium* spp. (Choudhary et al., 2016). The vast majority of cotton seeds planted in the United States were treated with *B. subtilis* strain GBo_3, registered as Kodiak (Gustafson, Inc, Plano, TX, USA). Response is typically a mixture of growth promotion (increased root mass) and disease suppression (*Rhizoctonia* and *Fusarium* spp.) (Brannen & Kenney, 1997). Similarly, *Bacillus toyonensis COPE52* and *B. amyloliquefaciens SQR9* isolated from rhizospheres exhibit antifungal activity (Cao et al., 2011; Dong et al., 2001; Rojas-Solis, Vences-Guzmán, Sohlenkamp, & Santoyo, 2020; Shen et al., 2013; Xiong et al., 2020; Xu et al., 2014; Yuan et al., 2015; Zhang et al., 2015, 2016).

11.2.3.3 PGPRs and viral disease management

Some PGPRs such as *Paenibacillus lentimorbus* act as a potent biocontrol agent against an economically important virus, cucumber mosaic virus (CMV), in the *Nicotiana tabacum* cv. Soil inoculation with *P. lentimorbus* B-30488 reduced the aggregation of CMV RNA in *Nicotiana tabacum* cv. White burley leaves by 91%. Soil inoculation of B-30488 increased the plant vigor while substantially decreasing the virulence and virus RNA accumulation by ~12 times (91%) in systemic leaves of CMV-infected tobacco plants as compared to the control ones. Besides, defense-related enzymes (guaiacol peroxidase, ascorbate peroxidase, superoxide dismutase, and catalase) induced due to CMV-infection have been ameliorated with the inoculation of B-30488, indicating systemic induced resistance-mediated protection against CMV in tobacco. The quantitative RT-PCR (Real time PCR) analyses of the genes linked to normal plant growth, stress, and pathogenesis also corroborate well with the biochemical data and showed the regulation (either up or down) of these genes in favor of plant to combat the CMV-mediated stress (Kumar et al., 2016). Strains of phosphate-solubilizing *Pseudomonas striata, B. polymyxa*, and *B. megaterium* have also been commercialized by AgriLife (India) (Chen et al., 2013).

11.3 Conclusion

The associations between plants and the rhizomicrobiome are ancient and represent the selective outcome of a coevolution. PGPR biofilms offer additional advantages in regulating bacterial, fungal, viral plant diseases, and plant growth promotion as compared to PGPR's planktonic lifestyle. However, the exact mechanism by which PGPRs inhibit plant diseases and improve disease resistance than other strains is not yet understood. Moreover, investigations directed toward the identification of key QS molecules, proteins, or other metabolites produced by PGPR strains and the understanding of the chemical complexity of root secretions will help to understand the molecular cues in the adhesion of PGPRs and development of PGPR biofilms. Furthermore, studies aligned with the in vitro production of polymicrobial PGPR biofilms and manipulation of microbial composition of PGPR biofilm based on the crop requirement will be helpful to understand the potential of PGPRs in plant disease management at large scale.

Acknowledgment

AP is supported by the ERASMUS + grant.

Funding

The work was supported by the Ramalingaswami fellowship program of the Department of Biotechnology, India under grant BT/RLF/Re-entry/41/2015 and ERASMUS + under grant 598515-EPP-1-2018-1-IN-EPPKA2-CBHE-JP.

References

Alavi, P., Starcher, M. R., Zachow, C., Müller, H., & Berg, G. (2013). Root-microbe systems: The effect and mode of interaction of stress protecting agent (SPA) *Stenotrophomonas rhizophila* DSM14405T. *Frontiers in Plant Science*, *4*, 1–11. Available from https://doi.org/10.3389/fpls.2013.00141.

Ansari, F. A., & Ahmad, I. (2019). Fluorescent *Pseudomonas* -FAP2 and *Bacillus licheniformis* interact positively in biofilm mode enhancing plant growth and photosynthetic attributes. *Scientific Reports*, *9*, 1–12. Available from https://doi.org/10.1038/s41598-019-408644.

Auger, S., Krin, E., Aymerich, S., & Gohar, M. (2006). Autoinducer 2 affects biofilm formation by *Bacillus cereus*. *Applied and Environmental Microbiology*, *72*, 937–941. Available from https://doi.org/10.1128/AEM.72.1.937-941.2006.

Bais, H. & Rudrapp, T. (2013). Inventors; University of Delaware, Assignee. Methods for promoting plant health. *United States patent, United States8551919B2*.

Bais, H. P., Fall, R., & Vivanco, J. M. (2004). Biocontrol of *Bacillus subtilis* against infection of arabidopsis roots by *Pseudomonas syringae* is facilitated by biofilm formation and surfactin production. *Plant Physiology*, *134*, 307–319. Available from https://doi.org/10.1104/pp.103.028712.

Bauer, J. S., Hauck, N., Christof, L., Mehnaz, S., Gust, B., & Gross, H. (2016). The systematic investigation of the quorum sensing system of the biocontrol strain *Pseudomonas chlororaphis* subsp. *aurantiaca* PB-St2 Unveils aurI to Be a biosynthetic origin for 3-Oxo-homoserine lactones. *PLoS One*, *11*. Available from https://doi.org/10.1371/journal.pone.0167002.

Bhattacharyya, P. N., & Jha, D. K. (2012). Plant growth-promoting rhizobacteria (PGPR): Emergence in agriculture. *World Journal of Microbiology and Biotechnology*, *28*, 1327–1350. Available from https://doi.org/10.1007/s11274-011-0979-9.

Brannen, P. M., & Kenney, D. S. (1997). A successful biological-control product for suppression of soil-borne plant pathogens of cotton. *Journal of Industrial Microbiology and Biotechnology*, *19*, 169–171. Available from https://doi.org/10.1038/sj.jim.2900439.

Bresson, J., Varoquaux, F., Bontpart, T., Touraine, B., & Vile, D. (2013). The PGPR strain *Phyllobacterium brassicacearum* STM196 induces a reproductive delay and physiological changes that result in improved drought tolerance in Arabidopsis. *New Phytologist*, *200*, 558–569. Available from https://doi.org/10.1111/nph.12383.

Camele, I., Elshafie, H. S., Caputo, L., Sakr, S. H., & De Feo, V. (2019). *Bacillus mojavensis*: Biofilm formation and biochemical investigation of its bioactive metabolites. *Journal of Biological Research (Italy)*, *92*, 39–45. Available from https://doi.org/10.4081/jbr.2019.8296.

Cao, Y., Zhang, Z., Ling, N., Yuan, Y., Zheng, X., Shen, B., et al. (2011). *Bacillus subtilis* SQR 9 can control Fusarium wilt in cucumber by colonizing plant roots. *Biology and Fertility of Soils*, *47*, 495–506. Available from https://doi.org/10.1007/s00374-011-0556-2.

Capita, R., & Alonso-Calleja, C. (2013). Antibiotic-resistant bacteria: A challenge for the food industry. *Critical Reviews in Food Science and Nutrition*, *53*, 11–48. Available from https://doi.org/10.1080/10408398.2010.519837.

Chen, Y., Yan, F., Chai, Y., Liu, H., Kolter, R., Losick, R., et al. (2013). Biocontrol of tomato wilt disease by *Bacillus subtilis* isolates from natural environments depends on conserved genes mediating biofilm formation. *Environmental Microbiology*, *15*, 848–864. Available from https://doi.org/10.1111/j.1462-2920.2012.02860.x.

Choudhary, D. K., Kasotia, A., Jain, S., Vaishnav, A., Kumari, S., Sharma, K. P., et al. (2016). Bacterial-mediated tolerance and resistance to plants under abiotic and biotic stresses. *Journal of Plant Growth Regulation*, *35*, 276–300. Available from https://doi.org/10.1007/s00344-015-9521-x.

Cornforth, D. M., Popat, R., McNally, L., Gurney, J., Scott-Phillips, T. C., Ivens, A., et al. (2014). Combinatorial quorum sensing allows bacteria to resolve their social and physical environment. *Proceedings of the National Academy of Sciences of the United States of America*, *111*, 4280–4284. Available from https://doi.org/10.1073/pnas.1319175111.

Dinesh, R., Anandaraj, M., Kumar, A., Bini, Y. K., Subila, K. P., & Aravind, R. (2015). Isolation, characterization, and evaluation of multi-trait plant growth promoting rhizobacteria for their growth promoting and disease suppressing effects on ginger. *Microbiological Research*, *173*, 34–43. Available from https://doi.org/10.1016/j.micres.2015.01.014.

Dogsa, I., Choudhary, K. S., Marsetic, Z., Hudaiberdiev, S., Vera, R., Pongor, S., et al. (2014). ComQXPA quorum sensing systems may not be unique to *Bacillus subtilis*: A census in prokaryotic genomes. *PLoS One*, *9*, 1–8. Available from https://doi.org/10.1371/journal.pone.0096122.

Domenech, J., Reddy, M. S., Kloepper, J. W., Ramos, B., & Gutierrez-Mañero, J. (2006). Combined application of the biological product LS213 with Bacillus, *Pseudomonas* or *Chryseobacterium* for growth promotion and biological control of soil-borne diseases in pepper and tomato. *BioControl*, *51*, 245–258. Available from https://doi.org/10.1007/s10526-005-2940-z.

Dong, Y. H., Wang, L. H., Xu, J. L., Zhang, H. B., Zhang, X. F., & Zhang, L. H. (2001). Quenching quorum-sensing-dependent bacterial infection by an *N*-acyl homoserine lactonase. *Nature*, *411*, 813–817. Available from https://doi.org/10.1038/35081101.

Dunlap, C. A., Bowman, M. J., & Schisler, D. A. (2013). Genomic analysis and secondary metabolite production in *Bacillus amyloliquefaciens* AS 43.3: A biocontrol antagonist of Fusarium head blight. *Biological Control*, *64*, 166–175. Available from https://doi.org/10.1016/j.biocontrol.2012.11.002.

Flaishman, M., Zahir, E., Zilberstein, A., Voisard, C., & Haas, D. (1996). Suppression of Septoria tritici Blotch and Leaf rust of wheat by recombinant cyanide producing strains of *Pseudomonas putida*. *Molecular Plant-Microbe Interaction*, *9*, 642–645.

Galelli, M. E., Sarti, G. C., & Miyazaki, S. S. (2015). *Lactuca sativa* biofertilization using biofilm from Bacillus with PGPR activity. *Journal of Applied Horticulture*, *17*, 186–191. Available from https://doi.org/10.37855/jah.2015.v17i03.35.

Gepts, P. (2002). A comparison between crop domestication, classical plant breeding, and genetic engineering. *Crop Science*, *42*, 1780–1790.

Giaouris, E., Chorianopoulos, N., Doulgeraki, A., & Nychas, G. J. (2013). Co-culture with listeria monocytogenes within a dual-species biofilm community strongly increases resistance of *Pseudomonas putida* to benzalkonium chloride. *PLoS One*, *8*, 1–14. Available from https://doi.org/10.1371/journal.pone.0077276.

Gupta, G., Parihar, S. S., Ahirwar, N. K., Snehi, S. K., & Singh, V. (2015). Plant growth promoting rhizobacteria (PGPR): Current and future prospects for development of sustainable agriculture. *Journal of Microbial & Biochemical Technology*, *07*, 96–102. Available from https://doi.org/10.4172/1948-5948.1000188.

Gupta, N., Debnath, S., Sharma, S., Sharma, P., & Purohit, J. (2017). Role of nutrients in controlling the plant diseases in sustainable agriculture. *Agriculturally Important Microbes for Sustainable Agriculture*, *2*, 217–262. Available from https://doi.org/10.1007/978-981-10-5343-6_8.

Holden, M. T. G., Chhabra, S. R., De Nys, R., Stead, P., Bainton, N. J., Hill, P. J., et al. (1999). Quorum-sensing cross talk: Isolation and chemical characterization of cyclic dipeptides from *Pseudomonas aeruginosa* and other Gram-negative bacteria. *Molecular Microbiology*, *33*, 1254–1266. Available from https://doi.org/10.1046/j.1365-2958.1999.01577.x.

Jayamohan, N. S., Patil, S. V., & Kumudini, B. S. (2018). Validation of molecular heterogeneity of Fluorescent *Pseudomonas* spp. and correlation with their potential biocontrol traits against fusarium wilt disease. *Agriculture and Natural Resources*, *52*, 317–324. Available from https://doi.org/10.1016/j.anres.2018.10.006.

Ji, S. H., Kim, J. S., Lee, C. H., Seo, H. S., Chun, S. C., Oh, J., et al. (2019). Enhancement of vitality and activity of a plant growth-promoting bacteria (PGPB) by atmospheric pressure non-thermal plasma. *Scientific Reports*, *9*, 1–16. Available from https://doi.org/10.1038/s41598-018-38026-z.

Jošić, D., Ćirić, A., Soković, M., Stanojković-Sebić, A., Pivić, R., Lepšanović, Z., et al. (2015). Antifungal activities of indigenous plant growth promoting *Pseudomonas* spp. from alfalfa and clover rhizosphere. *Frontiers in Life Science*, *8*, 131–138. Available from https://doi.org/10.1080/21553769.2014.998776.

Kasim, W. A., Gaafar, R. M., Abou-Ali, R. M., Omar, M. N., & Hewait, H. M. (2016). Effect of biofilm forming plant growth promoting rhizobacteria on salinity tolerance in barley. *Annals of Agricultural Sciences*, *61*, 217–227. Available from https://doi.org/10.1016/j.aoas.2016.07.003.

Kinsella, K., Schulthess, C. P., Morris, T. F., & Stuart, J. D. (2009). Rapid quantification of *Bacillus subtilis* antibiotics in the rhizosphere. *Soil Biology and Biochemistry*, *41*, 374–379. Available from https://doi.org/10.1016/j.soilbio.2008.11.019.

Krishnamurthy, K., & Gnanamanickam, S. S. (1998). Biological control of rice blast by *Pseudomonas fluorescens* strain Pf7–14: Evaluation of a marker gene and formulations. *Biological Control*, *13*, 158–165. Available from https://doi.org/10.1006/bcon.1998.0654.

Kumar, S., Chauhan, P. S., Agrawal, L., Raj, R., Srivastava, A., Gupta, S., et al. (2016). *Paenibacillus lentimorbus* inoculation enhances tobacco growth and extenuates the virulence of cucumber mosaic virus. *PLoS One*, *11*, 1–23. Available from https://doi.org/10.1371/journal.pone.0149980.

Li, T., Yang, B., Li, X., Li, J., Zhao, G., & Kan, J. (2018). Quorum sensing system and influence on food spoilage in *Pseudomonas fluorescens* from turbot. *Journal of Food Science and Technology*, *55*, 3016–3025. Available from https://doi.org/10.1007/s13197-018-3222-y.

Li, Z., Song, C., Yi, Y., & Kuipers, O. P. (2020). Characterization of plant growth-promoting rhizobacteria from perennial ryegrass and genome mining of novel antimicrobial gene clusters. *BMC Genomics*, *21*, 1–11. Available from https://doi.org/10.1186/s12864-020-6563-7.

Liu, K., Garrett, C., Fadamiro, H., & Kloepper, J. W. (2016). Induction of systemic resistance in Chinese cabbage against black rot by plant growth-promoting rhizobacteria. *Biological Control*, *99*, 8–13. Available from https://doi.org/10.1016/j.biocontrol.2016.04.007.

Luo, Y., Cheng, Y., Yi, J., Zhang, Z., Luo, Q., Zhang, D., et al. (2018). Complete genome sequence of industrial biocontrol strain *Paenibacillus polymyxa* HY96–2 and further analysis of its biocontrol mechanism. *Frontiers in Microbiology*, *9*, 1–14. Available from https://doi.org/10.3389/fmicb.2018.01520.

Ma, W., Guinel, F. C., & Glick, B. R. (2003). Rhizobium leguminosarum biovar viciae 1-aminocyclopropane-1-carboxylate deaminase promotes nodulation of pea plants.

Applied and Environmental Microbiology, *69*, 4396–4402. Available from https://doi.org/10.1128/AEM.69.8.4396-4402.2003.

Malešević, M., Di Lorenzo, F., Filipić, B., Stanisavljević, N., Novović, K., Senerovic, L., et al. (2019). *Pseudomonas aeruginosa* quorum sensing inhibition by clinical isolate *Delftia tsuruhatensis* 11304: Involvement of *N*-octadecanoylhomoserine lactones. *Scientific Reports*, *9*, 1–13. Available from https://doi.org/10.1038/s41598-019-52955-3.

Manjula, K., Kishore, G. K., Girish, A. G., & Singh, S. D. (2004). Combined application of *Pseudomonas fluorescens* and *Trichoderma viride* has an improved biocontrol activity against stem rot in groundnut. *Plant Pathology Journal*, *20*, 75–80. Available from https://doi.org/10.5423/PPJ.2004.20.1.075.

Medina-Martínez, M. S., Uyttendaele, M., Rajkovic, A., Nadal, P., & Debevere, J. (2007). Degradation of *N*-acyl-L-homoserine lactones by *Bacillus cereus* in culture media and pork extract. *Applied and Environmental Microbiology*, *73*, 2329–2332. Available from https://doi.org/10.1128/AEM.01993-06.

Meena, B. (2011). Effect of *Pseudomonas fluorescens* Pfl formulation application on rhizosphere and phyllosphere population in groundnut. *International Journal of Plant Protection*, *4*, 92–94.

Milner, J. L., Raffel, S. J., Lethbridge, B. J., & Handelsman, J. (1995). Culture conditions that influence accumulation of zwittermicin A by *Bacillus cereus* UW85. *Applied Microbiology and Biotechnology*, *43*, 685–UW91. Available from https://doi.org/10.1007/BF00164774.

Molina, L., Constantinescu, F., Michel, L., Reimmann, C., Duffy, B., & Défago, G. (2003). Degradation of pathogen quorum-sensing molecules by soil bacteria: A preventive and curative biological control mechanism. *FEMS Microbiology Ecology*, *45*, 71–81. Available from https://doi.org/10.1016/S0168-6496(03)00125-9.

Nadeem, S. M., Zahir, Z. A., Naveed, M., & Nawaz, S. (2013). Mitigation of salinity-induced negative impact on the growth and yield of wheat by plant growth-promoting rhizobacteria in naturally saline conditions. *Annals of Microbiology*, *63*, 225–232. Available from https://doi.org/10.1007/s13213-012-0465-0.

Naik, P. R., & Sakthivel, N. (2006). Functional characterization of a novel hydrocarbonoclastic *Pseudomonas* sp. strain PUP6 with plant-growth-promoting traits and antifungal potential. *Research in Microbiology*, *157*, 538–546. Available from https://doi.org/10.1016/j.resmic.2005.11.009.

Nissipaul, M., Triveni, S., Subhashreddy, R., & Suman, B. (2017). Novel biofilm biofertilizers for nutrient management and fusarium wilt control in Chickpea. *International Journal of Current Microbiology and Applied Sciences*, *6*, 1846–1852. Available from https://doi.org/10.20546/ijcmas.2017.606.215.

O'Brien, P. A. (2017). Biological control of plant diseases. *Australasian Plant Pathology*, *46*, 293–304. Available from https://doi.org/10.1007/s13313-017-0481-4.

Ortiz-Castro, R., Díaz-Pérez, C., Martínez-Trujillo, M., Del Río, R. E., Campos-García, J., & López-Bucio, J. (2011). Transkingdom signaling based on bacterial cyclodipeptides with auxin activity in plants. *Proceedings of the National Academy of Sciences of the United States of America*, *108*, 7253–7258. Available from https://doi.org/10.1073/pnas.1006740108.

Palmer, A. G., Senechal, A. C., Mukherjee, A., Ané, J. M., & Blackwell, H. E. (2014). Plant responses to bacterial *N*-acyl-L-homoserine lactones are dependent on enzymatic degradation to l-homoserine. *ACS Chemical Biology*, *9*, 1834–1845. Available from https://doi.org/10.1021/cb500191a.

Partida-Martínez, L. P., & Heil, M. (2011). The microbe-free plant: Fact or artifact? *Frontiers in Plant Science*, *2*, 1–16. Available from https://doi.org/10.3389/fpls.2011.00100.

Quiñones, B., Dulla, G., & Lindow, S. E. (2005). Quorum sensing regulates exopolysaccharide production, motility, and virulence in *Pseudomonas syringae*. *Molecular Plant-Microbe Interactions. Molecular Plant-Microbe Interactions*, *18*, 682–693. Available from https://doi.org/10.1094/MPMI-18-0682.

Raaijmakers, J. M., Paulitz, T. C., Steinberg, C., Alabouvette, C., & Moënne-Loccoz, Y. (2009). The rhizosphere: A playground and battlefield for soilborne pathogens and beneficial microorganisms. *Plant and Soil*, *321*, 341–361. Available from https://doi.org/10.1007/s11104-008-9568-6.

Ranjbariyan, A. R., Shams-Ghahfarokhi, M., Kalantari, S., & Razzaghi-Abyaneh, M. (2011). Molecular identification of antagonistic bacteria from Tehran soils and evaluation of their inhibitory activities toward pathogenic fungi. *Iranian Journal of Microbiology*, *3*, 140–146.

Redondo-nieto, M., Barret, M., Morrissey, J., Germaine, K., Martínez-granero, F., Barahona, E., et al. (2013). Genome sequence reveals that *Pseudomonas fluorescens* F113 possesses a large and diverse array of systems for rhizosphere function and host interaction. *BMC Genomics*, *14*, 1–17. Available from https://doi.org/10.1186/1471-2164-14-54.

Roberts, P. D., Momol, M. T., Ritchie, L., Olson, S. M., Jones, J. B., & Balogh, B. (2008). Evaluation of spray programs containing famoxadone plus cymoxanil, acibenzolar-*S*-methyl, and *Bacillus subtilis* compared to copper sprays for management of bacterial spot on tomato. *Crop Protection*, *27*, 1519–1526. Available from https://doi.org/10.1016/j.cropro.2008.06.007.

Rodríguez, M., Torres, M., Blanco, L., Béjar, V., Sampedro, I., & Llamas, I. (2020). Plant growth-promoting activity and quorum quenching-mediated biocontrol of bacterial phytopathogens by *Pseudomonas segetis* strain P6. *Scientific Reports*, *10*, 1–12. Available from https://doi.org/10.1038/s41598-020-61084-1.

Rojas-Solis, D., Vences-Guzmán, M. A., Sohlenkamp, C., & Santoyo, G. (2020). *Bacillus toyonensis* COPE52 modifies lipid and fatty acid composition, exhibits antifungal activity, and stimulates growth of tomato plants under saline conditions. *Current Microbiology*, *77*, 2735–2744. Available from https://doi.org/10.1007/s00284-020-02069-1.

Roy, T., Bandopadhyay, A., Sonawane, P. J., Majumdar, S., Mahapatra, N. R., Alam, S., et al. (2018). Bio-effective disease control and plant growth promotion in lentil by two pesticide degrading strains of Bacillus sp. *Biological Control*, *127*, 55–63. Available from https://doi.org/10.1016/j.biocontrol.2018.08.018.

Ryan, R. P., Fouhy, Y., Garcia, B. F., Watt, S. A., Niehaus, K., Yang, L., et al. (2008). Interspecies signalling via the *Stenotrophomonas maltophilia* diffusible signal factor influences biofilm formation and polymyxin tolerance in *Pseudomonas aeruginosa*. *Molecular Microbiology*, *68*, 75–86. Available from https://doi.org/10.1111/j.1365-2958.2008.06132.x.

Ryu, C. M., Choi, H. K., Lee, C. H., Murphy, J. F., Lee, J. K., & Kloepper, J. W. (2013). Modulation of quorum sensing in acyl-homoserine lactone-producing or -degrading tobacco plants leads to Alteration of induced systemic resistance elicited by the rhizobacterium Serratia marcescens 90–166. *Plant Pathology Journal*, *29*, 182–192. Available from https://doi.org/10.5423/PPJ.SI.11.2012.0173.

Ryu, C. M., Kim, J., Choi, O., Kim, S. H., & Park, C. S. (2006). Improvement of biological control capacity of *Paenibacillus polymyxa* E681 by seed pelleting on sesame. *Biological Control*, *39*, 282–289. Available from https://doi.org/10.1016/j.biocontrol.2006.04.014.

Shen, X., Hu, H., Peng, H., Wang, W., & Zhang, X. (2013). Comparative genomic analysis of four representative plant growth-promoting rhizobacteria in *Pseudomonas*. *BMC Genomics*, 14. Available from https://doi.org/10.1186/1471-2164-14-271.

Shi, S., Tian, L., Nasir, F., Li, X., Li, W., Tran, L. S. P., et al. (2018). Impact of domestication on the evolution of rhizomicrobiome of rice in response to the presence of *Magnaporthe oryzae*. *Plant Physiology and Biochemistry*, *132*, 156–165. Available from https://doi.org/10.1016/j.plaphy.2018.08.023.

Siddiqui, I. A., & Shaukat, S. S. (2002). Rhizobacteria-mediated induction of systemic resistance (ISR) in tomato against *Meloidogyne javanica*. *Journal of Phytopathology*, *150*, 469–473. Available from https://doi.org/10.1046/j.1439-0434.2002.00784.x.

Singh, H. B., Sarma, B. K., & Keswani, C. (Eds.), (2017). *Advances in PGPR research*. Varanasi: CABI. Available from https://doi.org/10.1079/9781786390325.0000.

Steindler, L., Bertani, I., De Sordi, L., Schwager, S., Eberl, L., & Venturi, V. (2009). LasI/R and RhlI/R quorum sensing in a strain of *Pseudomonas aeruginosa* beneficial to plants. *Applied and Environmental Microbiology*, *75*, 5131–5140. Available from https://doi.org/10.1128/AEM.02914-08.

Stone, V., Nowack, B., Baun, A., van den Brink, N., von der Kammer, F., Dusinska, M., et al. (2010). Nanomaterials for environmental studies: Classification, reference material issues, and strategies for physico-chemical characterisation. *Science of the Total Environment*, *408*, 1745–1754. Available from https://doi.org/10.1016/j.scitotenv.2009.10.035.

Syed Ab Rahman, S. F., Singh, E., Pieterse, C. M. J., & Schenk, P. M. (2018). Emerging microbial biocontrol strategies for plant pathogens. *Plant Science*, *267*, 102–111. Available from https://doi.org/10.1016/j.plantsci.2017.11.012.

Teplitski, M., Robinson, J. B., & Bauer, W. D. (2000). Plants secrete substances that mimic bacterial *N*-acyl homoserine lactone signal activities and affect population density-dependent behaviors in associated bacteria. *Molecular Plant-Microbe Interactions*, *13*, 637–648. Available from https://doi.org/10.1094/MPMI.2000.13.6.637.

Triveni, S., Prasanna, R., Kumar, A., Bidyarani, N., Singh, R., & Saxena, A. K. (2015). Evaluating the promise of Trichoderma and Anabaena based biofilms as multifunctional agents in *Macrophomina phaseolina*-infected cotton crop. *Biocontrol Science and Technology*, *25*, 656–670. Available from https://doi.org/10.1080/09583157.2015.1006171.

Vega, C., Rodríguez, M., Llamas, I., Béjar, V., & Sampedro, I. (2020). Silencing of phytopathogen communication by the halotolerant PGPR staphylococcus equorum strain EN21. *Microorganisms*, *8*. Available from https://doi.org/10.3390/microorganisms8010042.

Walsh, B., Ikeda, S. S., & Boland, G. J. (1999). Biology and management of dollar spot (*Sclerotinia homoeocarpa*); An important disease of turfgrass. *HortScience: A Publication of the American Society for Horticultural Science*, *34*, 13–21. Available from https://doi.org/10.21273/hortsci.34.1.13.

Wang, C., Knill, E., Glick, B. R., & Defago, G. (2000). Effect of transferring 1-aminocyclopropane-1-carboxylic acid (ACC) deaminase genes into *Pseudomonas fluorescens* strain CHA0 and its gacA derivative CHA96 on their growth-promoting and disease-suppressive capacities. *Canadian Journal of Microbiology*, *46*, 898–907. Available from https://doi.org/10.1139/cjm-46-10-898.

Weller, D. M., Landa, B. B., Mavrodi, O. V., Schroeder, K. L., De La Fuente, L., Blouin Bankhead, S., et al. (2007). Role of 2,4-diacetylphloroglucinol-producing fluorescent *Pseudomonas* spp. in the defense of plant roots. *Plant Biology*, *9*, 4–20. Available from https://doi.org/10.1055/s-2006-924473.

Weller, D. M., Mavrodi, D. V., Van Pelt, J. A., Pieterse, C. M. J., Van Loon, L. C., & Bakker, P. A. H. M. (2012). Induced systemic resistance in Arabidopsis thaliana against *Pseudomonas syringae* pv. tomato by 2,4-diacetylphloroglucinol-producing *Pseudomonas fluorescens*. *Phtopathology*, *102*, 403–412. Available from https://doi.org/10.1094/PHYTO-08-11-0222.

Xiong, Q., Liu, D., Zhang, H., Dong, X., Zhang, G., Liu, Y., et al. (2020). Quorum sensing signal autoinducer-2 promotes root colonization of *Bacillus velezensis* SQR9 by affecting biofilm formation and motility. *Applied Microbiology and Biotechnology*, *104*, 7177–7185. Available from https://doi.org/10.1007/s00253-020-10713-w.

Xu, Z., Zhang, R., Wang, D., Qiu, M., Feng, H., Zhang, N., et al. (2014). Enhanced control of cucumber wilt disease by bacillus amyloliquefaciens SQR9 by altering the regulation of its DegU phosphorylation. *Applied and Environmental Microbiology*, *80*, 2941–2950. Available from https://doi.org/10.1128/AEM.03943-13.

Yasmin, S., Hafeez, F. Y., Mirza, M. S., Rasul, M., Arshad, H. M. I., Zubair, M., et al. (2017). Biocontrol of Bacterial Leaf Blight of rice and profiling of secondary metabolites produced by rhizospheric *Pseudomonas aeruginosa* BRp3. *Frontiers in Microbiology*, 8. Available from https://doi.org/10.3389/fmicb.2017.01895.

Yuan, J., Zhang, N., Huang, Q., Raza, W., Li, R., Vivanco, J. M., et al. (2015). Organic acids from root exudates of banana help root colonization of PGPR strain *Bacillus amyloliquefaciens* NJN-6. *Scientific Reports*, *5*, 1–8. Available from https://doi.org/10.1038/srep13438.

Yuttavanichakul, W., Lawongsa, P., Wongkaew, S., Teaumroong, N., Boonkerd, N., Nomura, N., et al. (2012). Improvement of peanut rhizobial inoculant by incorporation of plant growth promoting rhizobacteria (PGPR) as biocontrol against the seed borne fungus, *Aspergillus niger*. *Biological Control*, *63*, 87–97. Available from https://doi.org/10.1016/j.biocontrol.2012.06.008.

Zhang, N., Yang, D., Wang, D., Miao, Y., Shao, J., Zhou, X., et al. (2015). Whole transcriptomic analysis of the plant-beneficial rhizobacterium *Bacillus amyloliquefaciens* SQR9 during enhanced biofilm formation regulated by maize root exudates. *BMC Genomics*, *16*, 1–20. Available from https://doi.org/10.1186/s12864-015-1825-5.

Zhang, T., Zhao, Y. L., Zhao, J. H., Wang, S., Jin, Y., Chen, Z. Q., et al. (2016). Cotton plants export microRNAs to inhibit virulence gene expression in a fungal pathogen. *Nature Plants*, *2*, 1–6. Available from https://doi.org/10.1038/nplants.2016.153.

Zheng, W., Zeng, S., Bais, H., LaManna, J. M., Hussey, D. S., Jacobson, D. L., et al. (2018). Plant growth-promoting rhizobacteria (PGPR) reduce evaporation and increase soil water retention. *Water Resources Research*, *54*, 3673–3687. Available from https://doi.org/10.1029/2018WR022656.

Zhu, M. L., Wu, X. Q., Wang, Y. H., & Dai, Y. (2020). Role of biofilm formation by *Bacillus pumilus* HR10 in biocontrol against pine seedling damping-off disease caused by *Rhizoctonia solani*. *Forests*, *11*, 1–17. Available from https://doi.org/10.3390/f11060652.

CHAPTER

Plant microbiota: a prospect to *Edge off* postharvest loss

12

Poonam Patel[1], Sushil Kumar[1] and Ajay Kumar[2]

[1]*Department of Biotechnology, Anand Agricultural University, Anand, India*

[2]*Postharvest Science, Agriculture Research Organization, Volcani Centre, Israel*

12.1 Introduction

The current scenario of today's world population is increasing at a pacing rate and is anticipated to reach 9.7 billion people by the year 2050. To meet this ever increasing demand and feed people, minimum of 60% rise in food production would be required. The produced food for consumption is wasted about one-third in the world per year which approximately costs around $680 billion and $310 billion (i.e., 670 and 630 million tons of food) for developed and developing countries, respectively. Recent statistics reveal that nearly 14% of world food production is lost before it even reaches the market, with the aided situation of the COVID-19 pandemic plushunger has made food security questionable due to reduced resources and transportation (FAO, 2020). The reported food loss during the year 2019 was around 40%–50% for root crops, vegetables, and fruits; 30% in cereals; 20% with oilseed crops, dairy products, and meat; and 35% accounted for fish (Sawicka, 2019).

12.2 Postharvest loss

Postharvest loss (PHL) refers to the loss of valuable food in terms of quantity and quality during or after harvest, this ultimately affects the farmers, small-scale food businesses, wastes limited resources and this all leads to climate-based consequences (Hodges, Buzby, & Bennett, 2010; Stathers et al., 2020). The menace of PHL is reported back from many years such as in 1975, Philippines faced a crisis in rice postharvest loss, estimated around 10%–37% (Bourne, 1977) are coherent with data of 2010 (Parfitt, Barthel, & Macnaughton, 2010) and 14% in 2019 (FAO, 2020). Willersinn, Mack, Mouron, Keiser, & Siegrist, 2015 reported 53%–55% initial fresh potato production and nearby 41%–46% initial processing potato production, losses caused by pathogen infestation, water loss of tubers, saccharification, and early sprouting in storage.

Biocontrol Mechanisms of Endophytic Microorganisms. DOI: https://doi.org/10.1016/B978-0-323-88478-5.00006-7

After the food crisis of the 1970s and 2007–08, PHL was considered an important critic in agriculture, to be resolved (Global Knowledge Initiative, 2014; World Bank, 2011).

Reasons behind PHL: The general causes of PHL are plenty which vary from crop to crop, use of agricultural practices, location and growing climatic factors, the field to food market supply chain, and economy of the country (developing or developed). The damage can occur at any stage from harvest to supply. The baseline fundamentals responsible for food loss are well explained by Bourne in 1977 in their "The food pipeline" (Bourne, 1977; Buchholz, Kostic, Sessitsch, & Mitter, 2018). The highest food loss is identified in storage conditions due to pathogen infestation (insect, mold, bacteria, etc.), environment in the vicinity (such as humidity, heat, rain, and frost), quality and sprouting (water loss, saccharification, and rancidity) or animals (rodents, birds) as shown in Fig. 12.1 (adapted from Bourne, 1977). Kiaya (2014) ascribed various internal and external factors responsible for PHL which are as follows:

Internal factors: The factors that affect the internal postharvest supply chain from harvesting to its marketing. They are : Harvesting (unfavorable conditions at maturity of crops and weather), precooling (high cost and inadequate precooling

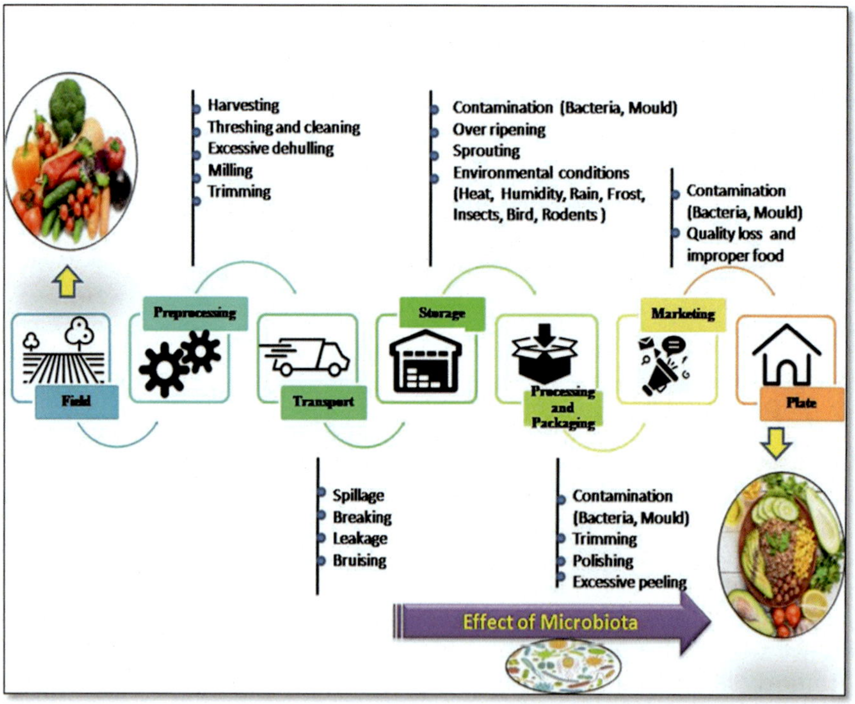

FIGURE 12.1

The food pipeline chain from field to plate and involved potential role of microbiota in it.

availability), transportation (lack of proper transport system, poor infrastructure, lack of refrigerated system for transport wherever needed), storage (a proper facility of storage, cleanliness, and temperature control are necessary factors), grading (lack of national standards and enforcement of laws, short of awareness, skill, and financial resources), packaging and labeling (fresh fruits and vegetables are more perishable, in general are sold unpacked in fresh or in wholesale markets, and reduces its shelf life), secondary processing (lack of infrastructure, technologies, and new commercialization agenda), factors such as biological (ethylene production, respiration rate, mechanical injuries, water stress, sprouting, rooting, pathological breakdown, etc.), microbiological (bacteria and fungi), and chemicals (chemical constituents within food, pesticides, or obnoxious chemicals).

External factors: Factors outside the food postharvest supply chain that contribute to PHL are environmental factors (temperature, humidity, altitude, and time), socioeconomic factors (urbanization, grain importation from other countries targets risk to native grain adulteration as well as chances of entry of new pathogens).

Key regulatory elements to overcome PHL: There have been various technical guidelines regarding common practices and techniques to be used in controlling PHL for crops to crops (World Bank, 2011). The strategies vary with the type of product and accordingly their management steps as needed to be followed are as follows:

1. A systemic approach of analysis should be carried out by commodity production and management system to identify proper strategies of PHL needed (Kitinoja & Gorny, 1999).
2. Strategies of PHL in cereals to be taken care of at different key stages: Harvesting, drying, threshing/shelling, winnowing/cleaning, and on storage farms.
3. Strategies of PHL in perishable crops (roots and tubers) to be taken care of at different key stages: Harvesting, handling, packing, transportation, and processing.
4. Strategies of PHL in perishable crops (fruits and vegetables) to be taken care of at different key stages: Harvesting, handling, sorting and cleaning, packaging, transportation, storage, and processing.
5. Making improvements in storage systems [mud silo (traditional) to metal silo (advanced system)].

Despite all these available strategies and guidelines, the system goes amiss. The scientific field and innovative minds have come up with a correlation of the postharvest loss–associated role of microbiota in reducing PHL. From Fig. 12.1, we can presume that a potential role of microorganisms is displayed in the food pipeline influencing the storage of crops. However, already quite work has been configured in the area of microbiota and their role in the growth and development of plants and increasing productivity. The potential role of microbiota in the postharvest pipeline is unmasked with few attempts carried out so far (Buchholz et al., 2018).

12.3 Microbiota

12.3.1 Plant microbiota

Plants have evolved with a surplus amount of various microbes during their growth and development. Microbes in the vicinity of plants are known for their established relationship and in counter effect display enormous benefits. A plethora of studies related to structural dynamics and functionalities of these microbiotas in association with plants are available. All the microbial assemblages' incognitos with plants surrounding the rhizosphere, endosphere, and phyllosphere are known as plant microbiota or plant microbiomes (all microbial genomes) (Compant, Samad, Faist, & Sessitsch, 2019). The term "microbiota" ("micro"—small and "biota"—living organism of a particular area) is originated from the ancient Greek language (Berg, Rybakova, Fischer, Cernava, & Vergès, 2020). Unraveling the interplay between the plant–microbe game can lead to enhanced perceptive to utilize them for targeted use in postharvest control as well as to combat the current challenges faced in postharvest loss during storage till usage, there is a demand of microbial innovations to performance level to deal with this situation.

12.3.2 Diversification in plant microbiota

There is wide diversification available within plant microbiotas among the bacteria (*Proteobacteria, Acidobacteria, Actinobacteria*), fungi (*Penicillium, Geotrichum, Aspergillus*), and archaea.

12.3.3 Bacteria

Plants actively employ the microbes from the vicinity of the rhizosphere, aerial parts of plants, the exterior of seeds and fruits. Compant et al. (2019) distinguished their bacterial microbes into two categories: below-ground plant microbiota, and above-ground microbiota.

Below-ground microbiota commonly includes root microbiota which is derived from soil microenvironment, include bacteria such as *Proteobacteria, Acidobacteria, Actinobacteria, Bacteriodetes, Planctomycetes*, and *Verrucomicrobia* (Fierer, 2017), and the below-ground environment is composed of bulk soil (devoid of plant roots), rhizosphere (associated with plant root) and endosphere (plant internal tissue), and microbiota from rhizosphere generally transmitted horizontally from soil to plants but are also transmitted vertically through seeds. The root system of plants proves to be a pleasant habitat for the colonization of microbiota, residing in the rhizosphere (a hotspot for microbial action) (Mendes, Garbeva, & Raaijmakers, 2013). The bacteria enter plant root via passive diffusion or active mechanism, through cracks or ruptured/opened part of lateral roots, then it colonizes and is transmitted in other plant parts depending on factors likeability of bacteria to colonize in specific part or allocation of plant resources to excess easily (Compant et al., 2005). Below ground the root exudates/rhizodeposits comprising fatty acids, amino acids, phenolics, plant growth regulators, sugars, nucleotides, vitamins, sterols, organic acids, etc. and influences microbial accumulation and its activity in relation to roots. This influential activity is called the rhizosphere effect (Driouich, Follet-Gueye, Vicré-Gibouin, & Hawes, 2013; Hartmann, Rothballer, & Schmid, 2003). Beirinckx, Viaene, and Haegeman (2020) reported the chilling responsive root microbiome of families the *Comamonadaceae* and the *Pseudomonadaceae* to be growth responsive and abundant in the root endosphere of maize under chilling stress. Kavamura, Robinson, and Hughes (2020) observed the change in soil bacterial communities in response to change in root traits from the use of dwarf varieties (tall to semidwarf wheat cultivars). Tall cultivars were abundant with *Actinobacteria, Bacteroidetes, and Proteobacteria*, whereas *Verrucomicrobia, Planctomycetes*, and *Acidobacteria* were found to be associated with semidwarf cultivars.

Above-ground microbiota commonly lies in leaves, vegetative foliar, and floral parts with great diversity; they enter through flowers and fruits and spread systemically through xylem tissue in various parts such as leaves, stems, and fruits (Compant, Clément, & Sessitsch, 2010). The compartmentalization of the phyllosphere/endosphere depends on resources allocated in plants and, thus, the presence of variable microbial communities. Dong, Wang, Li, and Shang (2019) represented bacterial diversity in different tissues such as *Acinetobacter, Enterobacter*, and *Pseudomonas* that were present enormously in stem, leaves, and root, *Enterobacter* was predominant in fruits especially in pericarp and seeds, *Acinetobacter* was found in placenta, whereas *Weissella* in the jelly. Venkatachalam et al. (2016) studied the diversity and abundance of phyllosphere microbiomes in rice var. Pusa Punjab Basmati 1509, results showed the supremacy of actinobacteria (38%) and α-proteobacteria (35%); *Pantoea, Exiguobacterium*, and Bacillus microbiotas. Mamphogoro, Maboko, Babalola, and Aiyegoro (2020) reported the dominance of bacterial phylum: *Proteobacteria, Firmicutes, Actinobacteria*, and *Bacteroidete* on the surface of fresh sweet pepper (*Capsicum annuum*).

12.3.4 Fungus and archaea

Most of the microbiome studies are well documented focusing on bacterial communities with plants, although there is significant microbial diversity. Fungi and archaea too have considerable biodiversity and reported friendly approaches toward plants with respect to growth and development. They are present in concentrations of 105−106 (fungi) and 107−108 (archaea) per g in the rhizosphere (Lee et al., 2019). Fungal strains such as *Trichoderma* sp., *Fusarium* sp., binucleate *Rhizoctonia solani*, and *Pythium* show biocontrol activity and promote plant growth (Whipps, 2001). Zhang, Gan, and Xu (2016) reported that fungi *Trichoderma longibrachiatum* T6 induces tolerance to salt stress and better growth in wheat seedlings. Poromarto, Nelson, and Freeman (1998) represented that the interaction of binucleate *Rhizoctonia* and soybean induced mechanism of biocontrol against *Rhizoctonia solani*, likewise Bacterial strain *Lysobacter enzymogenes* C3 and the binucleate *Rhizoctonia* strain BNR-8–2 in combination provoked biocontrol activity against common root rot of wheat (Eken & Yuen, 2014). Mycorrhizal fungi such as *Glomus intraradices* helps in providing mineral nutrition for plant development. Subramanian, Tenshia, Jayalakshmi, and Ramachandran (2009) showed the role of arbuscular mycorrhizal fungus—*G. intraradices* to enhance the Zn supply in maize by better root development and grains with high tryptophan and increased nutritional status. Ammonium oxidizing archaea that are predominately found in the rhizosphere (Chen, Zhu, Xia, Shen, & He, 2008) are key drivers that help in the conversion of nitrogen into a usable form for plants. The interaction among these three microbial domains (e.g., bacteria–fungus) plays an important role in providing various benefits to plants (Lee et al., 2019; Odelade & Babalola, 2019).

12.4 Prospective roles of microbiota in postharvest loss

Decay induced by microbial actions on food is the major reason for the postharvest loss. The role of microbiota effectively creates three relationships with plants: mutualism, commensalism, and parasitism/pathogenic association (Buchholz et al., 2018; Rodriguez et al., 2008). And this interaction is well affected by the biotic and abiotic stress, as well as the type microbiotic flora (pathogenic/beneficial microbes) (Brader et al., 2017). Pathogenic microbes or the change in the composition of the microbial community from "good to bad" microbes cause deterioration of food or diseases in plants. The role of microbiota in relation to postharvest loss is not a much-discussed topic but eye-catching importance is needed to be taken care of.

1. Seeds—Seeds represent an important pool of microbiomes, stipulated to pass onto progenies of plants and also are important in early growth and development during germination of plants (Nelson, 2018; Truyens, Weyens, Cuypers, & Vangronsveld, 2014). Seeds with microbiota (bacteria) can oppose the plant pathogens on the field as well as at postharvest (Furnkranz et al., 2012; Glassner et al., 2015). Rybakova et al. (2017) showed oilseed rape microbiome diversity shows resistance toward the colonization of pathogenic microorganisms. A novel approach "EndoSeedTM," that is, introducing new microbes in seeds and altering the plant microbiota and plant traits in a precise manner was reported recently (Mitter et al., 2017). All this can be done by introducing the microbial strain in a specific parent before the completion of seed development; this incorporates microbes into seeds. For this, scientists used the bacterium *Paraburkholderia phytofirmans* (PsJN), (a potential plant growth promoter) and induced stress resistance in various crops and vegetables (Mitter et al., 2013). Mitter et al. (2017) introduced PsJN into

seeds of maize, wheat, soy, and pepper, which stably integrated and after germination may proliferate and colonize in the subsequent plant progeny. This has more benefits compared to external applications such as seed coating. This can be a friendly approach to be applicable in terms of storage and quality of grains and cereals and protection in terms of postharvest losses and prevention of it from its own internal microbiota.

2. Role of microbiomes in sprouting, ripening, quality traits, and disease control.
3. Sprouting—in tuber crops such as potato and onion early sprouting is a key issue during storage conditions. Aksenova et al. (2013) explained sprouting as a complex physiological process that metabolizes complex proteins, lipids, and carbohydrates. In addition, there is a significant role of plant growth regulators, that is, cytokinin and indole acetic acid signaling, which triggers sprouting and gibberellins induce sprout growth, whereas abscisic acid and ethylene suppress sprouting (Sonnewald & Sonnewald, 2014). Slininger et al. (2007) showed the potential role of microbiota (bacteria) as a biocontrol agent for sprouting control and disease Fusarium dry rot. They showed that the mixture of strains−strain mix > *Pseudomonas fluorescens* S22: T: 04 > *P. fluorescens* S11: P: 12 have potential broad-spectrum suppression ability for postharvest sprouting and disease control. In previous studies, they had reported that bacterial strains: *P. fluorescens* S11: P: 12 (NRRL B-21133), *Enterobacter* sp., S11: T: 07 (NRRL B-21050), and S11: P:08 (NRRL B-21132) had potent activity in sprout control comparable with 16.6 ppm Chlorpropham (CIPC) thermal fog during 4−5 months of storage duration (Slininger, Burkhead, & Schisler, 2004). Weiss, Hertel, Grothe, Ha, & Hammes (2007) reported a strain of *Pseudomonas jessenii* (as protective culture) in different sprouts (adzuki, alfalfa, mung bean, radish, sesame, and wheat) which suppress the growth of pathogenic enterobacteria.
4. Ripening—A prolonged ripening attributes to nearby 20% postharvest loss in fruits and vegetables (Perry & Williams, 2014). During ripening food releases some gaseous compounds such as ethylene. The level of ethylene at later stages speeds up the ripening process in fruits. Few fruits produce ethylene at rapid rate results in fast ripening other fruit releases it gradually over the period of time and hence slow ripening. Overripening in fruits is an undesirable criteria at the consumer's end, therefore it is important to manage this condition at the postharvest stage. In general practices, postharvest ripening is prevented using cultivars with delayed ripening trait, for example, *Shiro* (plum variety) (Abdi, McGlasson, Holford, Williams, & Mizrahi, 1998) with slow ripening (as the production of ethylene is suppressed), Flavr savr genetically modified (GM) tomato (Bruening & Lyons, 2000), cold storage, application of 1-methyl cyclopropane, ethylene removal or scrubbing, and controlled atmospheric storage. The interrupted changes in ethylene levels can modulate the ripening and can also be achieved using microorganisms (bacteria, fungi) (Digiacomo et al., 2014). The production of ethylene in plants is mediated by 1-amino cyclopropane-1-carboxylate (ACC) oxidase

from ACC. There are plenty of bacterial strains present in plants as well as soil which cleave ACC and result in the production of 2-isobutyrate and ammonia via ACC deaminase activity, which leads to ethylene control in plants (Glick, 2005). Pierce et al. (2014) reported *Rhodococcus rhodochrous* DAP 96253 cells in close immediacy delay ripening in climacteric fruit (peaches, avocado, and banana). In peaches and banana, ripening could be delayed up to 7–14 days. The potential role of *Pseudomonas chlororaphis*, a nonpathogenic soil bacterium, codes for the enzyme that decreases ACC to different compounds and, thus, reduces ACC availability for ethylene production (ISAAA, 2020). Anderson, Staley, Challender, and Heuton (2018) reported a GM tomato with a gene from *P. chlororaphis* encoding ACC *deaminase* enzyme which delayed the ripening process by diminishing ethylene production. Thus we can presume *P. chlororaphis*, a safe alternative prospect in the future for delay ripening in fruits. Perry (2016) showed the efficacy of endophyte bacteria (*Bacillus* sp.) that were placed in sealed containers with organic bananas and showed delayed ripening postharvest. Banana showed more firmness, without visible spotting and nil ripened smell.

5. Improving quality trait and disease control.
6. Quality in food products is described as "the degree of excellence" (Kader, 2002) and meeting the demand at the consumer end regarding the desirability of the products. Various criteria describe quality such as appearance, freshness, flavor, nutritional value, and safety (Camelo, 2004). Controlling decay or diseases over postharvest stages maintains nutritional content and quality of food products. The microbiota displays diverse roles to attain the market demand. Barman, Patel, Sharma, & Singh (2017) showed that the dipping treatment of *P. fluorescens* at 108 CFU/mL (for 5 min) maintained the highest total carotenoids and total soluble solids content in mango; they also showed biocontrol efficacy against important postharvest diseases, anthracnose and stem-end rot caused by *Colletotrichum gloeosporioides* and *Botryodiplodia theobromae*, respectively, in mango during postharvest storage. Treatment caused a significant reduction of about 5.92% in the decay of mango fruits, lower respiration rate, ethylene production, loss of firmness, less total phenolics, and reduced acidity was also observed.

Postharvest losses could be prevented simply by using antagonist bacteria, fungi, or yeast. Yeast strains and bacterial extract (*Serratia* sp.) have shown that antagonistic activities against *C. gloeosporioides* (Granada et al., 2019), *Bacillus* sp. (from leaf and fruit surfaces of avocado itself) were found to be effective against anthracnose in avocado (Campos-Martínez et al., 2016). Bacteria antagonist, *P. fluorescens*, efficiently controlled green mold in oranges (Citrus sinensis Osbeck, cv. Jincheng) postharvest caused by *Penicillium digitatum*, via restricting spore germination and germ tube elongation as well as induced resistance in citrus peel (increased chitinases, β-1,3-glucanase, peroxidase, and phenylalanine ammonia-lyase activities) (Wang et al., 2018). *P. fluorescens* also were potent suppressors of blue mold caused by *Penicillium expansum*, in apples (McIntosh and Spartan variety) during commercial cold storage by inhibiting the conoidal germination and mycelial growth (Wallace, Hirkala, & Nelson, 2017). Zhang et al. (2008) investigated the role of yeast strain *Rhodotorula glutinis*, in reducing decay (gray mold decay and blue mold decay) caused by *Botrytis cinerea* and *P. expansum* in pear and observed good postharvest quality too. In the previous study, they reported that the combination of heat treatment and *Cryptococcus laurentii* was helpful in controlling *Rhizopus* and natural decay (40%–30%) in peaches on postharvest storage at 40°C (30 days) and 200°C (7 days) compared to control fruits, also firmness, total soluble sugar (TSS), ascorbic acid, or titratable acidities were unaffected in fruits (Zhang, Wang, Zheng, & Dong, 2007). Yeast *Metschnikowia fructicola* treatment controls postharvest rot in strawberry fruits in the greenhouse, opens field culture, and tunnels up to 70%, 64%, and 72%, respectively (Karabulut et al., 2004).

The recent study reported by Galsurker et al. (2020) pointed out that endophytic microbiota in short stem end of harvested mango fruit showed reduced stem-end rot disease during cold storage and shelf life. Koffi, Alloue-Boraud, Dadie, Koua, and Ongena (2017) showed the use of *Bacillus subtilis* GA1 for the inhibition of *Colletotrichum* and reducing spoilage in mango. Madhupani and Adikaram (2017) suggested that for avocado the disease incidence of stem-end rot can be delayed by thc application of cell suspension of *Aureobasidium pullulans* at the stem-end of unripe fruit The results showed the delay in disease incidence by 2 days in comparison to control by controlling conidia germination and shorter germ tubes.

12.4.1 Biocontrol products used to control postharvest losses: on way to commercialization

The research on biocontrol of postharvest losses is focused on the identification of microorganisms with potent biocontrol activities postharvest (Sharma, Singh, & Singh, 2009). There are different products in the market, registered with trade names and functionally active microbes used for control of postharvest losses. Some examples are enlisted in Table 12.1 and their applications (Spadaro & Droby, 2016). Although with various success stories in the marketplace, these products sometimes fall short to meet desired results such as variability in performance, problems in registration, costly in comparison to synthetic products, complicated processing and manufacturing, lack of industrial and farmer level acceptance (Droby, Wisniewski, Teixidó, Spadaro, & Jijakli, 2016; Wenneker & Thomma, 2020).

12.5 Mode of action of microbiota in postharvest

With the present biotechnological advancement and techniques involved, it is easier to dissect the microfloral diversity and its role in a precise manner.

Table 12.1 Different biocontrol products to control postharvest loss.

Product name	Active microbiota	Applications	References
Bio-Save Organic Decay-Control	Formulated *Pseudomonas syringae*	Diseases control in fruits and vegetables, seeds, and storage sweet potatoes and white potatoes	Janisiewicz and Peterson (2004)
Avogreen	*Bacillus subtilis*	Avocado	Korsten, Towsen, and Classens (1998)
Boni Protect	*Aureobasidium pullulans*	Apple	Weiss, Mogel, and Kunz (2006)
Candifruit	*Candida sake*	Pome fruits and grapevine	Calvo-Garrido et al. (2014)
Nexy	*Candida oleophila*	Pome fruits and banana	Janisiewicz and Jurick (2017)
Aspire	*C. oleophila*	Citrus and apples	Wisniewski et al. (2007)
Pantovital	*Pantoea agglomerans*	Citrus fruits	Janisiewicz and Jurick (2017)
Shemer	*Metschnikowia fructicola*	Peach, strawberry, and tomato	Droby, Wisniewski, Macarisin, and Wilson (2009)
Yield Plus	*Cryptococcus albidus*	Citrus and pome fruits	Janisiewicz and Korsten (2002)
Amylo-X	*Bacillus amyloliquefaciens*	Some vegetables	Usall, Torres, and Teixidó (2016)
NOLI	*M. fructicola*	Soft fruit, stone fruit, and grapes	Droby and Wisniewski (2019)

The unveiling of the details at the molecular, biochemical, microscopic level with the underlying mechanism of action can be obtained effortlessly. Spadaro and Droby (2016) explained a quadritrophic model for underlying interaction between the host, the antagonist (microbiota), the pathogen, and the endophytic microflora (Fig. 12.2 adapted from Spadaro & Droby, 2016). The different components of the quadritrophic system work with a complex network of interactions within each other. There are basically three different modules of microbes acting on a single host: the endophytes (that reside within-host and show beneficiary effects such as providing immunity, maintain physiology, growth, and development); the pathogen that imposes negative consequences, the invasion of pathogen starts decaying and causes diseases via the induction of pathogen-associated molecular pattern PAMPs, phytotoxins, cell wall degrading enzymes, effectors, pH alterations in the host. These diseases can be disastrous as the whole production sometimes could be at stake, exemplified by the well-known incidence of Panama

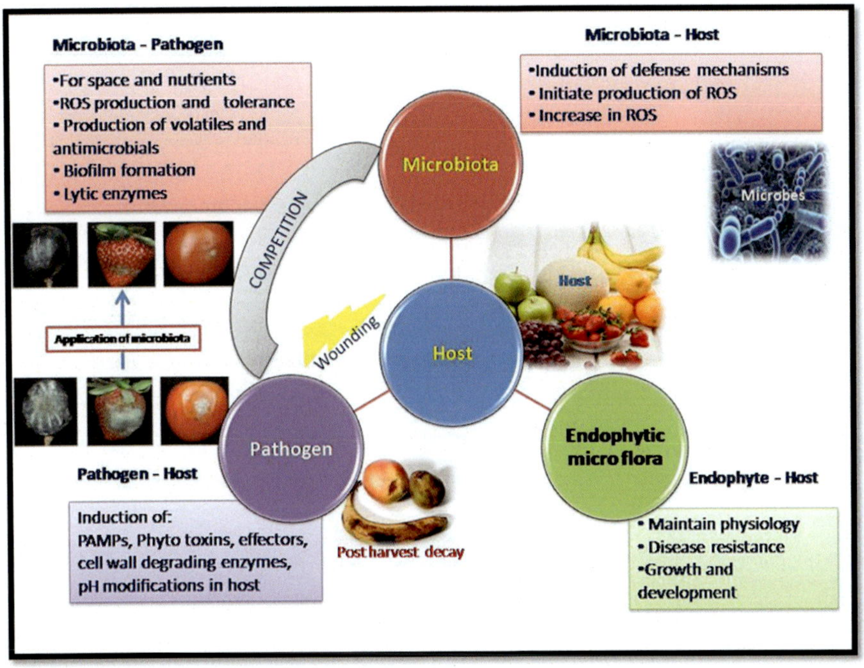

FIGURE 12.2

Quadritrophic model of underlying interaction between the host, the antagonist (microbiota), the pathogen, and the endophytic microflora.

disease (Fusarium wilt) in banana (Ploetz, 2005) and the microbiota involved helping host from invading pathogens and maintaining its quality attributes (Spadaro & Droby, 2016).

The key strategic mechanisms exerted by microbiota which behold antagonism against phytopathogen are briefly mentioned next (Carmona-Hernandez et al., 2019; Mamphogoro et al., 2020) (Fig. 12.3):

1. Nutrient and space (competition): The competition for nutrition (nitrogen, oxygen, and carbon sources) and space are key players to show efficacy as a biocontrol mechanism in microbiota against phytopathogen. Some bacteria and yeast species compete well with phytopathogen at the wounding site and restraining the nutritional inputs, in turn, inhibit the growth of competitors (Janisiewicz, Tworkoski, & Kurtzman, 2001). Zhang, Spadaro, Garibaldi, & Gullino (2011) reported that correlations among *Pichia guilliermondii* with *B. cinerea* in apple were regulated by sugars and nitrates as key resources.
2. Iron: Iron is considered an essential element for bacteria, fungi, yeast, and plants. The companionship of each other leads to a competitive situation, this condition determines the fecundity of host–pathogen interactions, which, in

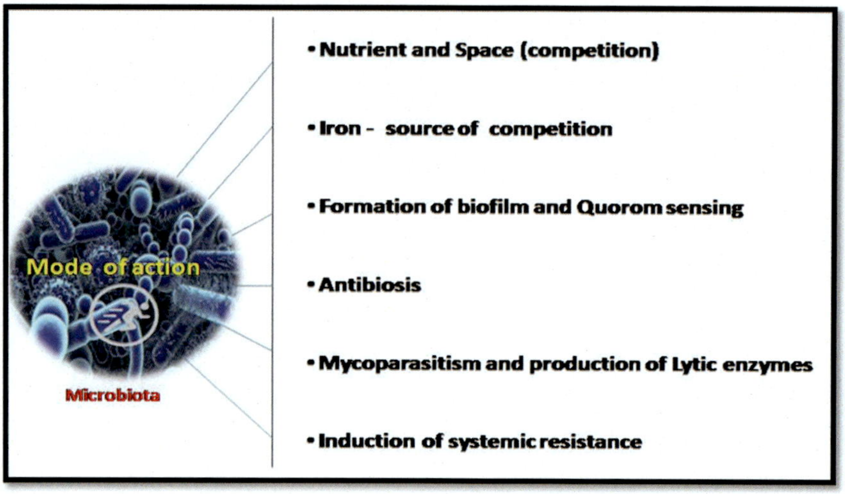

FIGURE 12.3

Key strategic mode of action by microbiota against pathogens.

turn, activate the immune response or can play a decisive role in resistance/susceptibility of the host (Naranjo-Arcos & Bauer, 2016; Payne, 1993). Iron is available in different forms such as iron—sulfur clusters (Fe/S), di-iron, heme (as a cofactor) containing proteins, for example, catalase, cytochromes, and various enzymes (cytochrome P450) pertaining to different roles. They play important role in fungal growth and pathogenesis, and competition for iron can be an effective strategic mechanism in the biocontrol of pathogens in postharvest conditions (Saravanakumar, Ciavorella, Spadaro, Garibaldi, & Gullino, 2008). Saravanakumar et al. (2008) in their studies on apple reported that iron depletion by *Metschnikowia pulcherrima* halts the conidial germination and mycelial growth of *B. cinerea, Alternaria alternata*, and *P. expansum*, also *M. fructicola* (in vitro) arrested the growth of *B. cinerea* and *P. digitatum*.

3. Biofilm formation and quorum sensing: The ability to adhere, colonize, and multiply over the fruit surface is important for antagonist bacteria in the formation of biofilms (microcolonies of bacteria produced protein hydrated matrix, nucleic, and polysaccharide acid) and this structural film is measured by quorum sensing (using regulators like farnesol, phenethyl alcohol, and tyrosol). The biofilm acts as an obstruction between invading pathogens over the bruised surface of the host. The key regulatory mechanism behind this is still not much elaborated (Beauregard, Chai, Vlamakis, Losick, & Kolter, 2013). Various microbes hold promising strategies via the formation of biofilm in association with plants (Thimmaraju, Biedrzycki, & Bais, 2008) of biocontrol activity and future directions to control postharvest loss; some

examples are *B. subtilis* (Lastochkina et al., 2019; Ostrowski, Meheter, Prescitt, Kiley, & Stanley-Wall, 2011), *Paenibacillus polymyxa* (Haggag & Timmusk, 2010), *P. fluorescens* (Silby & Levy, 2004), and *Microsphaeropsis* sp. (Carisse, El Bassam, & Benhamou, 2001).

4. Antibiosis: Antibiosis, that is, the production of antibiotics is one of the very important aspects of antagonist microbes. There are various antibiotic metabolites reported with potent antifungal, antibacterial properties. For example, *B. subtilis* (antifungal peptide) (Shankar & Shivakumar, 2013), *Saccharomyces cerevisiae* (killer toxins K1, K2, and K28) (Ferraz, Cássio, & Lucas, 2019), *Candida albicans* (farnesol) (Liu, Shi, Chen, & Long, 2014a), and *Kloeckera apiculata* (phenyl ethanol) (Liu et al., 2014b). Sometimes low molecular weight compounds are produced by microorganisms volatile organic compounds (VOCs); because of their volatile nature, they can reach out to distant places. VOCs are the modified products of fatty acid biosynthesis and shikimate pathways (Dickschat, Bode, Wenze, Müller, & Schulz, 2005). Raza, Ling, Yang, Huang, and Shen (2016) reported that volatile compounds produced by *Bacillus amyloliquefaciens* SQR-9 inhibited the growth of pathogen *Ralstonia solanacearum* in tomato wilt, with the production of antioxidant enzymes, exopolysaccharides, and the formation of biofilm.
5. Mycoparasitism: The antagonist directly destroys pathogens by feeding on them by releasing various lytic enzymes such as chitinases, glucanases, and proteases. Mycoparasitism occurs in a sequential event: secretion of lytic enzymes, mutual recognition by pathogen and antagonist, vigorous growth of antagonist in the host, and association of pathogen and host (Talibi, Boubaker, Boudyach, & Ait Ben Aoumar, 2014). Chanchaichaovivat, Panijpan, and Ruenwongsa (2008) studied the putative mode of action of *P. guilliermondii* R13 in controlling postharvest chilli anthracnose (*Colletotrichum capsici*) via the production of β-1,3-glucanase, and chitinases leading to the suppression of spore germination and germ tube length. Zhang, Spardo, Garibaldi, and Gullino (2010) suggested a possible mode of action based on the production of lytic enzymes (β-1,3-glucanase, exochitinase, and endochitinase) and competition of nutrients in antagonist *A. pullulans* PL5 against the postharvest pathogen (*Monilinia laxa*) in peach, plum, and *P. expansum* and *B. cinerea* on apples.
6. Induction of systemic resistance: Various molecular and biochemical elicited defense responses lead to acquired resistance on the application of microbiota over fruit skin (Spadaro & Droby, 2016). In normal conditions, plant generates defense response as a result of biotic/abiotic inducing factors that lead to induce infection or resistance to be localized or systemic (Bloemberg & Lugtenberg, 2001). Plant systemic resistance was reported by several bacteria such as *Saccharothrix algeriensis*, *Micromonospora*, and *P. fluorescens* against *B. cinerea* infections (Martínez-Hidalgo, García, & Pozo, 2015). Droby et al. (2002) reported *Candida oleophila* cell suspension

application onto grapefruit peel tissue—suppressed spore germination and germ tube growth, by increased activity of ethylene biosynthesis, phytoalexin accumulation, phenylalanine ammonia-lyase activity, chitinase, and endo-β-1,3-glucanase levels. Table 12.2 represents the few reported examples of fruits and antagonist with their mode of action against plant diseases.

Table 12.2 Few examples of fruit microbiota with their mode of action against pathogen.

Host	Antagonist	Pathogen	Mode of action	References
Apple	*Pseudomonas fluorescens*	*Botrytis cinerea* (gray mold)	Biofilm formation	Wallace, Hirkala, and Nelson (2018)
	Bacillus amyloliquefaciens PG12	*Botryosphaeria dothidea* (ring rot)	Lipopeptide production—"iturin"	Chen, Zhang, Fu, and Wang (2016)
Stone fruit	*Pseudomonas synxantha*	*Monilinia fructicola* and *Monilinia fructigena*	Production of antifungal compounds, VOCs, systemic resistance	Aiello, Restuccia, Stefani, Vitale, and Cirvilleri (2019)
Orange and lemon	Combination of *Pseudomonas* spp. and three *Trichoderma* spp.	*Penicillium digitatum* (Pers.) Sacc.	Competition for nutrients and space, production of cyclic lipodepsipeptides	Panebianco, Vitale, Polizzi, Scala, and Cirvilleri (2015)
Grapes	*Yeast: Meyerozyma guilliermondii and Candida membranifaciens*, Bacteria: *Bacillus* sp. and *Ralstonia* sp.	*B. cinerea*	Volatile and nonvolatile substances, competition for nutrient and space	Kasfi, Taheri, Jafarpour, and Tarighi (2018)
Peach, plum, and apple	*Aureobasidium pullulans PL5*	*Monilinia laxa*, *Penicillium expansum*, and *B. cinerea*	Production of lytic enzymes, competition of nutrients in antagonist	Zhang et al. (2010)
Tomato	*B. amyloliquefaciens* SQR-9	*Ralstonia solanacearum*	Production of antioxidant enzymes, VOCs, exopolysaccharides, and biofilm formation	Raza et al. (2016)

VOC, *Volatile organic compound.*

12.5.1 Biotechnological advancements aided to microbiota—postharvest loss

With advancements in biotechnological approaches, in the plethora of ways, this scientific development has uplifted the current scenario to obtain in-depth knowledge and unveil possibilities to overcome obstacles that we were facing so far. Genomics, transcriptomics, proteomics, bioinformatics, and metabolomics aided with gene editing/engineering tools have advanced the vision of "better than before." All these technologies may help to elucidate the changes that occurred due to environmental stress factors and physiological status of microbiota, gene interaction at host, pathogen, and microbiota level (Hershkovitz et al., 2013). Ke, Wang, and Yoshikuni (2020) reviewed potential applications of microbiome engineering with bottom-up and top-down strategies to remodel microbiomes to understand and promote plant−microbe interactions to address present problems of food safety and security. Saminathan et al. (2018) proposed the role of fruit-related microbiome in carbohydrate metabolism and ripening of mature fruits through metagenomics and transcriptome analysis in watermelon cultivars. Jiang, Chen, Miao, Krupinska, and Zheng (2009) assessed the interaction between cherry tomato fruit and *C. laurentii*, through microarray analysis. Data showed 194 (upregulated genes were involved in signal transduction, metabolism, and stress response) and 312 (downregulated were associated with energy metabolism and photosynthesis) genes. Digiacomo et al. (2014) reported engineered *Escherichia coli* that synthesizes and releases ethylene hormone to control fruit ripening. An enzyme [ethylene forming enzyme (EFE)/2-oxoglutarate oxygenase/decarboxylase] from *Pseudomonas syringae* pv. *phaseolicola* was inserted in *E. coli*, which catalyzes conversion of 2-oxoglutarate to ethylene, which consequently can seep across the bacterial membrane. This was helpful in speed up ripening process in tomatoes, apples, and kiwifruit. Lysoe et al. (2017) reported that three-way transcriptome interactions of *Clonostachys rosea* (biocontrol), *Helminthosporium solani* (pathogen), and potato (host) revealed that reductions in silver cruf were attributed to combinations of various mechanisms like plant defense responses, microbial competition for nutrient and space, antibiosis, and mycoparasitism.

Yang et al. (2020) carried out transcriptome and protein expression profiling of *P. expansum* (causal of blue mold decay in pear) induced by *Meyerozyma guilliermondii* to study fundamental molecular inhibitory response (*M. guilliermondii* against *P. expansum*). The transcriptome data showed that differentially expressed genes, *Cytochromes P450*, *Phosphatidate cytidylyltransferase*, and *Glutathione S-transferase*, were upregulated, while *Phosphoesterase*, *Polyketide synthase, HEAT*, *ATPase*, and *Ras-association* genes were downregulated. Whereas proteome data showed 66 differential proteins, 6 upregulated and 60 downregulated, it showed association with various pathways like ATP synthesis, basal metabolism, oxidative phosphorylation, and response regulation. Both downregulated genes and protein showed mutualistic association in response to the growth of *P. expansum* treated with *M. guilliermondii*.

12.6 Conclusion

In a nutshell, the role of microbiota to overcome postharvest losses in unexplored areas is slowly gaining attention with recent developments and more research. The problem can be dealt with efforts such as the use of microbiota but cannot be eradicated totally. A better understanding of the mechanism lying behind the interaction among host, pathogen, microbiota, and endophytes plays a crucial role in designing and toward a new perception of these microbiota to be used as a biocontrol agent in postharvest applications. Mining microbiota in association with plants, with biotechnological advancements, may help in exploring it in more depth to establish food safety and security in food chain link "field to plate."

References

Abdi, N., McGlasson, W. B., Holford, P., Williams, M., & Mizrahi, Y. (1998). Responses of climacteric and suppressed-climacteric plums to treatment with propylene and 1-methylcyclopropene. *Postharvest Biology and Technology*, *14*(1), 29–39.

Aiello, D., Restuccia, C., Stefani, E., Vitale, A., & Cirvilleri, G. (2019). Postharvest biocontrol ability of *Pseudomonas synxantha* against *Monilinia fructicola* and *Monilinia fructigena* on stone fruit. *Postharvest Biology and Technology*, *149*, 83–89.

Aksenova, N. P., Sergeeva, L. I., Konstantinova, T. N., Golyanovskaya, S. A., Kolachevskaya, O. O., & Romanov, G. A. (2013). Regulation of potato tuber dormancy and sprouting. *Russian Journal of Plant Physiology*, *60*, 301–312.

Anderson, J. A., Staley, J., Challender, M., & Heuton, J. (2018). Safety of *Pseudomonas chlororaphis* as a gene source for genetically modified crops. *Transgenic Research*, *27*, 103–113.

Barman, K., Patel, V. B., Sharma, S., & Singh, R. R. (2017). Effect of chitosan coating on postharvest diseases and fruit quality of mango (*Mangifera indica*). *Indian Journal of Agricultural Sciences*, *87*(5), 618–623.

Beauregard, P. B., Chai, Y., Vlamakis, H., Losick, R., & Kolter, R. (2013). *Bacillus subtilis* biofilm induction by plant polysaccharides. *Proceedings of the National Academy of Sciences of the United States of America*, *110*, E1621–E1630.

Beirinckx, S., Viaene, T., Haegeman, A., Debode, J., Amery, F., Vendenabeele, S., Goormachtig, S., et al. (2020). Tapping into the maize root microbiome to identify bacteria that promote growth under chilling conditions. *Microbiome*, *8*, 54.

Berg, G., Rybakova, D., Fischer, D., Cernava, T., Vergès, M., C., Charles, T., Schloter, M., et al. (2020). Microbiome definition re-visited: Old concepts and new challenges. *Microbiome*, *8*, 103.

Bloemberg, G. V., & Lugtenberg, B. J. (2001). Molecular basis of plant growth promotion and bio-control by rhizobacteria. *Current Opinion in Plant Biology*, *4*(4), 343–350.

Bourne, M. (1977). *Post harvest food losses—The neglected dimension in increasing the world food supply. Cornell international agriculture mimeograph* (p. 53) New York State College of Agriculture and Life Sciences, Cornell University.

Brader, G., Compant, S., Vescio, K., Mitter, B., Trognitz, F., Ma, L.-J., & Sessitsch, A. (2017). Ecology and genomic insights into plant-pathogenic and plant-nonpathogenic endophytes. *Annual Review of Phytopathology*, *55*, 61–83.

Bruening, G., & Lyons, J. M. (2000). The case of the FLAVR SAVR tomato. *California Agriculture*, *54*(4), 6–7.

Buchholz, F., Kostic, T., Sessitsch, A., & Mitter, B. (2018). The potential of plant microbiota in reducing postharvest food loss. *Microbial Biotechnology*, *11*, 971–975.

Calvo-Garrido, C., Vinas, I., Usall, J., Rodríguez-Romera, M., Ramos, M. C., & Teixido, N. (2014). Survival of the biological control agent *Candida sake* CPA-1 on grapes under the influence of abiotic factors. *Journal of Applied Microbiology*, *117*, 800e811.

Camelo, A. F. L. (2004). *Chapter 5: The quality in fruits and vegetables. Manual for the preparation and sale of fruits and vegetables from field to market, . FAO agricultural services bulletin* (Vol. 151). Food and Agriculture Organization o f the United Nations, ISSN 1010–1365.

Campos-Martínez, A., Velázquez-del, M. G., Flores-Moctezuma, H. E., Suárez-Rodríguez, R., Ramírez-Trujillo, J. A., & Hernández-Lauzardo, A. N. (2016). Antagonistic yeasts with potential to control *Colletotrichum gloeosporioides* (Penz.) Penz. & Sacc. and *Colletotrichum acutatum* on avocado fruits. *Crop Protection (Guildford, Surrey)*, *89*, 101–104, 2016.

Carisse, O., El Bassam, S., & Benhamou, N. (2001). Effect of *Microsphaeropsis* sp. strain P130A on germination and production of sclerotia of *Rhizoctonia solani* and interaction between the antagonist and the pathogen. *Phytopathology*, *91*, 782–791.

Carmona-Hernandez, S., Reyes-Pérez, J. J., Chiquito Contreras, R. J., Rincon-Enriquez, G., Cerdan-Cabrera, C. R., & Hernandez-Montiel, L. G. (2019). Biocontrol of postharvest fruit fungal diseases by bacterial antagonists: A review. *Agronomy*, *9*, 121.

Chanchaichaovivat, A., Panijpan, B., & Ruenwongsa, P. (2008). Putative modes of action of *Pichia guilliermondii* strain R13 in controlling chilli anthracnose after harvest. *Biological Control*, *47*(2), 207–215.

Chen, X., Zhang, Y., Fu, X., & Wang, Q. (2016). Isolation and characterization of *Bacillus amyloliquefaciens* PG12 for the biological control of apple ring rot. *Post Harvest Biology and Technology*, *115*, 113–121.

Chen, X. P., Zhu, Y. G., Xia, Y., Shen, J. P., & He, J. Z. (2008). Ammonia-oxidizing archaea: Important players in paddy rhizosphere soil? *Environmental Microbiology*, *10*, 1978–1987.

Compant, S., Clément, C., & Sessitsch, A. (2010). Plant growth-promoting bacteria in the rhizo- and endosphere of plants: Their role, colonization, mechanisms involved and prospects for utilization. *Soil Biology & Biochemistry*, *42*, 669–678.

Compant, S., Reiter, B., Sessitsch, A., Nowak, J., Clément, C., Ait., & Barka, E. (2005). Endophytic colonization of *Vitis vinifera* L. by plant growth-promoting bacterium *Burkholderia* sp. strain PsJN. *Applied and Environmental Microbiology*, *71*, 1685–1693.

Compant, S., Samad, A., Faist, H., & Sessitsch, A. (2019). A review on the plant microbiome: Ecology, functions, and emerging trends in microbial application. *Journal of Advanced Research*, *19*, 29–37.

Dickschat, J. S., Bode, H. B., Wenze, S. C., Müller, R., & Schulz, S. (2005). Biosynthesis and identification of volatiles released by the Myxobacterium *Stigmatella aurantiaca*. *Chembiochem: A European Journal of Chemical Biology*, *6*, 2023–2033.

Digiacomo, F., Girelli, G., Aor, B., Marchioretti, C., Pedrotti, M., Perli, T., Bianco, C., D., et al. (2014). Ethylene-producing bacteria that ripen fruit. *ACS Synthetic Biology*, *3*, 935–938.

Dong, C. J., Wang, L. L., Li, Q., & Shang, Q. M. (2019). Bacterial communities in the rhizosphere, phyllosphere and endosphere of tomato plants. *PLoS One*, *14*(11), e0223847.

Driouich, A., Follet-Gueye, M. L., Vicré-Gibouin, M., & Hawes, M. (2013). Root border cells and secretions as critical elements in plant host defense. *Current Opinion in Plant Biology*, *16*(4), 489–495.

Droby, S., Vinokur, V., Weiss, B., Cohen, L., Daus, A., Goldschmidt, E. E., & Porat, R. (2002). Induction of resistance to *Penicillium digitatum* in grapefruit by the yeast biocontrol agent *Candida oleophila*. *Phytopathology*, *92*, 393–399.

Droby, S., Wisniewski, M., Macarisin, D., & Wilson, C. (2009). Twenty years of postharvest biocontrol research: Is it time for a new paradigm? *Postharvest Biology and Technology*, *52*(2), 137–145.

Droby, S., Wisniewski, M., Teixidó, N., Spadaro, D., & Jijakli, M. H. (2016). The science, development, and commercialization of postharvest biocontrol products. *Postharvest Biology and Technology*, *122*, 22–29.

Droby., & Wisniewski. (2019). *International Workshop The Fruit Microbiome: A New Frontier.Innovative Fruit Production, Improvement, and Protection*. Kearneysville, WV: USDA. Available from https://www.ars.usda.gov/ARSUserFiles/80800505/International%20Workshop/Presentations/Thursday%20Morning/Samir%20Droby.pdf.

Eken, C., & Yuen, G. (2014). Biocontrol of common root rot of wheat with *Lysobacter enzymogenes* and *binucleate Rhizoctonia*. *Romanian Agricultural Research*, *31*, 309–314.

FAO. (2020). Food loss and waste must be reduced for greater food security and environmental sustainability. First International Day of Awareness of Food Loss and Waste. Food and Agriculture Organization of the United Nations. Available from http://www.fao.org/news/story/en/item/1310271/icode/.

Ferraz, P., Cássio, F., & Lucas, C. (2019). Potential of yeasts as biocontrol agents of the phytopathogen causing Cacao *Witches' Broom Disease*: Is microbial warfare a solution? *Frontiers in Microbiology*, *10*, 1766.

Fierer, N. (2017). Embracing the unknown: Disentangling the complexities of the soil microbiome. *Nature Reviews Microbiology*, *15*, 579–590.

Furnkranz, M., Lukesch, B., Muller, H., Huss, H., Grube, M., & Berg, G. (2012). Microbial diversity inside pumpkins: Microhabitat-specific communities display a high antagonistic potential against phytopathogens. *Microbial Ecology*, *63*, 418–428.

Galsurker, O., Diskin, S., Duanis-Assaf, D., Doron-Faigenboim, A., Maurer, D., Feygenberg, O., & Alkan, N. (2020). Harvesting mango fruit with a short stem-end altered endophytic microbiome and reduce stem-end rot. *Microorganisms*, *8*(558), 1–18.

Glassner, H., ZchoriFein, E., Compant, S., Sessitsch, A., Katzir, N., Portnoy, V., & Yaron, S. (2015). Characterization of endophytic bacteria from cucurbit fruits with potential benefits to agriculture in melons (*Cucumis melo* L.). *FEMS Microbiology Ecology*, *91*, fiv074.

Glick, B. R. (2005). Modulation of plant ethylene levels by the bacterial enzyme *ACC deaminase*. *FEMS Microbiology Letters*, *251*(1), 1–7.

Global Knowledge Initiative, 2014. *Reducing global food waste and spoilage: A rockefeller foundation initiative—Assessing resources needed and available to reduce postharvest food loss in Africa*. Available from https://go.nature.com/33hunKg.

Granada, D., López-lujan, L., Ramírez-restrepo, S., Morales, J., Peláez, C., Andrade, G., & Bedoya-pérez, J. (2019). Bacterial extracts and bioformulates as a promising control of fruit body rot and root rot in avocado cv. *Journal of Integrative Agriculture*, *18*, 2–12.

Haggag, W. M., & Timmusk, S. (2010). Colonization of peanut roots by biofilm-forming *Paenibacillus polymyxa* initiates biocontrol against crown rot disease. *Journal of Applied Microbiology*, *104*(4), 961–969.

Hartmann, A., Rothballer, M., & Schmid, M. (2003). Lorenz Hiltner, a pioneer in rhizosphere microbial ecology and soil bacteriology research. *Plant and Soil*, *312*, 7–14.

Hershkovitz, V., Sela, N., Taha-Salaime, L., Liu, J., Rafael, G., Kessler, C., Droby, S., et al. (2013). De-novo assemble and characterization of the transcriptome of *Metschnikowia fructicola* reveals differences in gene expression following interaction with *Penicillium digitatum* and grapefruit peel. *BMC Genomics*, *14*, 168.

Hodges, R. J., Buzby, J. C., & Bennett, B. (2010). Postharvest losses and waste in developed and less developed countries: Opportunities to improve resource use. *Journal of Agricultural Science*, *149*, 37–45.

ISAAA. (2020). *GM approval database*. Ithaca, NY: ISAAA. Available from https://www.isaaa.org/resources/publications/pocketk/12/default.asp.

Janisiewicz, W. J., & Jurick, W. M., II (2017). Sustainable approaches to control postharvest diseases of apples. In K. Evans (Ed.), *Achieving sustainable cultivation of Apples* (pp. 307–336). Cambridge: Burleigh Dodds Science Publishing.

Janisiewicz, W. J., & Korsten, L. (2002). Biological control of postharvest diseases of fruits. *Annual Review of Phytopathology*, *40*, 411e441.

Janisiewicz, W. J., & Peterson, D. L. (2004). Susceptibility of the stem pull area of mechanically harvested apples to blue mold decay and its control with a biocontrol agent. *Plant Disease*, *88*, 662e664.

Janisiewicz, W. J., Tworkoski, T. J., & Kurtzman, C. P. (2001). Biocontrol potential of *Metschnikowia pulcherrima* strains against blue mold of apple. *Phytopathology*, *91*, 1098–1108.

Jiang, F., Chen, J., Miao, Y., Krupinska, K. E., & Zheng, X. (2009). Identification of differentially expressed genes from cherry tomato fruit (*Lycopersicon esculentum*) after application of the biological control yeast *Cryptococcus laurentii*. *Postharvest Biology and Technology*, *53*(3), 131–137.

Kader, A. A. (2002). *Post-harvest technology of horticultural crops* (Vol. 3311, p. 535). Oakland, CA: University of California, Division of Agriculture and Natural Resources Publication.

Karabulut, O. A., Tezcan, H., Daus, A., Cohen, L., Wiess, B., & Dorby, S. (2004). Control of preharvest and postharvest fruit rot in Strawberry by *Metschnikowia fructicola*. *Biocontrol Science and Technology*, *14*(5), 513–521.

Kasfi, K., Taheri, P., Jafarpour, B., & Tarighi, S. (2018). Identification of epiphytic yeasts and bacteria with potential for biocontrol of grey mold disease on table grapes caused by *Botrytis cinerea*. *Spanish Journal of Agricultural Research*, *16*(1), e1002, 1–16.

Kavamura, V. N., Robinson, R. J., Hughes, D., Clark, I., Rossmann, M., de Melo, I., S., Mauchline, T., M., et al. (2020). Wheat dwarfing influences selection of the rhizosphere microbiome. *Scientific Reports*, *10*, 1452.

Ke, J., Wang, B., & Yoshikuni, Y. (2020). Microbiome engineering: Synthetic biology of plant-associated microbiomes in sustainable agriculture. *Trends in Biotechnology*, *1970*, 1–18.

Kiaya, V. (2014). *Post harvest losses and strategies to reduce them. Technical paper on Post-Harvest Losses* (pp. 1–25). Action Contre la Faim (ACF).

Kitinoja, L., & Gorny, J. R. (1999). Hort. Series No. 21. *Postharvest technology for small-scale produce marketers: Economic opportunities, quality and food safety*. Davis, USA: Department of Pomology, University of California.

Koffi, L. B., Alloue-Boraud, W. A. M., Dadie, A. T., Koua, S. H., & Ongena, M. (2017). Enhancement of mango fruit preservation by using antimicrobial properties of Bacillussubtilis GA1. *Cogent Food & Agriculture*, *3*(1), 1394249.

Korsten, L., Towsen, E., & Classens, V. (1998). Evaluation of Avogreen as post-harvest treatment for controlling anthracnose and stem-end rot on avocado fruit. *South African Avocado Growers' Association Yearbook*, *21*, 83–87.

Lastochkina, O., Seifikalhor, M., Aliniaeifard, S., Baymiev, A., Pusenkova, L., Garipova, S., ... Maksimov, I. (2019). *Bacillus* spp.: Efficient biotic strategy to control postharvest diseases of fruits and vegetables. *Plants (Basel, Switzerland)*, *8*(4), 97.

Lee, S. A., Kim, Y., Kim, J. M., Chu, B., Joa, J. H., Sang, M. K., ... Weon, H. Y. (2019). A preliminary examination of bacterial, archaeal, and fungal communities inhabiting different rhizocompartments of tomato plants under real-world environments. *Scientific Reports*, *9*, 9300.

Liu, P., Cheng, Y., Yang, M., Liu, Y., Chen, K., Long, C., Deng, X., et al. (2014b). Mechanisms of action for 2-phenylethanol isolated from *Kloeckera apiculatain* control of *Penicillium* molds of citrus fruits. *BMC Microbiology*, *14*, 242.

Liu, P., Shi, Y. Y., Chen, L., & Long, C. (2014a). Farnesol produced by the biocontrol agent *Candida ernobii* can be used in controlling the postharvest pathogen *Penicillium expansum*. *African Journal of Microbiology Research*, *8*(9), 922–928.

Lysoe, E., Dees, M. E., & Brurberg, M. B. (2017). A three way transcriptome interaction study of a biocontrol agent (*Clonostachys rosea*), a fungal pathogen (*Helminthosporium solani*) and a potato host (*Solanum tuberosum*). *Molecular Plant-Microbe interactions*, *30*(8), 646–655.

Madhupani, Y. D. S., & Adikaram, N. K. B. (2017). Delayed incidence of stem-end rot and enhanced defenses in *Aureobasidium pullulans*-treated avocado (*Persea americana* Mill.) fruit. *Journal of Plant Diseases and Protection*, *124*, 227–234.

Mamphogoro, T. P., Maboko, M. M., Babalola, O. O., & Aiyegoro, O. (2020). Bacterial communities associated with the surface of fresh sweet pepper (*Capsicum annuum*) and their potential as biocontrol. *Scientific Reports*, *10*, 8560.

Martínez-Hidalgo, P., García, J. M., & Pozo, M. J. (2015). Induced systemic resistance against *Botrytis cinerea* by Micromonospora strains isolated from root nodules. *Frontiers in Microbiology*, *6*, 922.

Mendes, R., Garbeva, P., & Raaijmakers, J. M. (2013). The rhizosphere microbiome: Significance of plant beneficial, plant pathogenic, and human pathogenic microorganisms. *FEMS Microbiology Reviews*, *37*, 634–663.

Mitter, B., Petric, A., Shin, M. W., Chain, P. S. G., Hauberg-Lotte, L., Reinhold-Hurek, B., Sessitsch, A., et al. (2013). Comparative genome analysis of *Burkholderia phytofirmans* PsJN reveals a wide spectrum of endophytic lifestyles based on interaction strategies with host plants. *Frontiers in Plant Science*, *4*, 120.

Mitter, B., Pfaffenbichler, N., Flavell, R., Compant, S., Antonielli, L., Petric, A., ... Sessitsch, A. (2017). A new approach to modify plant microbiomes and traits by introducing beneficial bacteria at flowering into progeny seeds. *Frontiers in Microbiology*, *8*, 11.

Naranjo-Arcos, M. A., & Bauer, P. (2016). Iron nutrition, oxidative stress, and pathogen defense. In P. Erkekoglu, & B. Kocer-Gumusel (Eds.), *Nutritional deficiency* (pp. 63–98). Available from http://doi.org/10.5772/63204.

Nelson, E. B. (2018). The seed microbiome: Origins, interactions, and impacts. *Plant and Soil*, *422*, 7–34.

Odelade, K. A., & Babalola, O. O. (2019). Bacteria, fungi and archaea domains in rhizospheric soil and their effects in enhancing agricultural productivity. *International Journal of Environmental Research and Public Health*, *16*(20), 3873.

Ostrowski, A., Meheter, A., Prescitt, A., Kiley, T. B., & Stanley-Wall, N. R. (2011). YuaB functions synergistically with the exopolysaccharide and Tas A amyloid fibers to allow biofilm formation by *Bacillus subtilis*. *Journal of Bacteriology*, *193*, 4821–4831.

Panebianco, S., Vitale, A., Polizzi, G., Scala, F., & Cirvilleri, G. (2015). Enhanced control of postharvest citrus fruit decay by means of the combined use of compatible biocontrol agents. *Biological Control*, *84*, 19–27.

Parfitt, J., Barthel, M., & Macnaughton, S. (2010). Food waste within food supply chains: Quantification and potential for change to 2050. *Philosophical Transactions of the Royal Society of London. Series B, Biological Sciences*, *365*, 3065–3081.

Payne, S. M. (1993). Iron acquisition in microbial pathogenesis. *Trends in Microbiology*, *1* (2), 66–69.

Perry, G. (2016). Ethylene induces endophyte bacteria to control early and late stage development in several plant species. *Peer J Preprints*, 1–9. Available from https://doi.org/10.7287/peerj.preprints.2611v1.

Perry, G., & Williams, D. (2014). Ethylene induces soil microbes to delay fruit ripening. *PeerJ Preprints*, *2*, e506v50.

Pierce, G. E., Tucker, T. A., Wang, C., Swensen, K., Sidney, A., & Crow, S. A. (2014). Delayed ripening of climacteric fruit by catalysts prepared from induced cells of *Rhodococcus rhodochrous* DAP 96253: A case for the biological modulation of yang-cycle driven processes by a prokaryote. *Industrial Biotechnology*, *10*(5), 354–362.

Ploetz, R. C. (2005). Panama disease, an old nemesis rears its ugly head: Part 1, The beginnings of the banana export trades. *Online. Plant Health Progress*. Available from https://doi.org/10.1094/PHP-2005-1221-01-RV.

Poromarto, H., Nelson, B. D., & Freeman, T. P. (1998). Association of binucleate *Rhizoctonia* with soybean and mechanism of biocontrol of *Rhizoctonia solani*. *Phytopathology*, *88*(10), 1056–1067.

Raza, W., Ling, N., Yang, L., Huang, Q., & Shen, Q. (2016). Response of tomato wilt pathogen *Ralstonia solanacearum* to the volatile organic compounds produced by a biocontrol strain *Bacillus amyloliquefaciens* SQR-9. *Scientific Reports*, *6*, 24856.

Rodriguez, R. J., Henson, J., Volkenburgh, E. V., Hoy, M., Wright, L., Beckwith, F., ... Redman, R. S. (2008). Stress tolerance in plants via habitat-adapted symbiosis. *The ISME Journal*, *2*, 404–416.

Rybakova, D., Mancinelli, R., Wikstr€om, M., Birch-Jensen, A., S., Postma, J., Ehlers, R., U., Berg, G., et al. (2017). The structure of the *Brassica napus* seed microbiome is cultivar-dependent and affects the interactions of symbionts and pathogens. *Microbiome*, *5*, 104.

Saminathan, T., García, M., Ghimire, B., Lopez, C., Bodunrin, A., Nimmakayala, P., ... Reddy, U. K. (2018). Metagenomic and metatranscriptomic analyses of diverse watermelon cultivars reveal the role of fruit associated microbiome in carbohydrate metabolism and ripening of mature fruits. *Frontiers in Plant Science*, *9*, 4.

Saravanakumar, D., Ciavorella, A., Spadaro, D., Garibaldi, A., & Gullino, M. L. (2008). *Metschnikowia pulcherrimastrain* MACH1 outcompetes *Botrytis cinerea*, *Alternaria alternata* and *Penicillium expansum* in apples through iron depletion. *Postharvest Biology and Technology*, *49*, 121e128.

Sawicka, B. (2019). Post-harvest losses of agricultural produce. In W. Leal Filho, et al. (Eds.), *Zero Hunger*. Springer Nature Switzerland AG. Available from https://doi.org/10.1007/978-3-319-69626-3_40-1.

Shankar, N., & Shivakumar, S. (2013). Potentiality of Bacillus subtilis as biocontrol agent for management of anthracnose disease of chilli caused by *Colletotrichum gloeosporioides* OGC1. *3 Biotech*, *4*(2), 1–10.

Sharma, R. R., Singh, D., & Singh, R. (2009). Biological control of postharvest diseases of fruits and vegetables by microbial antagonists: A review. *Biological Control*, *50*, 205–221.

Silby, M. W., & Levy, S. B. (2004). Use of *in vivo* expression technology to identify genes important in growth and survival of *Pseudomonas fluorescens* Pf0–1 in soil: Discovery of expressed sequences with novel genetic organization. *Journal of Bacteriology*, *186*, 7411–7419.

Slininger, P. J., Burkhead, K. D., & Schisler, D. A. (2004). Antifungal and sprout regulatory bioactivities of pheny-lacetic acid, indole-3-acetic acid, and tyrosol isolated from the potato dry rot suppressive bacterium *Enterobacter cloacae* S11: T:07. *Journal of Industrial Microbiology & Biotechnology*, *31*, 517–524.

Slininger, P. J., Schisler, D. A., Ericsson, L. D., Brandt, T. L., Frazier, M. J., Woodell, L. K., ... Kleinkopf, G. E. (2007). Biological control of post-harvest late blight of potatoes. *Biocontrol Science and Technology*, *17*(6), 47–663.

Sonnewald, S., & Sonnewald, U. (2014). Regulation of potato tuber sprouting. *Planta*, *239*, 27–38.

Spadaro, D., & Droby, S. (2016). Development of biocontrol products for postharvest diseases of fruit: The importance of elucidating the mechanisms of action of yeast antagonists. *Trends in Food Science and Technology*, *47*, 39–49.

Stathers, T., Holcroft, D., Kitinoja, L., Mvumi, B. M., English, A., Omotilewa, O., ... Torero, A. J. (2020). A scoping review of interventions for crop postharvest loss reduction in sub-Saharan Africa and South Asia. *Nature Sustainability*, *3*, 821–835.

Subramanian, K. S., Tenshia, V., Jayalakshmi, K., & Ramachandran, V. (2009). Role of arbuscular mycorrhizal fungus (*Glomus intraradices*)—(fungus aided) in zinc nutrition of maize. *Journal of Agricultural Biotechnology and Sustainable Development*, *1*, 029–038.

Talibi, I., Boubaker, H., Boudyach, E. H., & Ait Ben Aoumar, A. (2014). Alternative methods for the control of postharvest citrus diseases. *Journal of Applied Microbiology*, *117*(1), 1–17.

Thimmaraju, R., Biedrzycki, M. L., & Bais, H. P. (2008). Causes and consequences of plant-associated biofilms. *FEMS Microbiology Ecology*, *64*(2), 153–166.

Truyens, S., Weyens, N., Cuypers, A., & Vangronsveld, J. (2014). Bacterial seed endophytes: Genera, vertical transmission and interaction with plants. *Environmental Microbiology Reports*, *7*, 40–50.

Usall, T., Torres, R., & Teixidó, N. (2016). Biological control of postharvest diseases on fruit a suitable alternative? *Current Opinion In Food Science*, *11*, 51–55.

Venkatachalam, S., Ranjan, K., Prasanna, R., Ramakrishnan, B., Thapa, S., & Kanchan, A. (2016). Diversity and functional traits of culturable microbiome members, including cyanobacteria in the rice phyllosphere. *Plant Biology (Stuttgart)*, *18*(4), 627–637.

Wallace, R., Hirkala, D. L., & Nelson, L. M. (2017). Postharvest biological control of blue mold of apple by *Pseudomonas fluorescens* during commercial storage and potential modes of action. *Postharvest Biology and Technology*, *133*, 1–11.

Wallace, R. L., Hirkala, D. L., & Nelson, L. M. (2018). Mechanisms of action of three isolates of *Pseudomonas fluorescens* active against postharvest grey mold decay of apple during commercial storage. *Biological Control*, *117*, 13–20.

Wang, Z., Jiang, M., Chen, K., Wang, K., Du, M., Zalan, Z., ... Kan, J. (2018). Biocontrol of *Penicillium digitatum* on postharvest citrus fruits by *Pseudomonas fluorescens*. *Journal of Food Quality*, *2910481*, 1–10.

Weiss, A., Hertel, C., Grothe, S., Ha, D., & Hammes, W. P. (2007). Characterization of the microbiota of sprouts and their potential for application as protective cultures. *Systematic and Applied Microbiology*, *30*(6), 483–493.

Weiss, A., Mogel, G., Kunz, S. (2006). Development of "Boni-Protect"—a yeast preparation for use in the control of postharvest diseases of apples. In: Boos, Markus (Ed.) ecofruit—*12th International conference on cultivation technique and phytopathological problems in organic fruit-growing: Proceedings to the conference from 31st January to 2nd February 2006 at Weinsberg/Germany* (pp. 113–117), Fördergemeinschaft Ökologischer Obstbau e.V. (FÖKO), Weinsberg, Germany.

Wenneker, M., & Thomma, B. P. H. J. (2020). Latent postharvest pathogens of pome fruit and their management: From single measures to a systems intervention approach. *European Journal of Plant Pathology*, *156*(1), 663–681.

Whipps, J. M. (2001). Microbial interactions and biocontrol in the rhizosphere. *Journal of Experimental Botany*, *52*(1), 487–511.

Willersinn, C., Mack, G., Mouron, P., Keiser, A., & Siegrist, M. (2015). Quantity and quality of food losses along the Swiss potato supply chain: Stepwise investigation and the influence of quality standards on losses. *Waste Management*, *46*, 120–132.

Wisniewski, M., Wilson, C., Droby, S., Chalutz, E., El-Ghaouth, A., & Stevens, C. (2007). *Postharvest biocontrol: New concepts and applications. Biological control: A global perspective*. CAB International. Available from http://doi.org/10.1079/9781845932657.0262.

World Bank, 2011. *World Bank, NRI & FAO missing food: The case of postharvest grain losses in sub-Saharan Africa report no., 60371-AFR*; http://siteresources.worldbank.org/INTARD/Resources/MissingFoods10_web.pdf.

Yang, Q., Solairaj, D., Apaliya, M. T., Abdelhai, M., Zhu, M., Yan, Y., & Zhang, H. (2020). Protein expression profile and transcriptome characterization of *Penicillium expansum* Induced by *Meyerozyma guilliermondii. Journal of Food Quality, Article, 8056767*, 1–12.

Zhang, D., Spadaro, D., Garibaldi, A., & Gullino, M. L. (2011). Potential biocontrol activity of a strain of *Pichia guilliermondii* against grey mould of apples and its possible modes of action. *Biological Control*, *57*, 193e201.

Zhang, D., Spardo, D., Garibaldi, A., & Gullino, M. L. (2010). Efficacy of the antagonist *Aureobasidium pullulans* PL5 against postharvest pathogens of peach, apple and plum and its modes of action. *Biological Control*, *54*(3), 172–180.

Zhang, H., Wang, L., Dong, Y., Jiang, S., Zhang, H., & Zheng, X. (2008). Control of postharvest pear diseases using *Rhodotorula glutinis* and its effects on postharvest quality parameters. *International Journal of Food Microbiology*, *126*(1–2), 167–171.

Zhang, H., Wang, L., Zheng, X., & Dong, Y. (2007). Effect of yeast antagonist in combination with heat treatment on postharvest blue mold decay and *Rhizopus* decay of peaches. *International Journal of Food Microbiology*, *115*(1), 53–58.

Zhang, S., Gan, Y., & Xu, B. (2016). Application of plant-growth-promoting fungi *Trichoderma longibrachiatum* T6 enhances tolerance of wheat to salt stress through improvement of antioxidative defense system and gene expression. *Frontiers in Plant Science*, *7*, 1405.

CHAPTER

Endophytic microorganisms: utilization as a tool in present and future challenges in agriculture

13

Alisha Gupta[1], Meenakshi Raina[2] and Deepak Kumar[3]
[1]*Department of Genetics, University of Delhi, South Campus, India*
[2]*Government Degree College (Boys), India*
[3]*Department of Botany, Institute of Science, Banaras Hindu University, India*

13.1 Introduction

The world population is growing rapidly and has already touched 6.8 billion in the number directly resulting in an increase in food production by 70% globally (Alexandratos & Bruinsma, 2012). This is mainly achieved by bringing the new technology called conventional agriculture into practice that basically involves the application of chemical pesticides, mineral fertilizers, intensive tillage, and high irrigation to support the surplus population (Fasim & Uziar, 2019). Globally, the crops are majorly damaged due to several diseases (bacterial, fungal, and viral), insects, pests, and abiotic stresses, which are controlled by the regular use of the conventional agricultural practices that have an adverse effect on both the natural environment and the human health. The conventional agricultural practices adversely disturb the soil ecology by using surplus amount of fertilizers, put in high irrigation demands, contaminate the surface and groundwater and thus indirectly or directly affect the human health (Miliute, Buzaite, Baniulis, & Stanys, 2015). With the ever-increasing population, which is expected to reach 9.1 billion by 2050 (Alexandratos & Bruinsma, 2012), there will be an increase food demand. The major challenge for the government in this scenario would be to bring in sustainable method that should satisfy the increase food demand with the increasing population, should be environment- or ecosystem-friendly, should overcome the threats that lead to crop yield loss, and are reliable enough to maintain a long-term ecological balance (Fasim & Uziar, 2019). In this context, microorganisms prove to be of utmost importance to be used as biological control agents or biofertilizers such as endophytes, *Trichoderma* sp., arbuscular

Biocontrol Mechanisms of Endophytic Microorganisms. DOI: https://doi.org/10.1016/B978-0-323-88478-5.00013-4

mycorrhizal fungi, and plant growth-promoting rhizobacteria (PGPR) (Fasim & Uziar, 2019; Miliute, Buzaite, Baniulis, & Stanys, 2015).

A microorganism or microbial inoculants or biofertilizers are the live and latent cell of strains that plays an important role in nutrient cycling, energy flow, and decomposition. They make a close association with the plant and increase the nutrient availability that can be utilized by the plants efficiently. Microbial inoculants are efficient phosphate, potassium, nitrogen solubilizers, and siderophore producers are used for seed application, in solar composting areas where these microbe population increase incessantly and enhance the microbial processes thus boosting the nutrient availability to the plants, hence promoting the plant growth and productivity (Fasim & Uziar, 2019; Vyas, 2018). The knowledge of making and using a microbial inoculants or biofertilizers is passed down from one generation of farmer to the next generation. In making a biofertilizer, several things are considered such as growth profile of the selected microbe, organisms' optimum condition, and inoculum formulation. The crucial steps to make effective biofertilizers are inoculum formulation, application method, and product storage. Generally, there are six key steps involved in making biofertilizers, which are (1) organism selection, (2) isolation, (3) method selection, (4) propagation method selection, (5) prototype testing, and (6) large-scale testing (Fasim & Uziar, 2019). Endophytes are one such beneficial microorganism that shows a healthy and mutualistic relationship with their host plant. They inhabit within the plant tissue where they derive their nutrition from the plant and in return endophytes synthesize valuable compounds, including enzymes, organic acids, plant hormones, hydrogen cyanides, siderophore, and secondary metabolites that help in better growth of plants (Adetunji, Kumar, Raina, Arogundade, & Sarin, 2019; Vyas, 2018). Endophytes are beneficial microorganisms possessed by all plants and inhabits inside the plant tissue without causing any apparent sign of infection or disease. Of all the microbial endophytes, endophytic actinomycetes, bacteria, and fungi have received a great attention due to the numerous beneficial agriculturally important compounds they provide. They secrete innumerable and valuable compounds that provide several benefits to the plant such as suppression of the growth of weed or competitor plant species, reduction of oxidative stress of hosts, increase of the plant growth and development, providing protection from herbivore attack, diseases, insects, and pathogen attack, obtaining nutrients in the soil and transfers nutrients to the plant as in rhizophagy cycle and other nutrient transfer symbioses. They provide such effective functions important to agriculture, and thus they prove to be a better alternative to the agricultural conventional practices thus significantly reducing the use of agrochemicals such as fungicides, fertilizers, insecticides, and herbicides. Endophytes are mostly seed transmitted but can also be recruited from the soil providing the same benefits (White et al., 2019).

In this chapter, we will discuss the huge benefits of the endophytic microorganism that how they can be a good alternative to the chemical fertilizers that we use and help in combating the problem of huge crop loss, environmental damage, and human health deterioration and can be the sustainable approach for the global challenge of food security to the increasing population.

13.2 Biodiversity and distribution of endophytic microorganism

Microbes are the most diverse amongst all the life on the earth as compared to other organisms such as plants, vertebrates, and insects. They have adapted and can survive in any conditions even in the extreme ones and can feed on anything such as metals, acids, natural gases, petrol all of which are toxic to human beings and animals. Different groups of microbes associate to the different host plants in either as epiphytic, endophytic, or rhizospheric form (Rana et al., 2020). Endophytes are reported from different types of microbial groups such as bacteria, archaea, and fungi. Endophytic archaea and fungi form the major diverse endophytic microbes where the endophytic archaea belong to phylum Euryarchaeota and endophytic fungi belong to the phylum Ascomycota, Basidiomycota, and Mucoromycota. Out of all the phylum, Ascomycota was reported to be the most dominant (Rana et al., 2020; Suman, Yadav, & Verma, 2016). Endophytic bacteria belonging to 16 phyla of both culturable and unculturable bacteria have been reported (Arora & Ramawat, 2017) out of which the majorly of them belong to the phylum *Actinobacteria*, *Proteobacteria*, *Bacteroidetes*, *Firmicutes*, and *Deinococcus-Thermus*. Proteobacteria is the most dominant phylum and is further grouped as α-, β-, and γ-Proteobacteria. The least number of endophytic bacteria belonged to phylum *Deinococcus-Thermus* and Acidobacteria followed by Bacteroidetes (Rana et al., 2020; Suman et al., 2016). Endophytes are present almost on every plant whether cultivated or wild growing, herbaceous such as sorghum, rice, corn, and wheat; tree crops such as pear, spruce, and oak; leguminous plants such as common pea, bean, mungbean; and nonleguminous plants such as grape, Cannabis, rice, potato (Vasileva et al., 2019).

13.3 Plants and associated endophytes

Microbial endophytes include mainly actinomycetes, bacteria, and fungi and are found in wide variety of plant tissues such as fruits, seeds, pollen, leaves, buds, tubers, stems, roots, and flower tissues. The relation between the endophytes and their host plant ranges from latent phytopathogenesis to mutualism and they can be either facultative where they can grow and survive even outside the host plant or obligate where the endophytes depend completely on their hosts for growth and survival. Their colonization depends on the age, tissue type, and host genotype and in some plants, their colonization might be maximum in the root tissues (Vardharajula, SkZ, Shiva Krishna Prasad Vurukonda, & Shrivastava, 2017; Vyas, 2018). Endophytic microorganism originates from the rhizosphere or phyllosphere and enter into the plants either through the natural openings or wounding or even by the use of several enzymes such as cellulose, pectinase that dissolves the plant cell wall open the gates for endophytes to penetrate through roots and migrate and colonize into other tissues (Rana et al., 2020). They live in the intercellular spaces of plants, feed on the apoplastic nutrients as nonpathogens and promote the plant growth through both direct and indirect mechanisms (Yadav & Yadav, 2017). Direct mechanism benefits involve acting as iron chelators, nitrogen-fixing ability, antimicrobial metabolite production, and insecticidal by-products, induces systematic tolerance through the production of 1-aminocyclopropane-1-carboxylase deaminase, production of plant hormones and siderophore which influences the plant growth, secondary metabolite production providing plant defense, whereas indirect mechanism benefits include the mechanism of induced systemic resistance (ISR) that helps the plant in coping with unfavorable environment and biotic stresses like cold, drought, and hypersaline conditions or pathogenesis (Yadav & Yadav, 2017) (Fig. 13.1).

13.3.1 Endophytic bacteria

Bacterial endophytes are the class of endosymbiotic microorganism that is considered the subset of rhizospheric bacteria, commonly called PGPR. Endophytic

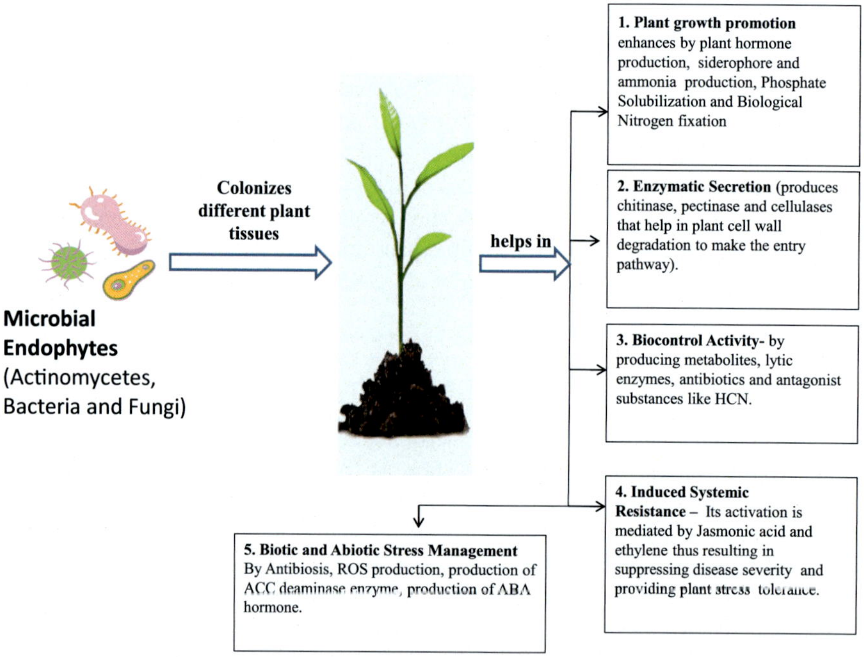

FIGURE 13.1

Mechanism and benefits endophytes provide to the plant.

bacterial growth has an added advantage over rhizospheric growth in respect to providing better protection to plant in stress condition (Afzal, Shinwari, Sikandar, & Shahzad, 2019; Miliute et al., 2015). The bacterial endophytes colonize both intercellular and intracellular spaces inside the all-plant compartments without causing any damage or disease or morphological change to the plants. The plant associates with a vast range of endophytes belonging to different phylum. The colonization process depends on several factors such as plant tissue type, plant genotype, the microbial taxon and strain type, and the abiotic and biotic environmental conditions. Endophytes colonize inside the host plants apoplast, intercellular and intracellular cell wall spaces, and xylem vessels of plant roots, leaves, and stems. Roots are the main entry point of the potential endophytes from the soil to the plant host (Miliute et al., 2015). The roots secrete root exudates such as amino acids, organic acids, and other components that provide nutrition to the bacteria thus determining the bacterial density and diversity of the colonizing bacteria. The bacteria then colonize on to the aerial plant parts and their final sink is the leaf tissue.

The bacterial cells first colonize the rhizosphere and the rhizoplane. Their colonization in the rhizosphere is a highly competitive process since they fight for the nutrients and space. Once the bacteria have colonized the rhizosphere, they then attach to the root surface-mediated by bacterial adhesions, pili, and polysaccharides forming a string of cells. From the root surface, they then colonize the whole root surface and some of the rhizodermal cells leading to the establishment of microfilms or biofilms. The endophytic bacteria reach the root entry site like lateral root emergence and wounds using type IV pili-mediated twitching mobility and thus penetrate inside the roots. The root colonization process depends on several factors such as root exudation patterns, bacterial attachment, and motility, bacterial quorum sensing, and bacterial growth rate.

The process of penetration can be active or passive. The active process is achieved by a dedicated machinery of attachment and proliferation that involves the use of pili, flagella, quorum sensing, lipopolysaccharides, and twitching mobility, whereas the passive process includes direct penetration at the cracks present at the root emergence area, root trips or wounds created by deleterious organisms. The bacteria then spread to the aerial parts of the plant systemically by degrading the plant cell wall with the help of cell wall degrading enzymes cellulase and pectinases secreted by them. In the xylem element, they move through perforated plates without requiring the use of cell wall degrading enzymes. The final sink of these bacteria is the leaf tissue where they can enter through the root pathway or from phyllosphere via leaf stomata (Afzal et al., 2019; Liu et al., 2017). Bacterial colonization can be categorized into obligate and facultative depending on whether they require plant tissues to live in and reproduce. Facultative bacteria widely live in soil and derive the process of colonization and infection when the conditions are suitable. They also live within the cortex but some can also enter the central phloem and xylem whereas the obligate endophytic bacteria totally originate from the seed and cannot survive in the soil (Liu et al., 2017).

Bacterial endophytes show beneficial effects to the plants such as biological nitrogen fixation, controlling phytopathogens, production of phytohormones, and enhancement of mineral uptake, inhibition of ethylene, and solubilization of phosphate (Ibáñez, Tonelli, Muñoz, Figueredo, & Fabra, 2017; Singh, Kumar, Singh, & Pandey, 2017). In addition to promoting plant growth, the bacterial endophytes also produce allelopathic effects against other competing plant species and help host tolerate stress conditions thus enabling host to survive better (Afzal et al., 2019). Gardner first observed the distribution of the endophytic bacteria in the roots of the rough lemon rootstock of the Florida citrus tree. The predominance of the endophytic phyla varies with the host plant species. Among the 13 bacterial genera found such as *Microbacterium, Micrococcus, Bacillus, Pseudomonas, Stenotrophomonas*, and *Burkholderia*, from which the bacteria most commonly isolate two genera, that is, *Pseudomonas* (40%) and *Bacillus*, which are regarded as the predominant genera and others are regarded as the rare one (Afzal et al., 2019; Lodewyckx et al., 2002). Some of the endophytic bacteria associated with the agriculturally important crop plants and found to be beneficial to the respective crop plant are mentioned in Table 13.1.

13.3.2 Endophytic fungi

Endophytic fungi are micromycetes that infect the living plant tissue internally without causing any disease. They form a mutualistic association with the plant's roots extending into rhizosphere for at least a part of their life cycle (Sharma et al., 2021). Endophytic fungi are majorly grouped into two major classes—Class 1 or clavicipitaceous endophytes (CEs) for example, *Epichloë/neotyphodium* and Class 2 or nonclavicipitaceous endophytes (NECs), for example,

Table 13.1 Some of endophytic bacteria associated with agriculturally important crop plants.

Plant type	Associated endophytic bacteria	Plant part	Reference
Alfalfa	*Bacillus, Erwinia, Microbacterium, Pseudomonas, Salmonella*	Root	Hallmann, Quadt-Hallmann, Mahaffee, and Kloepper (1997)
Carrot	*Agrobacterium, Staphylococcus, Pseudomonas, Klebsiella*	Crown tissues	Surette, Sturz, Lada, and Nowak (2003)
Cotton	*Clavibacter, Erwinia, Phyllobacterium*	Stem	Hallmann et al. (1997)
Maize	*Achromobacter, Arthrobacter, Enterobacter, Micrococcus, Rhizobium*	Kernels	Fisher, Petrini, and Scott (1992)
Tomato	*Brevibacillus, Escherichia, Pseudomonas, Salmonella*	Fruit	Hallmann et al. (1997)

Aspergillus niger, Aspergillus terreus, Aspergillus ochraceous, and *Trichoderma viride* (Sharma et al., 2021; Andrade-Linares & Franken, 2013). Class 1 or the CEs belong to the Clavicipitaceae family (phylum Ascomycota) that only infects the cool- and warm-grasses (Poaceae) and forms the mutualistic or symbiotic association with the host plant. The fungal hyphae colonize systemically all plant tissues but mainly shoot and rhizome and grow in intracellular spaces in the plant tissues. They are found in the seed coat and also in close association with the embryo and thus transmitted vertically via seeds (Sharma et al., 2021; Andrade-Linares & Franken, 2013; Stone, Polishook, & White, 2004). In contrast, Class 2 or NECs are inducible mutualists, phylogenetically diverse, belonging to different orders among the Ascomycota phylum infecting a wide range of plants and showing the horizontal mode of transmission (Sharma et al., 2021; Andrade-Linares & Franken, 2013). On the basis of host colonization patterns, transmission mechanism in the host generation, ecological function, and plant biodiversity levels, they are divided into three classes—Class 2, Class 3, and Class 4. Class 2 endophytes colonize extensively both above- and below-ground tissues, whereas Classes 3 and 4 both colonize, restrictively, above-ground tissues and roots. The mode of colonization in Class 3 is highly localized, whereas it is extensive in case of Class 4 (Rodriguez, White, Arnold, & Redman, 2009).

Endophytic fungi show beneficial effects to the plants by enhancing and improving the growth, yield, and plant fitness as they provide tolerance to plant in abiotic and biotic stress conditions. They further help in biodegradation of host plant litter, play major role in nitrogen and carbon cycling, release various secondary metabolites, and provide host defense against pathogenic microorganisms (Vyas, 2018). Some of the endophytic fungi associated with the plants that show benefits are mentioned in Table 13.2.

Table 13.2 Some of the endophytic fungi associated with the plants.

Plant type	Associated endophytic fungi	Plant part	References
Capsicum annuum	*Alternaria alternata*	Fruits	Sridhar (2019)
Ginkgo biloba	*Fusarium oxysporum*	Leaves	Sridhar (2019)
Allium sativum	*Trichoderma brevicompactum*	Bulbs	Rajamanikyam, Vadlapudi, and Upadhyayula (2017)
Musa acuminata	*Fusarium proliferatum, Trichoderma atroviride*	Leaf, corm, pseudostem, and fruit	Rajamanikyam et al. (2017)
Oryza meyeriana	*Xylaria striata*	Roots	Ortega, Torres-Mendoza, and Cubilla-Rios (2020)
Schima superba	*Botryosphaeria* sp.	Roots	Ortega et al. (2020)

13.3.3 Endophytic actinomycetes

Actinomycetes are Gram-positive, filamentous bacteria that forms a branching network of filaments and produce spores. They form one of the largest taxonomic groups of the known 18 lineages within the bacterial domain. They belong to the phylum Actinobacteria with 6 classes, 5 subclasses, 25 orders, 14 suborders, 52 families, and 232 genera and are widely distributed in the terrestrial and aquatic ecosystem (Vijayabharathi, Sathya, & Gopalakrishnan, 2016; Vyas, 2018). Endophytic actinomycetes harbor in the internal tissue of the plant without damaging or causing disease to the host plant. They have a diverse host range and harbor inside many plants such as barley, oats, rice, cowpea, tomato, and medicinal plants. Among the endophytic actinomycetes, Streptomyces sp. is the predominant, whereas Nocardia, *Micromonospora, Microbispora, Nocardioides*, and *Streptosporangium* are the common genera. Endophytic actinomycetes colonize mainly in the roots of the host (Vijayabharathi et al., 2016). Endophytic actinomycetes are promising producers of novel antibiotics in large amount; promote plant growth mainly showing the property of antagonism against the fungal pathogens; help in decomposing complex materials such as dead animals, fungi, algae; increase the fertility of soil by recycling nutrients, produces ammonia, siderophore, and auxin; and even show phosphate solubilization. They are the largest groups that produce lead compounds that help in developing new medicines and agrochemicals (Vyas, 2018). Some of the endophytic actinomycetes associated with the plants that show benefits are mentioned in Table 13.3.

13.4 Endophytes in sustainable agriculture

Agriculture is the major economic activity in most of the developing countries and is the major source of food for the growing population and to fulfill the food needs. Agriculture sector basically depends on the high yielding crop variety, intensive tillage, high irrigation, chemical fertilizers, and pesticides that cause detrimental effects to the environment as well as the human (Arora & Ramawat, 2017; Jain & Pundir, 2017). Endophytes are considered to be the promising alternative to be used as biofertilizers that enhance the plant growth, provide plant protection from pest and diseases, help in biological nitrogen fixation, siderophore production, antibiotic production, and systematic resistance induction (Jain & Pundir, 2017).

13.4.1 Endophytes as plant growth promoters

The production of phytohormones by endophytes is the most studied method. They utilize the nutrients secreted by the plant and in return release active metabolites that enhance the plant growth. Endophytes promote plant growth through the production of plant growth-enhancing substances such as plant hormones

Table 13.3 Some of the endophytic actinomycetes associated with the plants.

Plant type	Associated endophytic actinomycetes	Plant part	References
Triticum aestivum	*Streptomyces, Microbispora*	Roots	Shimizu (2011), Vijayabharathi et al. (2016)
Zea mays	*Streptosporangium, Streptomyces, Microbispora*	Roots and leaves	Vijayabharathi et al. (2016), Shimizu (2011)
Musa acuminata	*Actinomadura, Streptosporangium, Nocardia*	Root, stem, and leaves	Shimizu (2011), Vijayabharathi et al. (2016)
Rhododendron sp.	*Streptomyces*	Root, stem, and leaves	Vijayabharathi et al. (2016), Shimizu (2011)
Cannabis sativa	*Penicillium copticola*	Twigs, leaves, and apical and lateral buds	Shimizu (2011), Vijayabharathi et al. (2016)
Azadirachta indica	*Microbiospora, Nocardia, Streptoverticillium*	Root, stem, and leaves	Shimizu (2011), Vijayabharathi et al. (2016)
Monstera spp.	*Streptomyces*	Tree trunk	Shimizu (2011), Vijayabharathi et al. (2016)

[cytokines, indole acetic acid, gibberellins, ethylene, and abscisic acid (ABA)] that cause structural and morphological changes in the plant (Yadav & Yadav, 2017), siderophore production, phosphorus, potassium solubilization, atmospheric nitrogen fixation, supplying essential vitamins and enzymes and helping the plant in nutrient acquisition (Vardharajula et al., 2017). A study showed that diazotrophic endophytic bacterium—*Burkholderia vietnamiensis* isolated from cottonwood (*Populus trichocarpa*)—supported plant growth by secreting indole acetic acid (IAA). Another study showed that bioactive compounds GA_4, and GA_7 isolated from fungal strain, *Cladosporium sphaerospermum*, discovered in *Glycine max* roots helped in improving the growth of rice and soybean maximally (Fadiji & Babalola, 2020). *Azospirillum, Herbaspirillum, Pseudomonas, Pantoea, Gluconacetobacter* are the endophytes identified to be producing the plant hormone (Jain & Pundir, 2017). The mechanism for the production of plant hormones by endophytes is same as that of the plant growth-promoting rhizobacteria (Yadav & Yadav, 2017). It is reported that most of the endophytes such as *Acetobacter, Azotobacter, Herbaspirillum*, and *Rhizobium* produce more than one type of hormone. In the natural plant hormone IAA, the auxin is commonly produced by the endophytic bacteria that control the PGP processes. Another commonly produced hormone is cytokine involved in root development, cell division, and photosynthesis and chloroplast differentiations, which is available in small quantities in biological samples. The gaseous plant hormone ethylene, also known as the ripening hormone, plays an important role in inducing plant growth and development stimulates germination, root hair formation, breaks dormancy and senescence. All species of rhizobacteria produce this hormone (Prasad, Srinivasan, Chaudhary, Mahawer, & Jat, 2020).

Soil is rich in nutrients but it does lack sufficient quantity of one or more nutrient such as nitrogen, iron and phosphorus compound that is necessary for the plant growth and development. Endophyte plays a major role in fulfilling these nutrient requirements (Afzal et al., 2019). Nitrogen is the most important growth-limiting nutrient and plant absorbs it in the form of nitrate and ammonium ions. Endophytes make an association with the plants and increase the nutrient availability by three means. First type of endophytic microbes includes actinorhizal and rhizobial symbioses that fix atmospheric nitrogen in low oxygen nodules and transfer into root tissues. Another type of nutritional endophytic symbiosis involves dark septate endophytes and mycorrhizal fungi that inhabit both endophytic tissue and extend out into the soil. In these endophytes, hyphae grow endophytically in roots and mycelia extend into soil where it acquires nutrients and mobilize them back to plants. The third mechanism involves acquisition of those nutrients that cycle or extends between the plant and decaying insect showing symbiotic relationship with the plants. The endophytes decay the insect leading to the liberation of the nutrients thus resulting in their transfer back to the plant (Prasad et al., 2020; White, Kingsley et al., 2019).

Phosphorus is present both in organic and inorganic forms in the soil and plays important role in metabolic processes, namely, photosynthesis, macromolecular biosynthesis, respiration, and energy transfer thus promoting plant growth and development (Prasad et al., 2020). Despite having large reservoir and showing great importance, it is minutely available to the plants due to various fixation reactions occurring during phosphorus biogeochemical cycle (Mehta, Sharma, Putatunda, & Walia, 2019). A 95% of the available phosphorus is insoluble, immobilized, and precipitated in the form of mineral salts such as dicalcium and tricalcium phosphates, and hydroxyapatite rock phosphate (Prasad et al., 2020) and only 75% of the externally added phosphorus is absorbed by the plants and this phenomenon is strongly influenced by soil type and pH (Mehta et al., 2019) making it available in reduced amount to the plant and hampering the agricultural productivity. Phosphorus solubilization and mineralization are the important steps to increase the phosphorus productivity performed by rhizosphere colonizing and endophytic bacteria belonging to genera *Arthrobacter, Flavobacterium, Microbacterium, Enterobacter, Rhizobium, Bacillus, Pseudomonas*, etc. These bacteria release different types of organic acids such as carboxylic acid that breaks down the bound form of phosphorus particularly Ca-bonded phosphorus in calcareous soils (Prasad et al., 2020). Potassium helps in osmotic regulation, protein, and starch synthesis; improves resistance to pest and diseases; and plays role in energy relations. Potassium is minimally available to the plants or present in less amount in the soil (>90%) as insoluble silicate minerals in rocks. Endophytes or soil microbe such as bacteria such as *Bacillus mucilaginous*, *Acidithiobacillus ferrooxidans*, and *Bacillus circulans*, fungi; and actinomycetes solubilizes the insoluble potassium minerals (Prasad et al., 2020).

Iron is one of the most important micronutrients for the crop production that plays a major role in several enzymatic activities, photosynthesis, biosynthesis of chlorophyll, nitrogen accumulation, reduction of nitrates and sulfates and is available in low quantity particularly in the calcareous soils of many crops such as soybean, apple trees, peach, and peanut. To countercorrect the low availability of nitrogen, some endophytes produce siderophores that are iron-chelating compounds of low molecular weight and have high iron affinity (Yadav & Yadav, 2017). The mechanism goes as first the solubilization and acquiring of ferric ion takes place followed by the transfer of the iron to the plants and the cohabiting microorganism and also to the depriving pathogens. Based on the chemical mechanism involved in the iron chelation, there are three classes of siderophore, catechol/phenol, hydroxycarboxylic acid, and hydroxamate type (Prasad et al., 2020).

13.4.2 Endophytes as biocontrol agent

Diseases of fungal, bacterial, viral origin, and damage caused by insects and nematodes can be reduced following prior inoculation with endophytes. Endophytes live in the protected environment of the host having the advantage of survival as compared to rhizosphere due to their inability to survive and colonize in nonnative microclimates that are considered to be the effective biocontrol agent. They are mostly Gram-negative bacteria, the entire group of fluorescent pseudomonades and members of the family pseudomonadaceae (Yadav & Yadav, 2017). Endophytes reduce the plant pathogen severity and improve the host health by (1) competing for the nutrients in the same ecological niche—the endophytes rapidly colonized which ensures that the available nutrients are exhausted for the pathogens needed for their growth and survival (Vardharajula et al., 2017; Yadav & Yadav, 2017); (2) by producing antibiotics (microbial toxins) such as *pseudomycins, munumbicins, ecomycins*, and *xiamycins* in the rhizosphere (Yadav & Yadav, 2017) that either controls the growth of the harmful pathogen by either blocking the vascular tissues that in turn hinder the pathogen or by fighting with the pathogens (Vardharajula et al., 2017). The other method that involves the biocontrol mechanism is the secretion of wide variety of metabolites such as monoterpenes like limonene, myrcene, (E,Z)- allo-ocimene, linalool, other compounds such as indole and methyl salicylate that play crucial role in shaping the endospheric microflora (Yadav & Yadav, 2017). Some of the metabolites such as flavonoids and flavones are also signaling molecules that in addition to the metabolites are produced as signals when the microbe adheres to the root surface (phytoalexins). Plants secrete these signal molecules in the endosphere and tackle with the different types of pathogens such as symbionts, neutralist, pathogenic, and associative depending on how the endophytes associate with the plant (Yadav & Yadav, 2017). Endophytes can elicit the ISR in plants mediated by jasmonic acid or ethylene that causes great reduction in the disease severity and enhances the plant stress tolerance. The response is elicited not only against the pathogens but also the herbivores and insects (Vardharajula et al., 2017). ISR after activation induces different genes that are crucial and are involved in the plant immunization. These genes either alter the host physiology or metabolic responses that ultimately lead to increasing cell wall strength thus restricting the pathogens to the outer plant root cortex resulting plant immunization (Prasad et al., 2020). There are several steps involved in signal transduction leading to induced resistance to plants that is well studied in *Bacillus pumilus* SE34 that involves elaboration of structural barriers, toxic substance production such as phenolics and phytoalexins, molecules accumulation such as chitinase and hydrolytic enzyme such as β-1,3-glucanases that help in oligosaccharide releasing which directly stimulates other defense reactions. Endophytes also induce ISR to plants to reduce the impact of virus infection on host. There are several reports that show induction of ISR in tomato plants endophytes, *Bacillus subtilis* IN937b, *B. pumilus* SE34 against cucumber mosaic cucumovirus. In addition, endophytes protect the plant from root pathogen by preparing biofilm around the roots (Yadav & Yadav, 2017).

13.4.3 Endophytes in bioremediation and phytoremediation

Various anthropogenic activities such as rapid use of chemical fertilizers, climatic change, rapid industrialization, and opening of new mining sites have resulted in accumulation of various xenobiotic compounds in natural environment that has possessed serious risk to human health, soil, water, and plant natural environment as well as the natural microbial community (Gupta et al., 2020). In addition, intensive agricultural practices have led to heavy metal pollution, oil spills, polyaromatic hydrocarbons (PAHs), chemical fertilizers that impact the human health and the ecosystem productivity badly (Adetunji et al., 2019; Stępniewska & Kuźniar, 2013). Bioremediation is a technique that uses biological agents such as bacteria, yeast, fungi, plants, and their products (Gupta et al., 2020; Stępniewska & Kuźniar, 2013) to reduce heavy metal stress to plants, degrade xenobiotic compounds, remove greenhouse gases from air, and even control pest growth on the plant (Yadav & Yadav, 2017). The Biodegradation process is associated with microbial growth and metabolism and depends on the mobility, solubility, and bioavailability of contaminants (Stępniewska & Kuźniar, 2013).

Phytoremediation is a type of plant-based bioremediation strategy used to eliminate different type of pollutants such as organic like hydrocarbons, pesticides, or petroleum product or inorganic like heavy metals and metalloids particularly from the environment (Gupta et al., 2020). In phytoremediation process the plant utilizes the contaminant from the soil during water absorption and transports them to different parts of the plant where they are either turned into harmless forms or may be degraded. Here, the endophyte provides plants with the degradation pathway and metabolic abilities to reduce the phytotoxicity and enhance degradation. Thus endophyte helps in phytoremediation by decreasing metal phytotoxicity, enhancing plant growth, and affecting metal translocation and accumulation. Thus a number of wastelands and groundwater can be remediated (Yadav & Yadav, 2017). The phytoremediation technology seems to be environment friendly and cost-effective way of cleaning and even impacts the environment positively. Endophytes are helpful in PAH degradation. They increase their density in the endosphere and rhizosphere and catalysis the atmospheric oxygen into aromatic or aliphatic hydrocarbons producing corresponding alcohols (Gupta et al., 2020; Yadav & Yadav, 2017). It is found that such endophytic microflora that might be naturally present in the plant and grow in the contaminated site, for example, the halophytic plant, *Halonemum Strobilaceum* is found to have hydrocarbon utilizing microflora that helps in reducing oil load on coastal area (Yadav & Yadav, 2017).

13.4.4 Endophytic microorganism against for alleviation of biotic and abiotic stress

Plant faces a variety of biotic and abiotic stresses that show detrimental effects on their growth and productivity. A large number of various living organisms such as nematodes, viruses, fungi, bacteria, insects, arachnids, and weeds cause biotic stress by competing for the nutrients and thus making the host plant starve ultimately resulting in plant death. Abiotic stresses affecting the plants are salinity, heat, drought, cold, and heavy metals. Both the stresses majorly are responsible for reducing plant survival leading to agricultural yield loss and reduced biomass production thus resulting in food shortage and poor nutrient quality (Sharma et al., 2021).

Endophytic microorganism uses various mechanisms to overcome the biotic and abiotic stresses. One of them is that they discharge reactive oxygen species (ROS), like hydrogen peroxide (H_2O_2), superoxide anion (O_2), and hydroxyl radical that lead to the destruction of the cellular membrane of the host due to the pressure generated by these ROS. Destruction of the cellular membrane leads to protein, nucleic acid, and sugar leakage thus destructing their cytotoxicity and inhibiting the growth of these pests and pathogens (Adetunji et al., 2019). Antibiosis is another method to manage the stress; a method to release secondary metabolites like antibiotics and other volatile compounds, enzymes such as chitinase, cellulases, and 1.3-glucanase that hydrolyze the rigid cell wall and cell surface or even interrupt with the pathogen or pest metabolic pathway to check and lower down the pathogenesis of disease (Latha, Karthikeyan, & Rajeswari, 2019; Sharma et al., 2021). For example, antibiotic "surfactin" was found to be effective against the pathogen *Pseudomonas syringae* on *Arabidopsis*. A study shows that a combination of antibiotics fengycin and surfactin was found to suppress plant pathogen in bean and tomato plants (Latha et al., 2019). Microorganism enhances plants tolerance toward different abiotic stresses such as drought, chilling injury, metal toxicity, salinity stress by showing close interaction with their host plant tissue thus escaping competition with the rhizosphere microorganisms and other adverse environmental conditions. Endophytes have the ability to presensitize the plant cell metabolism to make them prepared to react quickly and efficiently on any exposure to the stress (Vardharajula et al., 2017). Endophytes alter the plant photosynthetic activity and carbohydrate metabolism that releases cold stress-related metabolites like proline, starch, and phenolics and thus endophytic microorganism enhances cold tolerance to the plant. Endophytes enhance drought stress to the plant by producing 1-aminocyclopropane-1-carboxylic acid (ACC) deaminase enzyme that cleaves the ACC, the precursor of ethylene thus decreasing the ethylene levels and in turn alleviating the drought stress. Endophytes belonging to genera *Arthrobacter*, *Bacillus*, and *Microbacterium* showed ACC deaminase activity in the plant *Capsicum annum* L thus alleviating the drought stress (Vardharajula et al., 2017; Yadav & Yadav, 2017). Plants accumulate endophyte-mediated glycine betaine compounds thus dealing with the salinity stress. ABA is shown to alleviate water stress in the plants by controlling stomata closure that reduces water loss. The major example of this stress is seen in maize plants where the ABA is secreted by endophytic microorganism *Azospirillum lipoferum* (Yadav & Yadav, 2017).

13.5 Conclusion

To support the agricultural demands of the growing population which is expected to reach to 9.1 billion by 2050, conventional agricultural practices that involve the use of chemical pesticides, mineral fertilizers, intensive tillage, and high irrigation are being continuously practiced worldwide. Though, this satiates the agricultural need for the surplus population growth but it is impacting negatively and adversely to the environment, ecosystem, and to the human health by disturbing the soil ecology, contaminating the surface and groundwater because of extensive use of chemical pesticides, insecticides, mineral fertilizers to fight against various pest and pathogen infections, plant diseases, etc. Endophytic microorganisms prove to be reliable to be used as biofertilizers or biocontrol agents that act as alternative method to conventional agricultural practices being ecosystem-friendly, ecosystem balancer, satisfying the surplus population demand and overcoming the threats that lead to crop yield loss. Microbial endophytes belong to three phyla bacteria, fungi, and actinomycetes found in wide variety of plant tissues such as fruits, seeds, pollen, leaves, buds, tubers, stems, roots, and flower tissues They inhabit the plant tissue internally, colonize them, enhance the plant growth, provide plant protection from pest and diseases, and help in biological nitrogen fixation, siderophore production, antibiotic production, and systematic resistance induction. To bring about a future where there is less dependency on the conventional agricultural practices, the use of endophytic microbes should be brought into practice in the agricultural sector as it is the main source that fulfills the major food demands. Thus, there is a need to develop better understanding, knowledge of how microbe functions in soil and plants and how their functions can be optimized to enhance crop protection and production.

Acknowledgements

DK acknowledges Science and Engineering Research Board (Grant No- EEQ/2016/000487), India for providing financial support to the laboratory and Department of Botany, Center of Advanced Study in Botany, Institute of Science, Banaras Hindu University for providing necessary infrastructure facilities.

References

Adetunji, C. O., Kumar, D., Raina, M., Arogundade, O., & Sarin, N. B. (2019). *Endophytic microorganisms as biological control agents for plant pathogens: A panacea for sustainable agriculture. Plant biotic interactions* (pp. 1–20). Springer.

Afzal, I., Shinwari, Z. K., Sikandar, S., & Shahzad, S. (2019). Plant beneficial endophytic bacteria: Mechanisms, diversity, host range and genetic determinants. *Microbiological Research*, *221*, 36–49.

Alexandratos, N. & J. Bruinsma (2012). "*World agriculture towards 2030/2050: The 2012 revision*."

Andrade-Linares, D. R., & Franken, P. (2013). *Fungal endophytes in plant roots: Taxonomy, colonization patterns, and functions. Symbiotic endophytes* (pp. 311–334). Springer.

Arora, J. and K.G. Ramawat (2017). An introduction to endophytes. Endophytes: Biology and biotechnology, Springer: 1–23.

Fadiji, A. E., & Babalola, O. O. (2020). Exploring the potentialities of beneficial endophytes for improved plant growth. *Saudi Journal of Biological Sciences*, *27*(12), 3622.

Fasim, F., & Uziar, B. (2019). *Applications of microorganisms in agriculture for nutrients availability. Soil microenvironment for bioremediation and polymer production* (pp. 1–16). John Wiley and Sons, USA. Available from https://doi.org/10.1002/9781119592129.

Fisher, P., Petrini, O., & Scott, H. L. (1992). The distribution of some fungal and bacterial endophytes in maize (*Zea mays* L.). *New Phytologist*, *122*(2), 299–305.

Gupta, A., Singh, S. K., Singh, V. K., Singh, M. K., Modi, A., Zhimo, V. Y., ... Kumar, A. (2020). *Endophytic microbe approaches in bioremediation of organic pollutants. Microbial endophytes* (pp. 157–174). Elsevier.

Hallmann, J., Quadt-Hallmann, A., Mahaffee, W., & Kloepper, J. (1997). Bacterial endophytes in agricultural crops. *Canadian Journal of Microbiology*, *43*(10), 895–914.

Ibáñez, F., Tonelli, M. L., Muñoz, V., Figueredo, M. S., & Fabra, A. (2017). *Bacterial endophytes of plants: Diversity, invasion mechanisms and effects on the host. Endophytes: Biology and biotechnology* (pp. 25–40). Springer.

Jain, P., & Pundir, R. K. (2017). *Potential role of endophytes in sustainable agriculture-recent developments and future prospects. Endophytes: Biology and biotechnology* (pp. 145–169). Springer.

Latha, P., Karthikeyan, M., & Rajeswari, E. (2019). *Endophytic bacteria: Prospects and applications for the plant disease management. Plant Health Under Biotic Stress* (pp. 1–50). Springer.

Liu, H., Carvalhais, L. C., Crawford, M., Singh, E., Dennis, P. G., Pieterse, C. M., & Schenk, P. M. (2017). Inner plant values: Diversity, colonization and benefits from endophytic bacteria. *Frontiers in Microbiology*, *8*, 2552.

Lodewyckx, C., Vangronsveld, J., Porteous, F., Moore, E. R., Taghavi, S., Mezgeay, M., & der Lelie, D. v (2002). Endophytic bacteria and their potential applications. *Critical Reviews in Plant Sciences*, *21*(6), 583–606.

Mehta, P., Sharma, R., Putatunda, C., & Walia, A. (2019). *Endophytic fungi: Role in phosphate solubilization. Advances in Endophytic Fungal Research* (pp. 183–209). Springer.

Miliute, I., Buzaite, O., Baniulis, D., & Stanys, V. (2015). Bacterial endophytes in agricultural crops and their role in stress tolerance: A review. *Zemdirbyste-Agriculture*, *102* (4), 465–478.

Ortega, H. E., Torres-Mendoza, D., & Cubilla-Rios, L. (2020). Patents on endophytic fungi for agriculture and bio-and phytoremediation applications. *Microorganisms*, *8*(8), 1237.

Prasad, M., Srinivasan, R., Chaudhary, M., Mahawer, S. K., & Jat, L. K. (2020). *Endophytic bacteria: Role in sustainable agriculture. Microbial endophytes* (pp. 37–60). Elsevier.

Rajamanikyam, M., Vadlapudi, V., & Upadhyayula, S. M. (2017). Endophytic fungi as novel resources of natural therapeutics. *Brazilian Archives of Biology and Technology*, *60*.

Rana, K. L., Kour, D., Kaur, T., Devi, R., Yadav, A. N., Yadav, N., ... Saxena, A. K. (2020). Endophytic microbes: Biodiversity, plant growth-promoting mechanisms and potential applications for agricultural sustainability. *Antonie Van Leeuwenhoek*, *113*, 1075–1107.

Rodriguez, R., White, J., Jr., Arnold, A., & Redman, Rs (2009). Fungal endophytes: Diversity and functional roles. *New Phytologist*, *182*(2), 314–330.

Sharma, A., Singh, A., Raina, M., & Kumar, D. (2021). Plant-Fungal Association: An Ideal Contrivance for Combating Plant Stress Tolerance. In Dr. R. Prasad, S. C. Nayak, R. N. Kharwar, & N. K. Dubey (Eds.), *Mycoremediation and Environmental Sustainability*. Switzerland: Springer Nature. Available from https://doi.org/10.1007/978-3-030-54422-5_13.

Shimizu, M. (2011). *"Endophytic actinomycetes: Biocontrol agents and growth promoters". Bacteria in Agrobiology: Plant Growth Responses* (pp. 201–220). Berlin, Heidelberg: Springer. Available from https://doi.org/10.1007/978-3-642-20332-9_10.

Singh, M., Kumar, A., Singh, R., & Pandey, K. D. (2017). Endophytic bacteria: A new source of bioactive compounds. *3 Biotech*, *7*(5), 1–14.

Sridhar, K. (2019). Diversity, ecology, and significance of fungal endophytes." Endophytes and secondary metabolites. In S. Jha (Ed.), *Reference series in phytochemistry*. Cham: Springer.

Stępniewska, Z., & Kuźniar, A. (2013). Endophytic microorganisms—promising applications in bioremediation of greenhouse gases. *Applied Microbiology and Biotechnology*, *97*(22), 9589–9596.

Stone, J. K., Polishook, J. D., & White, J. F. (2004). *Endophytic fungi. Biodiversity of fungi* (pp. 241–270). Burlington: Elsevier Academic Press.

Suman, A., Yadav, A. N., & Verma, P. (2016). *Endophytic microbes in crops: Diversity and beneficial impact for sustainable agriculture. Microbial inoculants in sustainable agricultural productivity* (pp. 117–143). Springer.

Surette, M. A., Sturz, A. V., Lada, R. R., & Nowak, J. (2003). Bacterial endophytes in processing carrots (*Daucus carota* L. var. sativus): Their localization, population density, biodiversity and their effects on plant growth. *Plant and Soil*, *253*(2), 381–390.

Vardharajula, S., SkZ, A., Shiva Krishna Prasad Vurukonda, S., & Shrivastava, M. (2017). Plant growth promoting endophytes and their interaction with plants to alleviate abiotic stress. *Current Biotechnology*, *6*(3), 252–263.

Vasileva, E. N., Akhtemova, G. A., Zhukov, V. A., & Tikhonovich, I. A. (2019). Endophytic microorganisms in fundamental research and agriculture. *Ecological genetics*, *17*(1), 19–32.

Vijayabharathi, R., Sathya, A., & Gopalakrishnan, S. (2016). *A renaissance in plant growth-promoting and biocontrol agents by endophytes. Microbial inoculants in sustainable agricultural productivity* (pp. 37–60). Springer.

Vyas, P. (2018). *Endophytic microorganisms as bio-inoculants for sustainable agriculture. Microbial bioprospecting for sustainable development* (pp. 41–60). Springer.

White, J. F., Kingsley, K. L., Zhang, Q., Verma, R., Obi, N., Dvinskikh, S., … Kowalski, K. P. (2019). Endophytic microbes and their potential applications in crop management. *Pest Management Science*, *75*(10), 2558–2565.

Yadav, A., & Yadav, K. (2017). Exploring the potential of endophytes in agriculture: A mini review. *Advances in Plants & Agriculture Research*, *6*(4), 00221.

CHAPTER

Microbially synthesized nanoparticles: aspect in plant disease management 14

Joorie Bhattacharya[1,2], Rahul Nitnavare[3,4], Aishwarya Shankhapal[1] and Sougata Ghosh[5]

[1]*Genetic Gains, International Crops Research Institute for the Semi-Arid Tropics, Hyderabad, India*

[2]*Department of Genetics, Osmania University, Hyderabad, India*

[3]*Division of Plant and Crop Sciences, School of Biosciences, University of Nottingham, Nottingham, United Kingdom*

[4]*Department of Plant Sciences, Rothamsted Research, Harpenden, United Kingdom*

[5]*Department of Microbiology, School of Science, RK University, Rajkot, India*

14.1 Introduction

A large part of the global agricultural production is hampered by the growing pest disease infestations as well as those of emerging pests every year. Various management methods for pest control have been employed through both physical and chemical methods. Out of these the utilization of pesticides is the most predominant. As has been established, the use of pesticides poses several human as well as environmental hazards, and thus their usage is not advisable. Researchers, therefore, have been looking for safer, more effective, and cost-intensive alternatives for pest disease management.

Nanotechnology has fast gained pace for its application and efficiency in the field of agriculture. Nanoparticles (NPs) are highly effective against pest-associated diseases, possess low toxicity and also have a broad pesticidal activity range. They are also extremely stable and are easily soluble in aqueous medium. The most commonly used NP in agriculture are silver nanoparticles (AgNPs) and are known to have strong antimicrobial effects and work against bacteria, fungus, and virus likewise (Gupta, Upadhyaya, Singh, Abd-Elsalam, & Prasad, 2018; Ghosh et al., 2016a,b,c; Ranpariya et al., 2021). NPs are usually generated through physical and chemical methods such as electrochemical method, electroreduction, photoreduction as well as oxidation. These methods, however, are not cost-effective and often are toxic for the environment (Bloch, Pardesi, Satriano, & Ghosh, 2021). Hereby, numerous biological methods are developed for synthesis of NPs employing bacteria, fungi, algae, and medicinal plants (Ghosh et al., 2016d; Rokade et al., 2017, 2018). A more environment-friendly, rapid, and

Biocontrol Mechanisms of Endophytic Microorganisms. DOI: https://doi.org/10.1016/B978-0-323-88478-5.00007-9

efficient alternative is the use of microorganisms for the purpose of NP generation (Ghosh, 2018).

Nanoagriculture, as a term and technology, has gained mileage and is essentially the use of NP in agriculture for various purposes such as disease management, enhanced yield, inducing seed germination, plant growth as well as seed vigor. Apart from metal NPs, biopolymer NPs are also being generated, such as chitosan. The synthesis of bionanoparticles, which are NPs derived from microorganisms ranging from bacteria to virus, is dependent on the choice of solvent medium, an environmentally stable and nontoxic reducing agent and finally a safe material for stabilization. Prior to this, research based on NPs production was based on organic solvents which cause severe environmental damage (Ghosh et al., 2015a; Shende et al., 2017, 2018).

Several bacterial species are able to synthesize metal NPs through extracellular and intracellular material production. Some of the most notable bacteria exhibiting this trait are *Escherichia coli, Bacillus* sp., *Pseudomonas aeruginosa, Salmonella typhi, Staphylococcus currens, Brevibacillus formosus*, and *Vibrio cholerae* (Bloch et al., 2021; Ghosh, 2018). These bacterial species have been found to effectively synthesize silver (Ag) and gold (Au) NPs.

Further, the term "myconanotechnology" was coined to define the use of fungus in nanotechnology. As compared to other microbes, synthesis of NPs using fungus is easier due to its ease in isolation, downstream processing, and production of large quantities of extracellular enzymes. *Aspergillus* and *Fusarium* sp. have been found to actively synthesize AgNPs and AuNPs (Ghosh, Shah, & Webster, 2021a).

The use of nanosized particles based on microbes for dealing with plant pathogenicity has been extensively reported. It is noteworthy that, to establish an efficient NP-based system against phytopathogen, requires information about antimicrobial activity of various NP compounds and the development of superior application strategy (Kaur, 2019). The most common approach for the application of NPs in plant–pathogen management is direct application of the nanomaterial to the seed, stem, root, and other such affected parts. The second approach is using nanomaterials as a carrier for pheromones, inhibitors, and other such bioactive components that act against the pathogen (Ghosh, Thongmee, & Kumar, 2021b). While applying the nanomaterial directly into the soil, the effect on nonspecific microorganisms especially those involved in vital pathways like mineral uptake and nitrogen fixation should be taken into consideration. Direct application of NPs is comparable to chemical method of pest management and hence can be used as nanopesticides (Khan & Rizvi, 2014). Metal NPs such as silver, gold, copper, and zinc have been known to resist bacterial and fungal pathogens such as *Bacillus subtilis, E. coli, Staphylococcus aureus, Aspergillus niger*, and *Fusarium oxysporum* amongst the many others (Ghosh et al., 2015b; Jamdade et al., 2019; Joshi, Ghosh, & Dhepe, 2019; Salunke et al., 2014; Shinde et al., 2018). The green synthesized NPs act against the phytopathogen through various mechanisms, namely, protein dysfunction, reactive oxygen species (ROS) generation, membrane function impairment, hindrance in nutrition uptake and genotoxicity. These mechanisms more often work collectively against the phytopathogen. The NPs adhere to the cell membrane of the microbe and cause physiological disruptions ultimately leading to cell death (Ali et al., 2020). The following chapter focuses on the various bacterial and fungal species employed for production of NPs against prominent plant diseases.

14.2 Nanoparticles for plant disease control

The ever-changing climatic condition is a major threat on the food quantity as well as quality. Change in the climatic condition mainly concerns the drought events caused due to increasing temperature. This affects the yield of various crops as well as makes crops more vulnerable to the plant pathogens and other stress conditions. In such a case, nanotechnology can be a potential sustainable solution, as it eliminates the usage of active chemicals such as pesticides that lead to the contamination of the groundwater ecosystems. Moreover, it is proven that nanotechnology provides an array of unique solutions, thereby increasing the solubility, delivery, and long-lasting residual activity for plant health management purpose. Among various sustainable solutions reported, the bacteriogenic and mycogenic NPs are most noteworthy emerging strategy for plant disease control (Alavi & Rai, 2019).

Microbe associated with green synthesis of NPs against the phytopathogenic strains of bacteria and fungi has gained popularity as it has widened the horizon of the NPs design and development. These biocontrol agents can be potentially cheaper and safer option compared to conventional biocidal and pesticidal agents. Microbial synthesis of NPs can be either extracellular or intracellular as seen in Tables 14.1 and 14.2. Usually, the metal salts are added to obtain NPs extracellularly and the synthesis is monitored by the color change of the mixture, whereas intracellularly synthesized NPs are recovered by disruption of cell membrane through sonication or heat shocks (Ali et al., 2020).

14.3 Bacteriogenic nanoparticles

Bacterial leaf blight (BLB) disease is widely renowned in the case of rice which is caused due to the pathogenic bacteria, *Xanthomonas oryzae* pv. oryzae (Xoo). One such study focuses on tackling this disease with the help of AgNPs to eliminate active chemical usage. In this study, *Bacillus cereus* strain SZT1 strain was used to obtain the NPs extracellularly. Fourier transform infrared spectroscopy (FTIR) revealed the existence of proteins and alcoholic functional groups, whereas X-ray diffraction (XRD) was used to obtain crystalline nature of the NPs. Moreover, the results obtained by scanning electron microscopy (SEM) and

Table 14.1 Bacteriogenic nanoparticles for plant disease management.

Sr. no	Name of organism	Nanoparticles	Size (nm)	Shape	Intracellular/ extracellular	Activity	Reference
1	*Aeromonas hydrophila*	ZnO	57–72	Crystalline	Extracellular	Antifungal against *Aspergillus flavus* in Maize	Jayaseelan et al. (2012)
2	*Bacillus cereus*	Ag	18–39	Spherical	Extracellular	Antibacterial against *Xanthomonas oryzae* pv. Oryzae in rice	Ahmed et al. (2020)
3	*Bacillus siamensis*	Ag	25–50	Spherical	Extracellular	Antibacterial against *X. oryzae* pv. oryzae in rice	Ibrahim et al. (2019)
4	*Bacillus* sp.	Ag	7–21	Spherical	Extracellular	Antifungal against *Fusarium oxysporum* in tomato	Gopinath and Velusamy (2013)
5	*Pseudomonas poae*	Ag	20–45	Spherical	Extracellular	Antifungal agent *Fusarium graminearum* in wheat	Ibrahim et al. (2020)
6	*Pseudomonas rhodesiae*	Ag	20–100	Spherical	Extracellular	Antibacterial against *Dickeya dadantii* in sweet potato	Hossain et al. (2019)
7	*Serratia* sp.	Ag	10–20	Spherical	Extracellular	Antifungal against *Bipolaris sorokiniana* in wheat	Mishra et al. (2014)
8	*Stenotrophomonas* sp.	Ag	12	Spherical	Extracellular	Antifungal against *Sclerotium rolfsii* in chickpea	Mishra et al. (2017)
9	*Streptomyces capillispiralis*	Cu	4–59	Spherical	Extracellular	Antifungal against various species in multiple crops	Hassan et al. (2018)
10	*Streptomyces griseus*	Cu	4–59	Spherical	Intracellular	Antifungal against various species in multiple crops	Ponmurugan et al. (2016)

Table 14.2 Mycogenic nanoparticles in plant disease management.

Sr. no	Name of organism	Nanoparticles	Size (nm)	Shape	Intracellular/ Extracellular	Activity	Reference
1	*Aspergillus niger*	Ag	10–100	–	Extracellular	Antimicrobial against *Aspergillus* spp. and *Penicillium digitatum* in various crop plants	Al-Zubaidi et al. (2019)
2	*Guignardia mangiferae*	Ag	5–30	Spherical	Extracellular	Antifungal against *Rhizoctonia solani* in rice	Balakumaran et al. (2015)
3	*Penicillium duclauxii*	Ag	3–32	Spherical	Extracellular	Antifungal against *Bipolaris sorghicola* in sorghum	Almaary et al. (2020)
4	*Setosphaeria rostrata*	Ag	2–50	Spherical	Extracellular	Antifungal agent *A. niger, R. solani, Fusarium graminearum*, and *Fusarium udum* in various crops	Akther and Hemalatha (2019)
5	*Streptomyces capillispiralis* Ca-1, *Streptomyces zaomyceticus* Oc-5, and *Streptomyces pseudogriseolus* Acv-11.	Ag	23.77–63.14, 11.32–36.72 11.70–44.73	Spherical	Extracellular	Antifungal against *Alternaria alternata, Fusarium oxysporum, Pythium ultimum*, and *A. niger* and larvicidal against *Culex pipiens* and *Musca domestica*	Fouda et al. (2020)
6	*Trichoderma harzianum*	Ag	12.7 ± 0.8	Spherical	Extracellular	Antifungal against *Sclerotinia sclerotiorum* in soybean	Guilger et al. (2017)

(*Continued*)

Table 14.2 Mycogenic nanoparticles in plant disease management. *Continued*

Sr. no	Name of organism	Nanoparticles	Size (nm)	Shape	Intracellular/ Extracellular	Activity	Reference
7	*T. harzianum*	Ag	12.7 ± 0.8	Spherical	Extracellular	Antimicrobial against *Helminthosporium* sp., *Alternaria alternate, Phytophthora arenaria* and *Botrytis* sp.	EL-Moslamy et al. (2017)
8	*Trichoderma longibrachiatum*	Ag	5–25	Spherical	Extracellular	Antifugal against *Fusarium verticillioides, Fusarium moniliforme, Penicillium brevicompactum, Helminthosporium oryzae*, and *Pyricularia grisea*	Elamawi et al. (2018)
9	*Trichoderma* sp. and *Cephalosporium* sp.	Ag	20–50	–	Extracellular	Antifugal against *F. oxysporum* f. sp. ciceri (FOC) in chickpea	Kaur et al. (2018)
10	*Xylaria acuta*	ZnO	40–55	Hexagonal	Extracellular	Antimicrobial against *Bacillus cereus, Staphylococcus aureus, and Pseudomonas aeruginosa*	Sumanth et al. (2020)

transmission electron microscopy (TEM) stated that the NPs are spherical and 18–39 nm in size as shown in Fig. 14.1. These characterized NPs were further checked for their antimicrobial activity against *(Xoo)* and gave around (24.21 ± 1.01 mm) zone of inhibition demonstrating its potential to protect paddy from phytopathogens. Application of the synthesized NPs was not lethal to the rice plants and showed increase in the biomass of the plants and reduction in the ROS of the plants. Application of NPs demonstrates a cost-effective as well as environmentally safe strategy in controlling BLB in rice plants (Ahmed et al., 2020).

On similar lines, AgNPs was used against *X. oryzae* pv. oryzae (Xoo) strain LND0005 and *Acidovorax oryzae* (Ao) strain RS-1, respectively. The NPs were isolated from *Bacillus siamensis*, endophytic bacteria from the medicinal plant, *Coriandrum sativum*. Ultraviolet (UV)–visible spectroscopy, SEM, TEM,

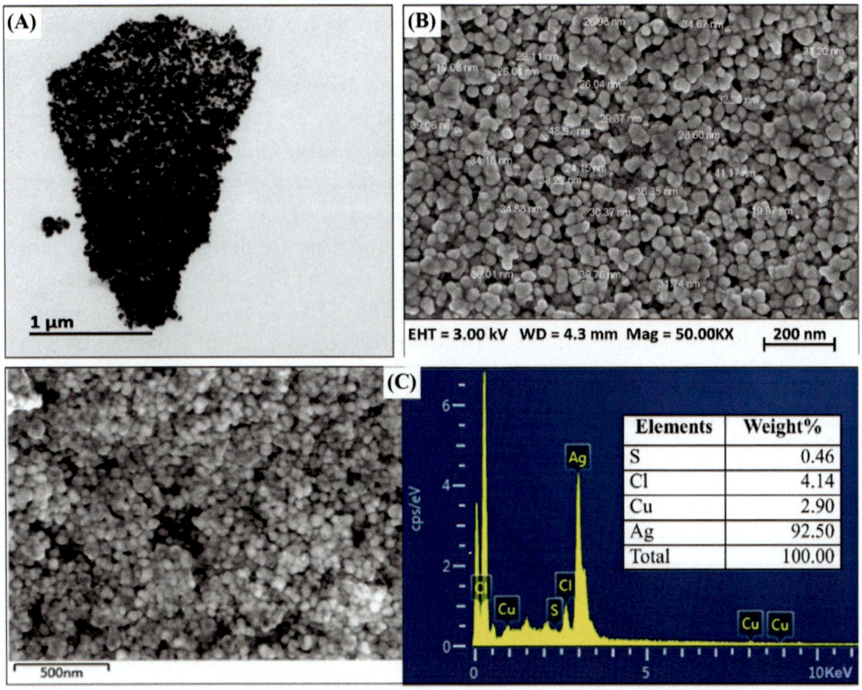

FIGURE 14.1

Characterization of the biogenic AgNPs synthesized from *Bacillus cereus* SZT1. (A) Transmission electron microscopy; (B) scanning electron microscopy; (C) EDS spectrum. *AgNPs*, Silver nanoparticles; *EDS*, energy dispersive spectroscopy.

Reprinted from Ahmed, T., Shahid, M., Noman, M., Niazi, M. B. K., Mahmood, F., Manzoor, I., Zhang, Y., Li, B., Yang, Y., Yan, C., & Chen, J. (2020). Silver nanoparticles synthesized by using Bacillus cereus SZT1 *ameliorated the damage of bacterial leaf blight pathogen in rice.* Pathogens, 9, *160. (Open access).*

infrared (IR) spectroscopy revealed the size of the NPs to be around 25–50 nm. An amount 20 mg/mL was confirmed to be the effective concentration as it showed maximum antibacterial activity against rice BLB and bacterial brown stripe. Overall the NPs showed inhibitory activity against the mentioned pathogenic strains while promoting healthy growth of the rice plant (Ibrahim et al., 2019).

Similarly, AgNPs were also used in controlling Wheat Fusarium head blight pathogen in wheat. The source synthesis of the NPs is *Pseudomonas poae* strain CO and the source of identification of said bacterium is garlic plant. Visible UV spectrometry was used to confirm the synthesis of AgNPs. Characterization of this novel, spherical AgNPs of 20–45 nm was done based on the FTIR, XRD, SEM, TEM, and energy dispersive X-ray spectroscopic (EDX) analyses. The AgNPs showed quite remarkable levels of antifungal activity against pathogenic *Fusarium graminearum* strain PH-1 to tackle Fusarium head blight disease in wheat. Application of the AgNPs attributed toward inhibition of spore germination, germ tube growth, mycotoxin production, and cell membrane disruption as illustrated in Fig. 14.2. Overall, the study unfolds the potential of *P. poae* in protecting the wheat from a fungal infection caused due to *F. graminearum* (Ibrahim et al., 2020).

Outbreaks of soft rot diseases have resulted in major losses of crop plants such as rice, maize, potato, sweet potato. This indicates the failure in management of soft rot *Pectobacterium* and *Dickeya*. Out of these two genus, *Dickeya dadantii* caused major breakout resulting in stem and root rot disease in sweet potato. Green synthesized AgNPs were used to counteract this problem. The *Bacillus amyloliquefaciens* strain A3, *Paenibacillus polymyxa* strain ShX304, and *Pseudomonas rhodesiae* strain G1 were used to synthesize the NPs in the culture-free supernatant. From the preliminary studies performed to gain insights into the antibacterial activity of the green synthesized NPs, *P. rhodesiae* was the only species which was able to show the highest inhibition activity independent of the components present in the cell-free supernatant. Source of isolation of *P. rhodesiae* was rhizospheric soil of cotton. The AgNPs when subjected to surface plasmon resonance (SPR) studies showed a characteristic peak at around 420–430 nm. SEM, TEM, and XRD studies revealed the size of green synthesized NPs of around 20–100 nm as shown in Fig. 14.3. Application of the NPs resulted in bactericidal activity in the form of stopping the swimming activity, biofilm formation, maceration of sweet potato tubers with the effective NPs concentration of 12 μg/mL. The results obtained from this study showed that the AgNPs synthesized from *P. rhodesiae* have a potential use in controlling the soft rot disease by controlling pathogen contamination of sweet potato seed tubers (Hossain et al., 2019).

AgNPs isolated from cell-free supernatant of *Bacillus sp*. GP-23 showed strong antifungal activity against plant pathogenic fungus, *F. oxysporum*. Source of isolation of the bacterium was soil samples. The strain of the bacterium was confirmed by 16S rRNA (ribosomal RNA) sequencing and maximum activity of these green synthesized NPs was found at the concentration of 8 μg/mL. SPR studies showed characteristic peak at 420 nm. FTIR, 10 XRD, high-resolution

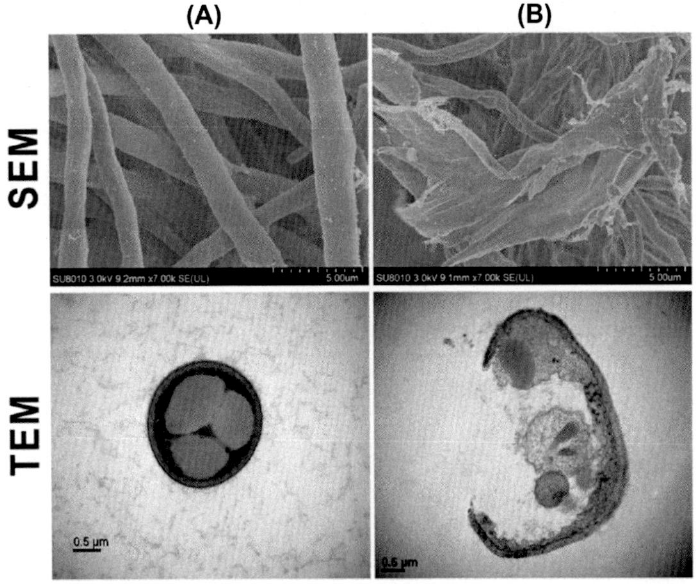

FIGURE 14.2

Microscopic images of SEM and TEM of *Fusarium graminearum* strain PH − 1 in the absence (A) and presence (B) of the synthesized AgNPs. *AgNPs*, Silver nanoparticles; *SEM*, scanning electron microscopy; *TEM*, transmission electron microscopy.

Reprinted from Ibrahim, E., Zhang, M., Zhang, Y., Hossain, A., Qiu, W., Chen, Y., Wang, Y., Wu, W., Sun, G., & Li, B. (2020) Green-synthesization of silver nanoparticles using endophytic bacteria isolated from garlic and its antifungal activity against wheat fusarium head blight pathogen Fusarium graminearum. Nanomaterials 10, *219. (Open access).*

TEM, EDX, and atomic force microscopy (AFM) studies suggested the shape of the AgNPs to be spherical with a size of 7–21 nm. This data also is an example of potential applicability of green synthesizes AgNPs in agriculture purpose (Gopinath & Velusamy, 2013).

The study presented by Mishra et al. (2014) suggested a means for effective management of spot blotch disease in wheat. Conidial germination on wheat leaves of *Bipolaris sorokiniana*, causative agent of spot blotch, was prevented significantly with an effective concentration of 2, 4, and 10 mg/mL of AgNPs synthesized from bacterium, *Serratia* sp. BHU-S4. Characterization studies such as SEM, TEM, FTIR, and AFM revealed the spherical morphology with a size of 10- to 20-nm range. The elemental analysis using EDX confirmed the presence of silver metals that are metallic and crystalline in nature. Histochemical studies showed the role of AgNPs in initiation of lignin deposition in the vascular bundles in the wheat. When application of AgNPs against this phytopathogen was checked, it was found that the NPs were able to reduce the infestation levels

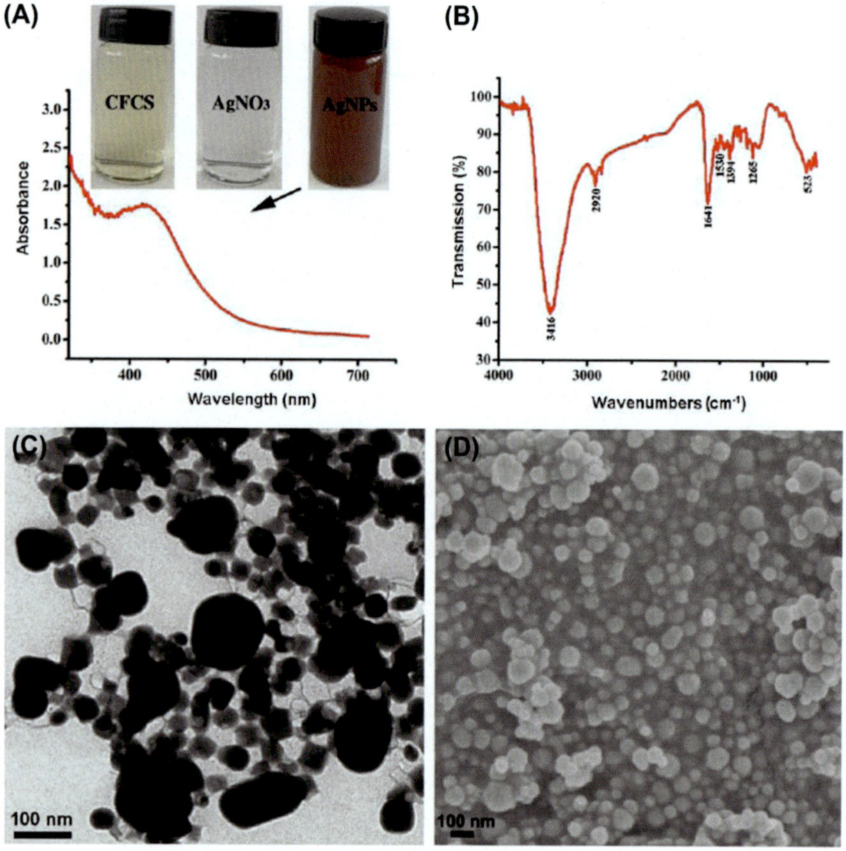

FIGURE 14.3

Characterization of AgNPs synthesized with CFCS of *Pseudomonas rhodesiae*. (A) Ultraviolet (UV)—visible absorption spectrum of the dark brown AgNP solution, which was formed by mixture of light yellow CFCS and colorless $AgNO_3$ solution. AgNPs display a clear surface plasmon resonance peak at 420–430 nm. (B) Fourier transform IR spectrum showing functional groups responsible for the synthesis and stabilization of AgNPs. (C) Transmission electron micrograph showing AgNPs in spherical forms about 20–100 nm in diameter. (D) Scanning electron micrograph showing AgNPs in spherical forms about 20–100 nm in diameter. *AgNPs*, Silver nanoparticles; *CFCS*, cell-free culture filtrate; *IR*, infrared.

Reprinted from Hossain, A., Hong, X., Ibrahim, E., Li, B., Sun, G., Meng, Y., Wang, Y., & An, Q. (2019). Green synthesis of silver nanoparticles with culture supernatant of a bacterium Pseudomonas rhodesiae *and their antibacterial activity against soft rot pathogen* Dickeya dadantii. Molecules 24*, 2303. (Open access).*

significantly. This study indicated exciting possibilities in the application of AgNPs as a fungicidal agent managing the spot blotch disease in wheat (Mishra et al., 2014). Taking inference from the previous study, Mishra, Singh, Naqvi, and Singh (2017) checked the applicability of AgNPs against *Sclerotium rolfsii. S. rolfsii* is identified as a soilborne pathogen causing collar rot in chickpea. The AgNPs were synthesized extracellularly from a soilborne organism, *Stenotrophomonas* sp. BHU-S7 (MTCC 5978). Maximum inhibitory concentration of the NPs was checked at 2, 4, 10 μg/mL and the rate of conidial germination was recorded under in vitro conditions. Results obtained from these studies suggested exposure of *S. rolfsii* to *Stenotrophomonas* sp. AgNPs failed to germinate when grown on potato dextrose agar medium and soil system. The biosynthesized NPs were spherical in shape and having particle size of ~12 nm. AgNPs have played important role not only in inhibiting the conidial growth of *S. rolfsii* but also in triggering the defense mechanism of chickpea thereby inducing phenolic acids, altered lignification, and hydrogen peroxide (H_2O_2) production. This suggests the adverse effects of the AgNPs on phytopathogens raising a possibility of *S. rolfsii* reduced infestations in chickpea. This study fully depicted the protection mechanism of the AgNPs with respect to *S. rolfsii* infection in chickpea. It has given a new perspective in the usage of *Stenotrophomonas* sp. BHU-S7 (MTCC 5978) against phytopathogens (Mishra et al., 2017).

In this way, NPs have received attention through biotechnological applications to regulate various processes due to benefits such as cost-effectiveness, low energy consumption, clean, eco-friendly, and time-efficient methods for synthesis. Out of various other NPs, copper and their oxides, NPs are considered noble and most promising due to their high surface-to-volume ratio (Bhagwat et al., 2018). Copper nanoparticles (CuNPs) are biocompatible and are reported to be highly antioxidant and can also ameliorate the diabetes-associated complications (Ghosh et al., 2015c).

Application of cupric oxide (CuO)NPs resulted in the resistance of phytopathogenic bacterial and fungal strains such as *Alternaria alternata, F. oxysporum, Pythium ultimum*, and *A. niger*. Endophytic actinomycete *Streptomyces capillispiralis* Ca-1 was used for the synthesis of CuONPs which was isolated from healthy medicinal plant (*Convolvulus arvensis*) (L.). UV–vis spectroscopy at 600 nm demonstrated the conversion of biomass filtrate from light blue to greenish brown color. TEM and XRD analysis revealed the size of the NPs to be 3.6–59 nm and the presence of functional groups in the NPs, respectively. Moreover, FTIR analysis of the synthesized NPs resulted into the presence of bioactive compounds, which may possibly be responsible for antimicrobial activity. The data represent promising use of CuONPs in various biomedical applications (Hassan et al., 2018).

Similarly, the application potential of CuNPs isolated from *Streptomyces griseus* was assayed on tea plants infested with *Poria hypolateritia. P. hypolateritia* Berk. ex Cooke isolated from soil in the rhizospheric zone of tea plants was found to be the major causative agent of red rot root disease leading to sudden death of tea bushes and subsequently resulting in major capital loss. Characterization studies of the CuNPs resulted into the characteristic spectrum at 592 nm through UV-spectrometric studies. Further studies revealed the shape of the NPs to be spherical and size, around 30–50 nm. FTIR analysis confirmed that the formation of the CuNPs in *S. griseus* is solely due to the protein or enzymatic induced/reduction reactions. When the impact of the CuNPs was assessed with respect to the commercially available fungicide, carbendazim, the results were comparable as the maximum inhibitory concentration was noted to be 2.5 ppm that ably increased the leaf yield. In this way, this investigation has opened a channel for green synthesis of CuNPs, which can effectively work in controlling the population density of the pathogenic *P. hypolateritia* around the rhizospheric region of the tea crop (Ponmurugan, Manjukarunambika, Elango, & Gnanamangai, 2016).

Apart from CuONPs, zinc oxide–based NPs ZnONPs have also been studied for their antibacterial properties, antioxidant and antidiabetic (Adersh et al., 2015; Robkhob et al., 2020). Advantage of using ZnONPs as an antibacterial compound in agriculture purpose is that their application does not hinder the soil fertility compared to the traditional active antifungal compound (Karmakar, Ghosh, & Kumbhakar, 2020; Kitture et al., 2015). Moreover, ZnONPs facilitate prolonged contact between the bacterium cell membrane and the NPs that results in the bactericidal activity of the NPs. In the study performed by Jayaseelan et al. (2012), ZnO NPs were successfully isolated from *Aeromonas hydrophila*. Synthesis of ZnONPs was confirmed by 374-nm characteristic peak using UV spectrometry. Further, the characterization of the synthesized NPs was done using XRD, FTIR, AFM, noncontact AFM, and field emission scanning electron microscopy with EDX analyses. This characterization confirmed the crystalline nature as well as 57.72 nm spherical/oval nature of the NPs. Maximum inhibition concentration was found to be 25 g/mL, when tested against pathogenic strains of bacteria and fungi, *P. aeruginosa* and *Aspergillus flavus*, respectively (Jayaseelan et al., 2012). *A. flavus* was found to be an opportunistic pathogen against many oilseed crops such as maize, peanut, cottonseed, and tree nuts, as it produces aflatoxin as a secondary metabolite in these crops before and after harvest (Klich, 2007). Therefore application of ZnONPs could potentially prevent growth of *A. flavus* in the oilseed crops without causing major side effects to healthy growth of plant.

Taking advantage of good penetration power of NPs inside the pathogen cell membrane, researchers have created more reliable ways to tackle phytopathogens through synthesizing metal composites. Strayer-Scherer et al. (2018) applied Cu composites to suppress the infection levels of *Xanthomonas* spp. responsible for causing bacterial spot disease in tomato. Composites are more sophisticated form of NPs as they prevent particles agglomeration and provide long-term stability. Core–shell copper (CS-Cu), multivalent copper (MV-Cu), and fixed quaternary ammonium copper (FQ-Cu) were used to check the efficacy levels against the said pathogen. Greenhouse studies of these composites suggested significantly reduced levels of bacterial spot disease severity when the results were compared with copper–mancozeb.

14.4 Mycogenic nanoparticles

In recent years, it has been documented that the diverse metal NPs isolated from various fungal isolates have an inhibitory effect on different phytopathogenic strains of bacteria as well as fungi as listed in Table 14.2. Moreover, it has been found that endophytic fungi are excellent sources of secondary bioactive compounds, the potential of which has been explored for impressive antifungal, antimicrobial, and other biological activities. Filamentous fungi such as ascomycetes and imperfect fungi possess high level of tolerance toward the heavy metals as well as capability of internalizing and accumulating a high concentration of metals. Fungi are considered to be the most convenient source for synthesis of NPs due to uniformity in their shape and size (Ghosh & Webster, 2021c). The culture conditions for synthesis of NPs are same as that of bacterial cultures and are discussed in the bacteriogenic NPs section.

Concerning the green synthesis of NPs using fungal culture filtrates, Fouda, Hassan, Abdo, and El-Gamal (2020) reported synthesis of AgNPs from endophytic actinomycetes of *S. capillispiralis Ca-1, Streptomyces zaomyceticus Oc-5, and Streptomyces pseudogriseolus Acv-11*. The source of isolation of these strains was the healthy leaves of *C. arvensis L.* and *Oxalis corniculata L.* After characterization studies were performed, these green synthesized NPs were assessed for the minimum inhibitory concentration (MIC) against the phytopathogenic strains such as *A. alternata, F. oxysporum, P. ultimum*, and *A. niger*. Further, antioxidant and larvicidal activity against *Culex pipiens* and *Musca domestica* was also checked. The results obtained from these tests suggested an effective biocontrol reagent against phytopathogenic fungal species to successively achieve healthy growth of the plant.

It has been reported that the commercially available fungicides not only kill the fungi infecting the plants, but also it affects the normal flora and fauna of the niche. This makes NP-based delivery systems a promising solution to achieve more sustainable source of resistance. Concerning this issue, synthesized mycosilver NPs were tested and its efficacy against phytopathogens without majorly disturbing the microbiota around was studied. The mycosilver NPs were synthesized from endophytic fungi isolated from *Solanum nigrum*. The bioactivity of AgNPs was tested against *F. graminearum, Fusarium udum, Rhizoctonia solani, and A. niger*. The characteristic SPR peak at 350 nm confirmed the shape of the NPs to be spherical shape, whereas other characterization studies of these NPs were found to be of size ranging from 2 to 50 nm. The antifungal mechanism of the mycosilver NPs could be because of the induction of ROS in the phytopathogens which may lead to the degradation of protein, lipids, and nucleic acids and eventually damage of the cells. The results give a way to eradicate the food/grain contamination that leads to the health-related issues (Akther & Hemalatha, 2019).

Similarly, AgNPs synthesized from a cosmopolitan endophytic fungus, *Guignardia mangiferae* isolated from the leaves of *Citrus* sp. gave a way forward for applying simple, faster, and cost-effective way to tackle phytopathogens.

Extracellular synthesis gave an upper hand for large-scale bioproduction of the AgNPs. Characterization studies of the NPs confirmed that the NPs are spherical and of 5- to 30-nm diameter shape as depicted in Fig. 14.4. Moreover, Energy dispersive X-ray spectrum confirmed that the synthesis of the AgNPs is due to *G. mangiferae*. Antimicrobial activity of *P. aeruginosa* and *S. aureus* with a clear zone of inhibition of 16 mm with an effective concentration, 1 mg/mL was observed. This could be the clear evidence of the biocompatibility to use these green NPs as a biocontrol reagent (Balakumaran, Ramachandran, & Kalaichelvan, 2015).

AgNPs synthesized from *Trichoderma* sp. and *Cephalosporium* sp. were also explored for their antibacterial potential in controlling wilt disease caused by *F. oxysporum* f. sp. ciceri in chickpea, a rich source of iron and vitamin B. The

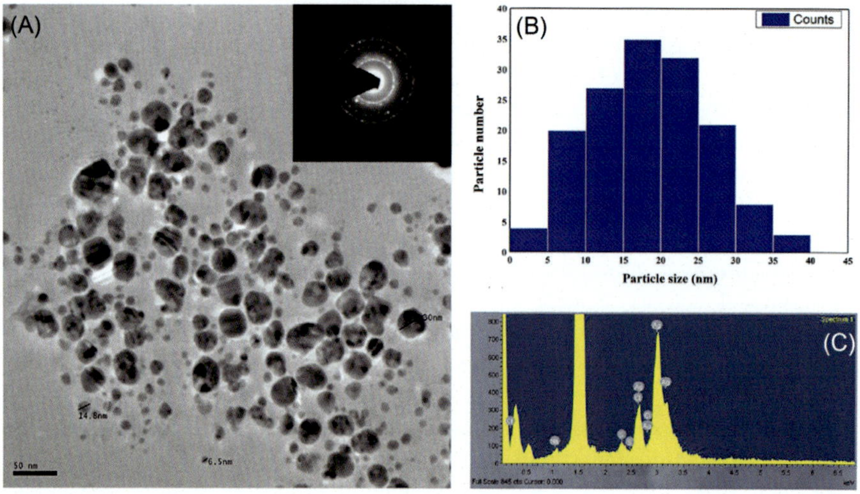

FIGURE 14.4

Characterization of silver nanoparticles synthesized from *Guignardia mangiferae*. (A) High-resolution transmission electron microscopic image of spherical-shaped silver nanoparticles; the size of the nanoparticles was 8–20 nm (scale bar = 50 nm) and the inset shows the SAED pattern; (B) histogram analysis of the particle size distribution of silver nanoparticles. It reveals an average size of 5- to 30-nm diameters by calculating 150 randomly selected silver nanoparticles in HR-TEM images; (C) energy dispersive X-ray spectrum shows strong signal in the silver region and confirms the formation of silver nanoparticles from *G. mangiferae*. Other elemental signals were also recorded possibly due to elements from enzymes or proteins present in the fungal extract. *HR-TEM*, High-resolution transmission electron microscopy.

Reprinted with permission from Balakumaran, M. D., Ramachandran, R., Kalaichelvan, & P. T. (2015). Exploitation of endophytic fungus, Guignardia mangiferae *for extracellular synthesis of silver nanoparticles and their in vitro biological activities.* Microbiological Research, 178, *9–17. Copyright © 2015 Elsevier GmbH.*

AgNPs synthesized extracellularly by *Trichoderma* sp. were 22 nm in size. MIC of the NPs was tested through pot experiment held at the glasshouse maintained chickpea seeds infected with the phytopathogen. The germination efficiency was increased comparatively as the AgNP coating might have increased the permeability of the seed capsule thereby also enhancing the availability of dioxygen molecules from the soil. This process might have accelerated the overall metabolism germination rate of the seed. The maximum percentage of survival of the chickpea plant was reported to be 73.33%. This work indicates the potential of mycogenic AgNPs to control Fusarium wilt of chickpea in vitro and the disease control of the plant using its own rhizospheric microflora. Also there were no ill effects found on the normal microbiota of the soil in the peripheral region of the plant (Kaur, Thakur, Duhan, & Chaudhury, 2018).

Aspergillus spp. and *Penicillium digitatum* are found to infest various crop plants such as orange, lemon, tomatoes, grapes, strawberries, cucurbit, cucumbers, eggplants, bell pepper, soft dates, and onion. Due to the low, moderated, and high levels of infestation in this crop plants, approximately 20% of the crops are lost every year. Generation of higher amount of pectinases and cellulases and thereby deterioration of the cell membrane of fruits and vegetable is the main cause of the spoilage of the crop plants. Therefore management of the fungal infection and growth is really important. Al-Zubaidi, Al-Ayafi, and Abdelkader (2019) synthesized AgNPs from *A. niger* and checked their pathogenicity against the abovementioned phytopathogens. SPR studies of synthesized NPs showed peak at 430 nm, which corresponds to AgNPs. Further characterization studies deduced the size of these synthesized AgNPs to be around 10–100 nm. The MIC of AgNPs of phytopathogenic strains such as *P. digitatum, A. flavus*, and *F. oxysporum* was observed to be 6.75 ± 0.24, 7.45 ± 0.18, and 9.62 ± 0.14, respectively. Therefore mycosynthesized AgNPs represent a powerful solution against *P. digitatum, A. flavus, F. oxysporum*. So from the application point of view, appropriate formulation by using AgNPs can be commercialized for controlling the fungal growth on the vegetable and fruit crops (Al-Zubaidi et al., 2019).

Similarly, in the series of mass production of NPs from mycogenic sources, *Trichoderma longibrachiatum*'s potential was studied through the extracellular synthesis of AgNP and used effectively against *Fusarium verticillioides, Fusarium moniliforme, Penicillium brevicompactum, Helminthosporium oryzae*, and *Pyricularia grisea*. UV–visible spectroscopy was performed to confirm the presence of AgNPs. TEM studies revealed the spherical nature of the NPs with a size range of 5 and 25 nm. The synthesis studies are yet to be optimized to achieve same size of the NPs along with polydispersity. The exact mechanism of action of these mycogenic AgNPs is not yet known. So it is imperative to focus on these points along with designing a potential strategy toward commercialization of this bioactive compound for crop protection (Elamawi, Al-Harbi, & Hendi, 2018).

Target leaf spot disease of sorghum is caused by the necrotrophic fungus *Bipolaris sorghicola* affects the grain yield in sorghum by causing premature drying of leaves and defoliation. The infestation of this pest is majorly found in

India, Japan, and the United States at various levels. Almaary et al. (2020) suggested the use of mycosynthesized AgNPs to this disease. The fungal species used for synthesis is *Penicillium duclauxii*. Crystallographic analysis revealed the polycrystalline face-centered cubic symmetry of the NPs. Synthesis of 3–32 nm, spherical AgNPs was confirmed by TEM analysis studies. Antifungal activity tests confirmed effective monitoring of the growth of *Sclerotium cepivorum, B. sorghicola*, and *Botrytis cinereal* in onions, sorghum, and strawberries, respectively. Thus this study underlines the effective use of the seed-borne fungus *P. duclauxii* in synthesis of AgNPs and their bioactivity against other fungal species that are pathogenic to the various crop species (Almaary et al., 2020).

Apart from increasing the yields of the crop plants, there are several other parameters that need to be attended such as climate change, pest infestation, and disease control. Among all the amenable crops, soybean stands at the second position worldwide. In spite of being a rapidly growing important part of the food supply, this crop has been drastically affected by a strain of pathogenic fungus, *Sclerotinia sclerotiorum*, which causes the disease known as white mold. This leads to reduction in the yields and contamination of the soil in the infected region. The organism can remain viable for as long as 11 years in the soil. Therefore Guilger et al. (2017) gave an alternative to eradicate *S. sclerotiorum*, by synthesizing AgNPs from *Trichoderma harzianum*. Physicochemical characterization was performed by dynamic light scattering, zeta potential using microelectrophoresis, NP tracking analysis, and SEM coupled with energy dispersive spectroscopy. The AgNPs showed cytotoxic as well as genotoxic effects on *S. sclerotiorum*. There were some evidences of effects of the AgNPs on the bacterial population, which positively could be recovered over a period of few years, although no ill effects were observed on the healthy germination and growth of the soybean (Guilger et al., 2017).

In further studies, these mycosynthesized NPs were used to check their bactericidal/fungicidal activity against *Helminthosporium* sp., *Alternaria alternate, Phytophthora arenaria*, and *Botrytis* sp., by agar well diffusion method. In this study the spherical nano-Ag mycosynthesized particles were of size of around 12.7 ± 0.8 nm. The MIC was found to be 100ug/mL against *A. alternate, Helminthosporium* sp., *Botrytis* sp., and *P. arenaria* with zone of clearance of 43, 35, 32, and 28 mm, respectively. The possible mechanism for interaction between the AgNPs and the phytopathogens could be the dissolution of the lipid bilayers, thus resulting in penetration into the cell organelles and nuclei causing metal accumulation. This could further lead to the death of the cells. This study is an attempt to develop reliable and low cost, large-scale bioactive formula with an eco-friendly approach.

ZnONPs have an advantage over other metal NPs due to their broad-spectrum antimicrobial activity at minimum concentrations. It has been previously demonstrated that ZnO is biocompatible and available in plenty ecologically. Moreover, green amalgamation of NPs is a quite eco-friendly approach that involves fewer efforts in the synthesis as well as having less or no harmful effects on the ecosystem when compared with the physical and synthetic chemical. *Xylaria acuta*, an endophytic fungus, was used to synthesize the ZnONPs extracellularly. *Millingtonia hortensis* L. f. was the source of isolation of endophytic fungi. Zinc nitrate hexahydrate [Zn $(NO_3)_2 0.6H_2O$] was used as a precursor for the synthesis process. SEM and TEM micrographs revealed the structure of green synthesized ZnONPs to be hexagonal in shape and possessing a diameter of around 40–55 nm on average. Antibacterial activity of the ZnONPs was observed by zone of inhibition against *B. cereus, S. aureus*, and *P. aeruginosa*. The test results showed that *S. aureus* and *B. cereus* are highly susceptible to ZnONPs with MIC concentration 15.6 μg/mL, followed by *E. coli* and *P. aeruginosa* with MIC concentration of 31.3 μg/mL. Mechanism of action responsible for this is could be increase of H_2O_2 (highly reactive species), thereby increasing penetration power of ZnONPs and causing high cellular damage to the pathogens (Sumanth et al., 2020).

Almost all of the methods are quite cost-effective and can be used liberally, instead of synthetic commercially available pesticides, fungicides, and antimicrobial agents (El-Moslamy, Elkady, Rezk, & Abdel-Fattah, 2017). Among all the microbial entities, bacteria and fungi have attracted more researchers due to their high accumulation capacity, biocompatibility in nature and easy and fast growth patterns. The discoveries show mass scale production of the metal NPs from bacterial or fungal source and thereby widening the array of nanobiotechnological applications. In spite of being a quite new field, pest or disease control through application of biogenic NPs has demonstrated quite remarkable discoveries at impressive levels. The technology also holds some pitfalls such as lack of the data of the effects of the metal NPs on the microbiota or if the NPs are used in the soil. Likewise, the effect of NPs on the soil fertility is unknown and whether metal accumulation in the form of NPs may harm the ecosystems is needed to be investigated (Ismail, Prasad, Ibrahim, & Ahmed, 2017). Relevant experiments and generation of significant results will lead to the commercial formulation inclusive of enhancers and stabilizers to fight against phytopathogenic strains of bacteria and fungi. This strategy will increase the productivity of the crop plants thereby enhancing the healthy growth pattern of these plants. Therefore the overall discoveries in this field represent a positive collaboration of nanobiotechnology and agricultural plant sciences and appear to offer quite high number of promising solutions.

14.5 Future prospects

The development of green routes for NPs synthesis is an essential element for application of nanotechnology in agriculture. The management of pests and pathogens using biogenic NPs includes microbially derived NPs and also NP obtained from plants. However, microbe synthesized NPs possess an upper hand over phytogenic NPs as they also contain inherent biocontrol properties. Also, they can be used for delivery of genes and nucleic acid into the plant along with holding the ability to undergo genomic- and proteomic-level modifications. Moreover, microbes encompass the ability to synthesize NPs through both intracellular and extracellular means. The extracellular mode of NPs synthesis allows for an easier route for further purification as compared to intracellular NPs (Naik, 2020). The direct utilization of microbially derived NPs is as biopesticide. A NP is considered to be effective if it is able to withstand extreme conditions such as high temperature and still remain stable. Microbially derived NPs also provide the added advantage of improved biological efficacy, high dispersion rate, easy adherence, reduced toxicity, and high mass delivery. In addition to utilization of NPs for plant protection, nanosensors are also being actively used for crop monitoring and detection of pathogens. Nanosensors planted in fields provide a timely analysis of the real-time crop status and infestation that allows for management prior to heavy infestation and damage. Microbially derived NPs from various microorganisms have been used for this purpose. Selenium NPs derived from *B. subtilis* was employed for developing horseradish peroxidase biosensors. Further, two microbes were employed for development of nanosensors that allowed rapid-diagnosis of a soilborne disease. Such biosensors, while not directly protecting the plant against pathogens, provide a timely report before severe damage occurs within the plant system.

Although there are several successful studies for the utilization of microbial NP against phytopathogens, a few drawbacks and issues are needed to be addressed. The uptake of the NPs by the plant, effect of size, stability of the NPs in the plant and the ultimate consequences of the NPs on the plant are a few of the factors that need to be kept in mind to determine the efficacy of NPs in inhibition of pathogenic activities. Attention should also be paid to the development and regulation of hybrid carriers for delivery of active agents such as nucleic acid for crop improvement. While the biosensors have futuristic implications in terms of efficacy and applicability, their commercialization is a major concern.

The exact mechanism of NP should be determined to enhance specificity and avoid targeting nonspecific organisms. The compounds responsible for reduction of NPs can be characterized to reduce the response time and increase biosynthesis. Finally, the toxic implications of NPs on the environment are crucial to comprehend and should be extensively unveiled for various forms of nanomaterials (Kaushal, 2018). The use of superparamagnetic iron oxide NPs for phytopathogen detection has recently been pursued. The magnetic NPs (MNPs) bind to biological tissues and facilitate the detection. This technique has been utilized in the detection of *F. oxysporum* where the MNPs adhere to the surface of the fungus and allow easy visualization. MNPs can therefore be exploited for other plant pathogens and thus can prevent severe infections through early detection (Pandey, 2018; Worrall, Hamid, Mody, Mitter, & Pappu, 2018). Additionally, nanoenabled biosensor mentioned earlier can be collectively used with Global Positioning System to detect and monitor specific areas prior to pest infestation (Elmer & White, 2018).

14.6 Conclusion

Bionanoparticles for pest management represent a revolutionary advancement due to their advantages related to environment and human health. However, its reach into the market is not high owing to its emergence in the recent past. The NP research is till at preliminary stages with lack of long-term trials along with little existing detail about its efficacy and exact toxicity mechanism. Additionally, the clarity of regulations regarding application of NPs is limited as compared to more prevalent pesticides. It is essential to establish robust data with thorough analysis regarding the safety in long-term in-field trials, production cost, regulations and policies, and the general public outlook. An integrative approach collaborating different techniques would provide a more in-depth assessment of the pros and cons of using microbe synthesized NPs for plant disease management. Additionally, determination of the structural properties like size, shape, active functional groups, and binding capacity would be the starting point for the development of a rationale in generating appropriate NPs. A comprehensive idea about the permissible dosage of NPs along with the trans-generational effects also is crucial to elucidate NPs as a safe alternative to pesticides and insecticides. The use of metallic and carbon-based NPs has the potential to replace other genetic delivery systems. Further, compared to conventional metallic fungicides, NP-based pesticides provide a reduced requirement of metals such as Ag, Cu, and Zn as they exist in the nanoform. The implementation of new tools like biosensors has been huge in the field of medicine; however, its application in agriculture and subsequently, pest detection, is yet to be explored. A small section of research is directed toward this, which calls for a new approach to discover and adapt nanotechnology-driven solutions in agriculture.

References

Adersh, A., Kulkarni, A. R., Ghos, S., More, P., Chopade, A., & Gandhi, M. N. (2015). Surface defect rich ZnO quantum dots as antioxidant inhibiting α-amylase and α-glucosidase: A potential anti-diabetic nanomedicine. *Journal of Materials Chemistry B*, *3* (22), 4597–4606.

Ahmed, T., Shahid, M., Noman, M., Niazi, M. B. K., Mahmood, F., Manzoor, I., ... Chen, J. (2020). Silver nanoparticles synthesized by using *Bacillus cereus SZT1* ameliorated the damage of bacterial leaf blight pathogen in rice. *Pathogen*, *9*(3), 160.

Akther, T., & Hemalatha, S. (2019). Mycosilver nanoparticles: Synthesis, characterization and its efficacy against plant pathogenic fungi. *Bionanoscience*, *9*(2), 296–301.

Alavi, M., & Rai, M. (2019). Recent advances in antibacterial applications of metal nanoparticles (MNPs) and metal nanocomposites (MNCs) against multidrug-resistant (MDR) bacteria. *Expert Review of Anti-Infective Therapy*, *17*(6), 419–428.

Ali, A., Ahmed, T., Wu, W., Hossain, A., Hafeez, R., Masum, M. I., ... Li, B. (2020). Advancements in plant and microbe-based synthesis of metallic nanoparticles and their antimicrobial activity against plant pathogens. *Nanomaterials*, *10*(6), 1146.

Almaary, K. S., Sayed, S. R. M., Abd-Elkader, O. H., Dawoud, T. M., El Orabi, N. F., & Elgorban, A. M. (2020). Complete green synthesis of silver-nanoparticles applying seed-borne *Penicillium duclauxii. Saudi Journal of Biological Sciences, 27*(5), 1333–1339.

Al-Zubaidi, S., Al-Ayafi, A., & Abdelkader, H. (2019). Biosynthesis, characterization and antifungal activity of silver nanoparticles by *Aspergillus Niger* isolate. *Journal of Nanoparticle Research, 1*(1), 23–36.

Balakumaran, M. D., Ramachandran, R., & Kalaichelvan, P. T. (2015). Exploitation of endophytic fungus, *Guignardia mangiferae* for extracellular synthesis of silver nanoparticles and their in vitro biological activities. *Microbiological Research, 178*, 9–17.

Bhagwat, T. R., Joshi, K. A., Parihar, V., Asok, A., Bellare, J., & Ghosh, S. (2018). Biogenic copper nanoparticles from medicinal plants as novel antidiabetic nanomedicine. *World Journal of Pharmaceutical Research, 7*(4), 183–196.

Bloch, K., Pardesi, K., Satriano, C., & Ghosh, S. (2021). Bacteriogenic platinum nanoparticles for application in nanomedicine. *Frontiers in Chemistry, 9*, 624344.

Elamawi, R. M., Al-Harbi, R. E., & Hendi, A. A. (2018). Biosynthesis and characterization of silver nanoparticles using *Trichoderma longibrachiatum* and their effect on phytopathogenic fungi. *Egyptian Journal of Biological Pest Control, 28*(1), 1–11.

Elmer, W., & White, J. C. (2018). The future of nanotechnology in plant pathology. *Annual Review of Phytopathology, 56*, 111–133.

El-Moslamy, S. H., Elkady, M. F., Rezk, A. H., & Abdel-Fattah, Y. R. (2017). Applying Taguchi design and large-scale strategy for mycosynthesis of nano-silver from endophytic *Trichoderma harzianum SYA.F4* and its application against phytopathogens. *Scientific Reports, 7*(1), 1–22.

Fouda, A., Hassan, S. E. D., Abdo, A. M., & El-Gamal, M. S. (2020). Antimicrobial, antioxidant and larvicidal activities of spherical silver nanoparticles synthesized by endophytic *Streptomyces spp. Biological Trace Element Research, 195*(2), 707–724.

Ghosh, S. (2018). Copper and palladium nanostructures: A bacteriogenic approach. *Applied Microbiology and Biotechnology, 101*(18), 7693–7701.

Ghosh, S., Harke, A. N., Chacko, M. J., Gurav, S. P., Joshi, K. A., Dhepe, A., ... Chopade, B. A. (2016a). *Gloriosa superba* mediated synthesis of silver and gold nanoparticles for anticancer applications. *Journal of Nanomedicine and Nanotechnology, 7* (390), 2.

Ghosh, S., Chacko, M. J., Harke, A. N., Gurav, S. P., Joshi, K. A., Dhepe, A., ... Chopade, B. A. (2016b). *Barleria prionitis* leaf mediated synthesis of silver and gold nanocatalysts. *Journal of Nanomedicine and Nanotechnology, 7*(394), 4.

Ghosh, S., Gurav, S. P., Harke, A. N., Chacko, M. J., Joshi, K. A., Dhepe, A., ... Chopade, B. A. (2016c). *Dioscorea oppositifolia* mediated synthesis of gold and silver nanoparticles with catalytic activity. *Journal of Nanomedicine and Nanotechnology, 7* (398), 2.

Ghosh, S., More, P., Derle, A., Kitture, R., Kale, T., Gorain, M., ... Chopade, B. A. (2015a). Diosgenin functionalized iron oxide nanoparticles as novel nanomaterial against breast cancer. *Journal of Nanoscience and Nanotechnology, 15*(12), 9464–9472.

Ghosh, S., Jagtap, S., More, P., Shete, U. J., Maheshwari, N. O., Rao, S. J., ... Chopade, B. A. (2015b). *Dioscorea bulbifera* mediated synthesis of novel $Au_{core}Ag_{shell}$ nanoparticles with potent antibiofilm and antileishmanial activity. *Journal of Nanomaterials, 16*.

Ghosh, S., More, P., Nitnavare, R., Jagtap, S., Chippalkatti, R., Derle, A., . . . Chopade, B. A. (2015c). Antidiabetic and antioxidant properties of copper nanoparticles synthesized by medicinal plant *Dioscorea bulbifera*. *Journal of Nanomedicine and Nanotechnology*, *S6*, 1.

Ghosh, S., Patil, S., Chopade, N. B., Luikham, S., Kitture, R., Gurav, D. D., . . . Chopade, B. A. (2016d). *Gnidia glauca* leaf and stem extract mediated synthesis of gold nanocatalysts with free radical scavenging potential. *Journal of Nanomedicine and Nanotechnology*, *7*(358), 2.

Ghosh, S., Shah, S., & Webster, T. J. (2021a). *Recent trends in fungal biosynthesis of nanoparticles. Fungi bio-prospects in sustainable agriculture, environment and nanotechnology* (pp. 403–452). Academic Press.

Ghosh, S., Thongmee, S., & Kumar, A. (2021b). *Agricultural nanobiotechnology: Biogenic nanoparticles, nanofertilizers and nanoscale biocontrol agents*. Elsevier Inc. (In Press).

Ghosh, S., & Webster, T. J. (2021c). *Nanobiotechnology: Microbes and plant assisted synthesis of nanoparticles, mechanisms and applications*. Elsevier Inc.

Gopinath, V., & Velusamy, P. (2013). Extracellular biosynthesis of silver nanoparticles using *Bacillus* sp. GP-23 and evaluation of their antifungal activity towards *Fusarium oxysporum*. *Spectrochimica Acta, Part A: Molecular and Biomolecular Spectroscopy*, *106*, 170–174.

Guilger, M., Pasquoto-Stigliani, T., Bilesky-Jose, N., Grillo, R., Abhilash, P. C., Fraceto, L. F., & De Lima, R. (2017). Biogenic silver nanoparticles based on *Trichoderma harzianum*: Synthesis, characterization, toxicity evaluation and biological activity. *Scientific Reports*, *7*(1), 1–13.

Gupta, N., Upadhyaya, C. P., Singh, A., Abd-Elsalam, K. A., & Prasad, R. (2018). *Applications of silver nanoparticles in plant protection. Nanobiotechnology applications in plant protection* (pp. 247–265). Cham: Springer.

Hassan, S. E. D., Salem, S. S., Fouda, A., Awad, M. A., El-Gamal, M. S., & Abdo, A. M. (2018). New approach for antimicrobial activity and bio-control of various pathogens by biosynthesized copper nanoparticles using endophytic actinomycetes. *Journal of Radiation Research and Applied Sciences*, *11*(3), 262–270.

Hossain, A., Hong, X., Ibrahim, E., Li, B., Sun, G., Meng, Y., . . . An, Q. (2019). Green synthesis of silver nanoparticles with culture supernatant of a bacterium *Pseudomonas rhodesiae* and their antibacterial activity against soft rot pathogen *Dickeya dadantii*. *Molecules (Basel, Switzerland)*, *24*(12), 2303.

Ibrahim, E., Fouad, H., Zhang, M., Zhang, Y., Qiu, W., Yan, C., . . . Chen, J. (2019). Biosynthesis of silver nanoparticles using endophytic bacteria and their role in inhibition of rice pathogenic bacteria and plant growth promotion. *RSC Advances*, *9*(50), 29293–29299.

Ibrahim, E., Zhang, M., Zhang, Y., Hossain, A., Qiu, W., Chen, Y., . . . Li, B. (2020). Green-synthesization of silver nanoparticles using endophytic bacteria isolated from garlic and its antifungal activity against wheat Fusarium head blight pathogen *Fusarium graminearum*. *Nanomaterials*, *10*(2), 219.

Ismail, M., Prasad, R., Ibrahim, A. I., & Ahmed, A. I. (2017). *Modern prospects of nanotechnology in plant pathology. Nanotechnology* (pp. 305–317). Singapore: Springer.

Jamdade, D. A., Rajpali, D., Joshi, K. A., Kitture, R., Kulkarni, A. S., Shinde, V. S., . . . Ghosh, S. (2019). *Gnidia glauca* and *Plumbago zeylanica* mediated synthesis ofnovel copper nanoparticles as promising antidiabetic agents. *Advances in Pharmacological Sciences*, *2019*, 9080279.

Jayaseelan, C., Rahuman, A. A., Kirthi, A. V., Marimuthu, S., Santhoshkumar, T., Bagavan, A., ... Rao, K. V. B. (2012). Novel microbial route to synthesize ZnO nanoparticles using *Aeromonas hydrophila* and their activity against pathogenic bacteria and fungi. *Spectrochimica Acta, Part A: Molecular and Biomolecular Spectroscopy*, *90*, 78–84.

Joshi, K. A., Ghosh, S., & Dhepe, A. (2019). Green synthesis of antimicrobial nanosilver using *in-vitro* cultured *Dioscorea bulbifera*. *Asian Journal of Organic & Medicinal Chemistry*, *4*(4), 222–227.

Karmakar, S., Ghosh, S., & Kumbhakar, P. (2020). Enhanced sunlight driven photocatalytic and antibacterial activity of flower-like ZnO@MoS_2 nanocomposite. *Journal of Nanoparticle Research*, *22*(1), 1–20.

Kaur, P. (2019). Biosynthesis of nanoparticles using eco-friendly factories and their role in plant pathogenicity. *Biotechnology Research and Innovation*, *2*(1), 63–73.

Kaur, P., Thakur, R., Duhan, J. S., & Chaudhury, A. (2018). Management of wilt disease of chickpea in vivo by silver nanoparticles biosynthesized by rhizospheric microflora of chickpea (*Cicer arietinum*). *Journal of Chemical Technology and Biotechnology (Oxford, Oxfordshire: 1986)*, *93*(11), 3233–3243.

Kaushal, M. (2018). *Role of microbes in plant protection using intersection of nanotechnology and biology. Nanobiotechnology applications in plant protection* (pp. 111–135). Cham: Springer.

Khan, M. R., & Rizvi, T. F. (2014). Nanotechnology: Scope and application in plant disease management. *Plant Pathology Journal*, *13*(3), 214–231.

Kitture, R., Chordiya, K., Gaware, S., Ghosh, S., More, A., Kulkarni, P., ... Kale, S. N. (2015). ZnO nanoparticles-red sandalwood conjugate: A promising anti-diabetic agent. *Journal of Nanoscience and Nanotechnology*, *15*(6), 4046–4051.

Klich, M. A. (2007). *Aspergillus flavus*: The major producer of aflatoxin. *Molecular Plant Pathology*, *8*(6), 713–722.

Mishra, S., Singh, B. R., Naqvi, A. H., & Singh, H. B. (2017). Potential of biosynthesized silver nanoparticles using *Stenotrophomonas* sp. BHU-S7 (MTCC 5978) for management of soil-borne and foliar phytopathogens. *Scientific Reports*, *7*(1), 1–15.

Mishra, S., Singh, B. R., Singh, A., Keswani, C., Naqvi, A. H., & Singh, H. B. (2014). Biofabricated silver nanoparticles act as a strong fungicide against *Bipolaris sorokiniana* causing spot blotch disease in wheat. *PLoS One*, *9*(5), e97881.

Naik, B. S. (2020). Biosynthesis of silver nanoparticles from endophytic fungi and their role in plant disease management. *Microbial Endophytes*, 307–321.

Pandey, G. (2018). Challenges and future prospects of agri-nanotechnology for sustainable agriculture in India. *Environmental Technology & Innovation*, *11*, 299–307.

Ponmurugan, P., Manjukarunambika, K., Elango, V., & Gnanamangai, B. M. (2016). Antifungal activity of biosynthesised copper nanoparticles evaluated against red root-rot disease in tea plants. *Journal of Experimental Nanoscience*, *11*(13), 1019–1031.

Ranpariya, B., Salunke, G., Karmakar, S., Babiya, K., Sutar, S., Kadoo, N., ... Ghosh, S. (2021). Antimicrobial synergy of silver-platinum nanohybrids with antibiotics. *Frontiers in Microbiology*, *11*, 610968.

Robkhob, P., Ghosh, S., Bellare, J., Jamdade, D., Tang, I. M., & Thongmee, S. (2020). Effect of silver doping on antidiabetic and antioxidant potential of ZnO nanorods. *Journal of Trace Elements in Medicine and Biology: Organ of the Society for Minerals and Trace Elements (GMS)*, *58*, 126448.

Rokade, S., Joshi, K., Mahajan, K., Patil, S., Tomar, G., Dubal, D., ... Ghosh, S. (2018). *Gloriosa superba* mediated synthesis of platinum and palladium nanoparticles for induction of apoptosis in breast cancer. *Bioinorganic Chemistry and Applications, 2018*, 4924186.

Rokade, S. S., Joshi, K. A., Mahajan, K., Tomar, G., Dubal, D. S., Parihar, V. S., ... Ghosh, S. (2017). Novel anticancer platinum and palladium nanoparticles from *Barleria prionitis*. *Global Journal of Nanomedicine*, *2*(5), 555600.

Salunke, G. R., Ghosh, S., Santosh, R. J., Khade, S., Vashisth, P., Kale, T., ... Chopade, B. A. (2014). Rapid efficient synthesis and characterization of AgNPs, AuNPs and AgAuNPs from a medicinal plant, *Plumbago zeylanica* and their application in biofilm control. *International Journal of Nanomedicine*, *9*, 2635–2653.

Shende, S., Joshi, K. A., Kulkarni, A. S., Charolkar, C., Shinde, V. S., Parihar, V. S., ... Ghosh, S. (2018). *Platanus orientalis* leaf mediated rapid synthesis of catalytic gold and silver nanoparticles. *Journal of Nanomedicine and Nanotechnology*, *9*(494), 2.

Shende, S., Joshi, K. A., Kulkarni, A. S., Shinde, V. S., Parihar, V. S., Kitture, R., ... Ghosh, S. (2017). *Litchi chinensis* peel: A novel source for synthesis of gold and silver nanocatalysts. *Global Journal of Nanomedicine*, *3*(1), 555603.

Shinde, S. S., Joshi, K. A., Patil, S., Singh, S., Kitture, R., Bellare, J., & Ghosh, S. (2018). Green synthesis of silver nanoparticles using *Gnidia glauca* and computational evaluation of synergistic potential with antimicrobial drugs. *World Journal of Pharmaceutical Research*, *7*(4), 156–171.

Strayer-Scherer, A., Liao, Y. Y., Young, M., Ritchie, L., Vallad, G. E., Santra, S., ... Paret, M. L. (2018). Advanced copper composites against copper-tolerant *Xanthomonas perforans* and tomato bacterial spot. *Phytopathology*, *108*(2), 196–205.

Sumanth, B., Lakshmeesha, T. R., Ansari, M. A., Alzohairy, M. A., Udayashankar, A. C., Shobha, B., ... Almatroudi, A. (2020). Mycogenic synthesis of extracellular zinc oxide nanoparticles from *Xylaria acuta* and its nanoantibiotic potential. *International Journal of Nanomedicine*, *15*, 8519–8536.

Worrall, E. A., Hamid, A., Mody, K. T., Mitter, N., & Pappu, H. R. (2018). Nanotechnology for plant disease management. *Agronomy*, *8*(12), 285.

Index

Note: Page numbers followed by "*f*" and "*t*" refer to figures and tables, respectively.

C

D

E

F

G

Q

R

Printed in the United States
by Baker & Taylor Publisher Services